Physics

Phase 2

NOTION PRESS

NOTION PRESS

India. Singapore. Malaysia.

ISBN xxx-x-xxxxx-xx-x

Dedicated to our Beloved Prime Minister Shree Narendra Modi ji, whose life gives the Inspiration to every Individual

*If a Man **Decides** to achieve something in his life nothing is Impossible*

CHAPTERS

1. WORK, ENERGY AND POWER

1. INTRODUCTION

This chapter explains the concepts of work and energy and how these quantities are related to each other. The law of conservation of energy is an important tool in physics, for the analysis of motion of a system of particles or bodies, and in understanding various phenomena in nature. When the nature of forces involved in a process are not exactly known, or when we want to avoid complicated calculations, then the law of conservation of energy proves to be an indispensable tool in solving many problems. The importance of energy cannot be explained in words. The progress of science and civilization is based on finding new ways to efficiently use the energy available in nature in various forms. Energy is required by a person to perform his/her daily activities, as well as to run our automobiles and machines. Depletion of natural energy resources is a major concern these days. The efficiency of energy utilization processes and quantity of energy sources harnessed by a country determines the pace of its economic development.

2. WORK

2.1 Work

In physics, a force is said to do work only when it acts on a body, and if there is a consequential displacement of the point of application in the direction of the force.

For example, say if a constant force F displaces a body through displacement s then the work done, W, is given by

$$W = Fs\cos\theta = \vec{F}.\vec{s}$$

where s is magnitude of displacement and θ is angle between force and displacement. The SI unit of work is Joule or Newton-metre.

Sign Convention of Work

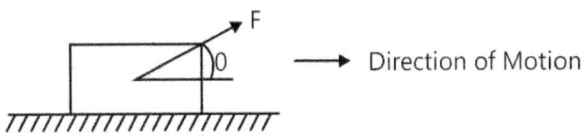

Figure 5.1: Motion of block in directon of applied force

We now define the sign convention of work as follows:

When $0 < \theta < 90^0$,

then $W = Fs\cos\theta$ is positive

i.e., when the force constantly supports the motion of a body, work done by that force is said to be positive.

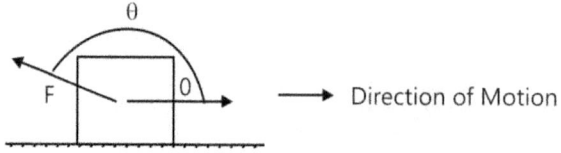

Figure 5.2: Motion of block

then $W = Fs \cos\theta = -\text{ve}$

i.e., in this case force is not truly supporting the motion of the body and hence the work done by that force is said to be negative.

2.2 Nature of Work

Work done is signified by the equation: $\vec{F}.\vec{S}$

Based on this equation, three possible situations are possible regarding the nature or sign of the work done as listed here under:

(a) To begin with, the work done is said to be positive if the angle between the force and the displacement vectors is an acute angle.

E.g., when a horse pulls a cart on a level road, the work done by the horse is positive.

(b) Second, the work done is zero if the force and the displacement vectors are perpendicular to each other.

E.g., when a body is moved along a circular path by a string, then the work done due to the string is zero.

(c) The last possible situation is that the work done is said to be negative if the angle between the force and the displacement vectors is an obtuse angle.

E.g., when a body slides over a rough surface, the resultant work done due to the frictional force is negative. (It is pertinent here to remember the fact that the angle between the force and the displacement is 180 degrees.)

> **NOMORECLASS CONCEPTS**
>
> Students should be able to deduce that by positive work, force is actually doing what it is meant for, i.e. force wants to move a body in certain direction and if it moves in that direction then it's positive work.

Illustration 1: Assume that a body is displaced from $\vec{r}_A = (2m, 4m, -6m)$ to $\vec{r}_B = (6i - 4j + 2k)m$ under a constant force $\vec{F} = (2i + 3j - k)N$. Now, calculate the total work done. **(JEE MAIN)**

Sol: The work done by the constant force $\vec{F}$ during displacement $\vec{S}$ of a particle is scalar product of force and displacement and is given by $W = \vec{F} \cdot \vec{S}$

$$\vec{r}_A = \left(2\hat{i} + 4\hat{j} - 6\hat{k}\right)m\vec{S} = \vec{r}_B - \vec{r}_A = \left(6\hat{i} - 4\hat{j} + 2\hat{k}\right) - \left(2\hat{i} + 4\hat{j} - 6\hat{k}\right) = 4\hat{i} - 8\hat{j} + 8\hat{k}$$

$$W = \vec{F}.\vec{S} = \left(2\hat{i} + 3\hat{j} - \hat{k}\right).\left(4\hat{i} - 8\hat{j} + 8\hat{k}\right) = 8 - 24 - 8 = (-24\hat{j})$$

Illustration 2: A block of total mass 5 kg is being raised vertically upwards with the help of a string attached to it and it rises with an acceleration of 2 m/s². Find the work done due to the tension in the string if the block rises by 2.5 m. Also, calculate the work done due to the gravity and the net work done. **(JEE ADVANCED)**

1.2

Sol: The tension in the string is acting vertically upwards and the block is also moving vertically upwards, so the work done by the tension will be positive. The force of gravity is acting vertically downwards so the work done by gravity will be negative.

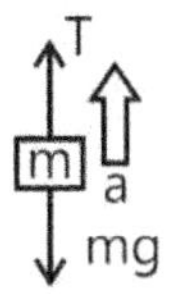

Let us first calculate the tension T.

From the force diagram T-mg = 5a; T = 5(9.8 + 2) = 59 N.

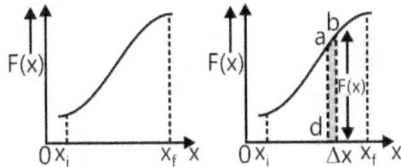

Figure 5.3

As it is clear that both T and displacement S are in the same direction (upwards), then work done by the tension T is W based on which we calculate that $W = Ts = 59(2.5) = 147.5\,J$.

Now, work done due to gravity = $-mgs = -5(9.8)(2.5) = -122.5$ J

Therefore, net work done on the block = work done by T + work done by mg = 147.5 + (–122.5) = 25 J.

> **NOMORECLASS CONCEPTS**
>
> Point of application of force also plays a major role.
>
> Zero work is done by a force in following cases: -If the point of application of force is not changed in space but the body moves. If body doesn't move but the point of application of force moves.

3. WORK DONE BY A VARIABLE FORCE

We need to be aware of the fact that when the force is an arbitrary function of position, then we need the principles of calculus to evaluate the work done by it. The Fig. 5.4 given here under shows F (x) as some function x. We now begin our evaluation in this regard by replacing the actual variation of the force by a series of small steps. In the Fig. 5.4 provided, the area under each segment of the curve is approximately equal to the area of a rectangle. Based on the height of the rectangle, the amount of work done is given by the relation, $\Delta W_n = F_n \Delta X_n$. Therefore, the total work done is approximately given by the summation of the areas of both the recangles: $W \approx \sum F_n \Delta X_n$. As the number of the steps is reduced, the tops portions of the rectangle more closely resemble the actual curve shown in the Fig.5.4. In limit $\Delta x \rightarrow 0$, which is equivalent to letting the number of steps to be infinite, the discrete sum is replaced by a continuous integral. $W = \int_{x_1}^{x_2} F(x)\, dx =$ area under the F – x curve and the x – axis

Figure 5.4: Work done on particle by variable force

Illustration 3: A force $F = (10 + 0.50X)$ is observed to act on a particle in the x direction, where F is in newton and x in meter. Find the actual work done by this force during a displacement from x=0 to x=2.0 m.　　　**(JEE MAIN)**

Sol: If a particle is being displaced under action of variable force, the work done by this force is calculated as $W = \int_{s_1}^{s_2} \vec{F} \cdot d\vec{s}$.

As we know that the force is a variable quantity, we shall find the work done in a small displacement from x to x +

dx and then integrate the resultant value to calcuate the total work done. The work done in this small displacement is calculated as

$$dW = \vec{F}.\vec{dx} = (10 + 0.50x)dx. \text{ Thus, } W = \int_0^{2.0} (10 + 0.50x)dx = \left[10x + 0.50\frac{x^2}{2}\right]_0^{2.0} = 21 \text{ J.}$$

4. CONSERVATIVE AND NON-CONSERVATIVE FORCES

A force is said to be of the conservative category if the work done by it in moving a particle from one point to another does not depend upon the path taken but depends only upon the initial and final positions. The work done by a conservative force around a closed path calculated to be zero. Gravitational force, electric force, spring force, etc. are some of the examples of this category. Basically, all central forces are conservative forces. In contrast, if the work done by a force in moving a body from one point to another depends upon the path followed, then the force is said to be of the nonconservative category. The work done by such a force around a closed path cannot be zero. For example, both the frictional and viscous forces work in an irreversible manner and hence a definite part of energy is lost in overcoming these frictional forces. (Mechanical energy is converted to other energy forms such as heat, sound, etc.). Therefore, these forces are of the nonconservative category.

5. WORK DONE AGAINST FRICTION

We know that the frictional force always acts opposite to the direction of motion (and hence direction of the displacement); therefore, the work done by the frictional force is always on the negative side. Further, the work done by the frictional force is invariably lost in the form of heat and sound energy and thus it is a nonconservative force.

NOMORECLASS CONCEPTS

The work done by the frictional force is either negative or zero, but never positive. The frictional force always resists the attempted work done along a horizontal surface. Work done along a horizontal surface is given by: $-\mu mgl$, where

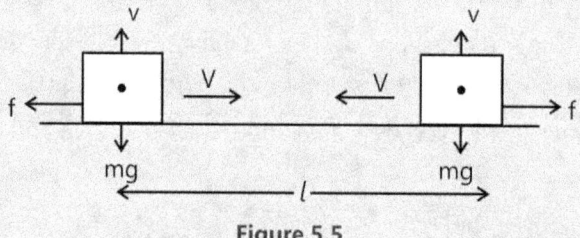

Figure 5.5

m is the mass of the object ;

μ is the coefficient of friction

g is the acceleration due to gravity (9.8m/s²)

l is the distance traveled by the block along the rough surface

Similarly, work done along an inclined surface with an angle θ from horizontal is given by $-\mu mgl\cos\theta$

Illustration 4: It is observed that a block of mass 4 kg slides down a plane inclined at 37° with the horizontal. The length of the plane is calculated to be of 3 m. The value of the coefficient of sliding friction between the block and the plane is 0.2. Based on the above, find the work done due to the gravity, the frictional force, and the normal reaction between the block and the plane.

(JEE MAIN)

Sol: Normal reaction is always perpendicular to the inclined plane hence it is perpendicular to the displacement and thus the work done by it is zero. Whereas the frictional force is in opposite direction to the displacement and hence the work done by the firctional force is negative. The work done by the component of gravitational force along the inclined plane will be positive.

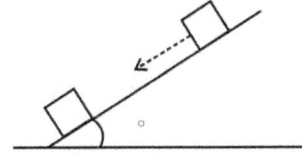

Figure 5.6

Total force acting on the block moving on inclined plane constitutes frictional force, normal reaction due to ground and gravitational force acting on wire. The work done on block is given as $W = Fs\cos\theta$

As the normal reaction is perpendicular to the point of displacement, work done by the normal reaction $R = R\,s\cos 90° = 0$. The magnitude of displacement $s = 3$ m and the angle between force of gravity (mg) and displacement is equal to $(90°-37°)$.

Therefore, work done by gravity $= mgs\cos(90°-37°)$

$= mgs\sin 37^0 = 4 \times 9.8 \times 3 \times 3/5 = 70.56\,J$

Work done by friction $= -(\mu R)s = -(\mu\,mg\cos 37°)s = -0.2 \times 4 \times 9.8 \times 4/5 \times 3 = -18.816\,J$.

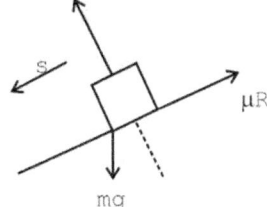

Figure 5.7

6. POWER

Power is defined as the rate at which the actual work is done. If an amount of work ΔW is done in time Δt, then

average power, $P_n = \dfrac{\Delta W}{\Delta t}$ and instantaneous power, $P = \underset{\Delta t \to 0}{\text{Lim}}\left(\dfrac{\Delta W}{\Delta t}\right) = \dfrac{dW}{dt}$.

It is a well-known fact that work done by a force F on an object that has infinitesimally small displacement ds is

$dw = F.ds$. Then, instantaneous power, $P = \dfrac{dW}{dt} = \dfrac{\vec{F}d\vec{s}}{dt} = \vec{F}.\vec{v}$.

The S I unit of power is Watt (W) or Joule/second (J/s) and it is a scalar quantity. Dimensions of power is $M^1L^2T^{-3}$.

Illustration 5: A block of mass m is allowed to slide down a fixed smooth inclined plane of angle θ and length ℓ. Calculate the magnitude of power developed by the gravitational force when the block reaches the bottom.

(JEE ADVANCED)

Sol: The power dlivered by the force $\vec{F}$ is the scalar product of the force and velocity i.e. $P = \vec{F}.\vec{v}$

When body reaches bottom of the inclined plane the velocity of of body is $v = \sqrt{2gh} = \sqrt{2g\cdot \ell\sin\theta}$ and the angle between velocity and vertical will be $(90-\theta)°$. $P = \vec{F}.\vec{V} = mg\sin\theta\sqrt{2g\ell\sin\theta} = \sqrt{2m^2g^3\ell\sin^3\theta}$.

Illustration 6: A particle of mass m is moving in a circular path of constant radius r such that its centripetal accelecration a_c is varying with time t as $a_c = k^2rt^2$, where k is a constant. The power delivered to the particle by the force acting on it is

(JEE MAIN)

(A) $2\pi mk^2r^2$ (B) mk^2r^2t (C) $\dfrac{\left(mk^4r^2t^5\right)}{3}$ (D) Zero

Sol: (B) As the centripetal force is perpendicular to the direction of the velocity, the work done and power delivered by the centripetal force will be zero, whereas the tangential force is in the direction of the velocity so the power delivered to the particle of mass m is $P = F_t \cdot v$

Here $a_c = k^2rt^2$ or $\dfrac{v^2}{r} = k^2rt^2$ or $v = krt$

Therefore, tangential acceleration, $a_t = \dfrac{dv}{dt} = kr$ or tangential force, $F_t = m\,a_t = m\,kr$

However, only tangential force does work. Power $= F_t v = (mkr)(krt)$ or Power $= mk^2r^2t$

1.5

7. ENERGY

Generally, the energy of a body is signified by the body's capacity to do work. It is a scalar quantity and shares the same unit as that of work (Joule in SI unit). In mechanics, both kinetic and potential energies are involved with dynamics of the body.

7.1 Potential Energy

7.1.1. Potential Energy

Potential energy of a body is the energy possessed by virtue of its position or due to its state. It is independent of the way in which the body is transformed to this state. Although it is a relative parameter, it depends upon its value at reference level. We can define the change in potential energy as the negative of work done by the conservative force in operation in carrying a body from a reference position to the position under consideration.

7.1.2 Definition

$\Delta U = -W_{AB}$ where A is the initial state, B is the final state, and W_{AB} is the total work done by conservative forces. We know that potential energy depends upon the work done by conservative force only. Hence, it cannot be defined for the nonconservative force (s). This is because of the proven fact that in this type work done depends upon the path followed alone.

7.1.3 Gravitational Potential Energy (GPE)

Suppose if we lift a block through some height (h) from A to B, then the work is done defying the gravity. The work done in such a case is stored normally in the form of gravitational potential energy of the block-energy system. Therefore, we can write that work done in raising the block = (mg)h. This is exactly equal to the increase in gravitational potential energy (GPE) of the block.

If the center of a body of mass m is raised by a height h, then increase in GPE = mgh

If the center of a body of mass m is lowered by a distance h, decrease in GPE = mgh

7.1.4 Elastic Potential Energy

Suppose when a spring is elongated (or compressed), then work is done against the restoring force of the spring. This resultant work done is stored in the spring in the form of elastic potential energy.

7.1.5 Nature of Restoring Force

Suppose if a spring is extended or compressed by a distance x, the spring then exerts a restoring force so as to oppose this change.

NOMORECLASS CONCEPTS

GPE is always thought of as only of block. But to be more specific it is the energy of block-earth system. Potential energy never comes in context of a single particle. It is always for a configuration. In the case of GPE, writers however generally skip writing "Earth" each time.

7.1.6 Spring

In case of a spring, natural length of the spring is assumed to be the reference point and correspondingly is always assigned zero potential energy (This is a universal assumption.). However, in gravity, we can choose any point as our reference and hence assign it any value of potential energy.

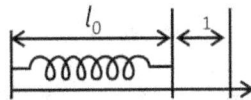

Figure 5.8: Energy stored in stretched spring

For Stretching

$$U_f - U_i = -\int_i^f \vec{F}.d\vec{S} \; ; U_f - 0 = -\int_0^{x_i} kx(-i)(dx)i \; ; U = \frac{1}{2}kx_1^2$$

For Compression

$$U_f - U_i = -\int_i^f \vec{F}.d\vec{S} = -\int_0^{x_i} kxi(dx)(-i) = U = \frac{1}{2}kx^2$$

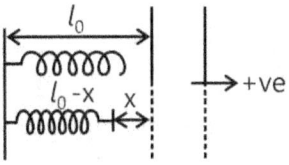

Figure 5.9: Energy stored in compressed spring

Thus, if the spring is either stretched or compressed from natural length by x the corresponding potential energy is $1/2kx^2$

7.1.7 Relationship between Force and Potential Energy

Now, let us discuss the relationship between force and potential energy.

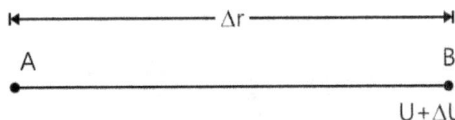

Figure 5.10

Let us assume that a body is taken from A to B in such away that there is no net change in its kinetic engery. Then

$\Rightarrow$ Work done = $-$change in P.E. ; $\quad F\Delta r = U - (U + \Delta U) = -\Delta U$

$\Rightarrow \quad F_{avg} = -\left(\dfrac{\Delta U}{\Delta r}\right)$ if $\Delta r \to 0;$ $\quad F = -\lim_{\Delta r \to 0} \dfrac{\Delta u}{\Delta r} = -\dfrac{\partial U}{\partial r}$

7.2 Kinetic Energy

Kinetic energy (KE) is the energy of a body possessed by virtue of its motion alone. Therefore, a body of mass m and moving with a velocity v has a kinetic energy $E_k = \frac{1}{2}mv^2$.

We already know that velocity is a relative parameter; therefore, KE is also a relative parameter.

We provide a detailed account on kinetic energy after presenting the concept of conservation of mechanical energy.

8. EQUILIBRIUM

We have already studied in the chapter on "Laws of Motion" that a body is said to be in translatory equilibrium only if net force acting on the body is zero, i.e., $\vec{F}_{net} = 0$

However, if the forces are conservative, then $F = -\dfrac{dU}{dr}$; for equilibrium, then

$F = 0$; Thus, $-\dfrac{dU}{dr}=0,$ or $\dfrac{dU}{dr}=0$

i.e., exactly at the equilibrium position the slope of U-r graph is zero or the potential energy is optimum (maximum or minimum or constant). Equilibria are of three types, i.e., stable equilibrium, unstable equilibrium, and neutral equilibrium. Further, the situations where $F = 0$ and $dU/dr = 0$ can be obtained only under three conditions as specified hereunder.

(a) If $\dfrac{d^2U}{dr^2} > 0$, then it is stable equilibrium;

(b) If $\dfrac{d^2U}{dr^2} < 0$, then it is unstable equilibrium; and

(c) If $\dfrac{d^2U}{dr^2} = 0$, then it is neutral equilibrium.

NOMORECLASS CONCEPTS

A system always wants to minimize its energy. The above equilibriums are categorized only on this basis. Stable indicates that if system is disturbed slightly, from these configuration, it would try to come back to its original state (position of energy minima). For unstable equilibrium, a slight disturbance would cause the system to find some other suitable configuration (position of energy maxima). A neutral equilibrium is generally found when U becomes constant and each position is a state of equilibrium. A slight disturbance has no after reactions and the new state is also an equilibrium position.

Illustration 7: The potential energy of a particle of mass 5 kg, moving in xy plane, is given by $U = (-7x + 24y)J$

where x and y being in meters. Initially (at t=0), the particle is at the origin and has velocity $\vec{v} = \left(14.4\,\hat{i} + 4.2\,\hat{j}\right) m/s$.

Then Calculate (a) the acceleration of the particle and (b) the direction of acceleration of the particle. (c) The speed of the particle at t = 4 s. **(JEE MAIN)**

Sol: If particle has potential energy U then corresponding conservative force, is $F = -\dfrac{dU}{dr}$ and according to the

Newton's second law of motion $\vec{F} = m\vec{a}$. The direction of acceleration is calculated as $\tan\theta = \dfrac{a_y}{a_x}$.

(a) Acceleration,

$F_x = \dfrac{\delta U}{\delta x}, F_y = -\dfrac{\delta U}{\delta y}$ $\Rightarrow F_x = 7N,$ $\quad F_y = -24N;$ $\quad\Rightarrow a_x = 7/5,\ a_y = -24/5$

(b) Direction of acceleration $\theta = \tan^{-1}\left(\dfrac{a_y}{a_x}\right)$;

(c) $\vec{v} = \vec{u} + \vec{a}\,t$; $v_x = 14.4 + \dfrac{7}{5}\times 4 = 20$; $v_y = 4.2 - \dfrac{24}{5}\times 4 = (-15)$

Illustration 8: The potential energy of a particle in a certain field has the form $U = a/r^2 - b/r$, where a and b are positive constants and r is the distance from the center of the field. Find the value of r_0 corresponding to equilibrium position of the particles and hence examine whether this position is stable. **(JEE ADVANCED)**

Sol: Conservative force acting on the particle is $F = -\dfrac{dU}{dr}$. Under stable equilibrium particle has minimum potential energy while potential energy is maximum in case of unstable equilibrium.

$U(r) = a/r^2 - b/r$

$\text{Force} = F = -\dfrac{dU}{dr} = -\left(\dfrac{-2a}{r^3} + \dfrac{b}{r^2}\right); \qquad F = -\dfrac{(br - 2a)}{r^3}$

At equilibrium, then $F = \dfrac{dU}{dr} = 0$

Hence, $br - 2a = 0$ at equilibrium.

Further, $r = r_0 = 2a/b$ corresponds to equilibrium.

At stable equilibrium, the potential energy of a particle is at its minimum, whereas at unstable equilibrium, it is the maximum. From the principles of calculus, we know that for minimum value around a point $r = r_0$, the first derivative should be zero and the second derivative should be invariably positive.

For minimum potential energy, the applicable conditions are

$\dfrac{dU}{dr} = 0 \quad \text{and} \quad \dfrac{d^2U}{dr^2} > 0 \quad \text{at} \quad r = r_0$

However, we have already used $dU/dr = 0$ to obtain $r = r_0 = 2a/b$.

Now, in a similar way let us investigate the second derivative.

$\dfrac{d^2U}{dr^2} = \dfrac{d}{dr}\left(\dfrac{dU}{dr}\right) = \dfrac{d}{dr}\left(-\dfrac{2a}{r^3} + \dfrac{b}{r^2}\right) = \dfrac{6a}{r^4} - \dfrac{2b}{r^3}$

At $r = r_0 = 2a/b$, $\dfrac{d^2U}{dr^2} = \dfrac{6a - 2br_0}{r_0^4} = \dfrac{2a}{r_0^4} > 0$.

Based on our calculations, the potential energy function U(r) has a minimum value only when $r_0 = 2a/b$. Therefore, we conclude that the system has stable equilibrium only at the minimum potential energy state.

9. WORK ENERGY THEOREM

Suppose that a particle is acted upon by various forces and consequently undergoes a displacement. Then there is a change in its kinetic energy by an amount equal to the total (net) work (W_{net}) done on the particle by all the forces.

i.e., $\quad W_{net} = K_f - K_i = \Delta K$ ... (i)

We call the above expression as the work-energy theorem.

Expression (i) is valid irrespective of the fact that whether the forces are constant or varying and whether the path followed by the particle is straight or curved.

We further elaborate expression (i) as follows:

$W_c + W_{NC} + W_{Oth} = \Delta K$ ... (ii)

where W_c is the work done by conservative forces

W_{sc} is the work done by nonconservative forces

W_{oth} is the work done by all other forces which are not included in the category of conservative, nonconservative, and pseudo forces.

'Since $W_c = \Delta U$' (based on definition of potential energy), therefore, expression (ii) can be accordingly modified as

$W_{NC} + W_{oth} = \Delta K + \Delta U = \Delta(K + U) = \Delta E$ ... (iii)

In expression (iii), the term K + U = E is known as the mechanical energy of the system.

Illustration 9: Find how much will mass "m" rise if 4 m falls away. Block are at rest and in equilibrium **(JEE MAIN)**

Sol: Initially the block is at rest. When the block rises to the maximum height, it again comes to rest momentarily. So, by work energy theorem the total work done on the block by force of gravity and spring force is zero.

Applying work energy theorem (WET) on a block of mass m

$$W_g + W_{sp} = K.E._f - K.E._i$$

Let the final displacement of the block from the initial equilibrium is x. Then

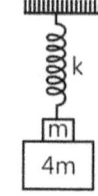

$$-mg\left(\frac{5mg}{k} + X\right) + \frac{1}{2}k\left(\frac{25m^2g^2}{k^2}\right) - \frac{1}{2}kx^2 = 0; \frac{1}{2}kx^2 + mgx - \frac{15m^2g^2}{2k} = 0 ; \qquad x = \frac{3mg}{k}$$

Figure 5.11

NOMORECLASS CONCEPTS

Whenever there is frictional force, energy is dissipated which is equal to work done by frictional force and the dissipated energy converts into heat. Practically, machine handlers do a lot of things to minimize friction and reduce energy losses by applying lubricants and rollers in their parts.

Illustration 10: A body of mass m was slowly hauled up the hill as shown in the Fig. 5.12 provided by a force F which at each point was directed along a tangent to the trajectory. Find the work done due to this force if the height of the hill is h, the length of its base is l, and the coefficient of friction is μ. **(JEE ADVANCED)**

Sol: As block hauls slowly, the kinetic energy will not change throughout the motion. And the sum of the work done by applied force, gravitational force, normal reaction and frictional force will be zero as per work energy theorem.

The four forces that are acting on the body are listed hereunder.

(a) Weight (mg),

(b) Normal reaction (N),

(c) Friction (f), and

(d) The applied force (F)

According to the principle of work-energy theorem

$$W_{net} = \Delta KE \text{ or } W_{mg} + W_N + W_f + W_F = 0 \qquad \text{... (i)}$$

Here, $\Delta KE = 0$, because $K_i = 0 = K_f$ ∴ $W_{mg} = -mgh$; $W_N = 0$

(This is because the normal reaction is perpendicular to displacement at all the points.)

Figure 5.12

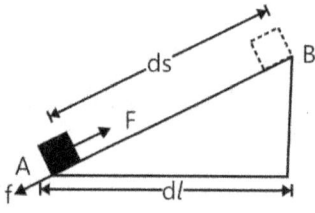

W_f can be calculated as $f = \mu mg \cos \theta$

∴ $(dW_{AB})_f = -fds = -(\mu mg \cos \theta)ds = -\mu mg(dl)$ (as ds cos θ = dl)

∴ $f = -\mu mg \sum dl = -\mu mgl$

Figure 5.13

Substituting these values in Eq. (i), we obtain the expression $W_F = mgh + \mu mgl$.

Note: Here again, if we desire to solve this problem without using the concept of work-energy theorem, then we will first evaluate magnitude of applied force $\vec{F}$ at different locations following which we will then integrate $\left(= \vec{F}.\vec{dr}\right)$ with proper limits.

10. KINETIC ENERGY

Now, let us attempt to develop a relationship between the work done and the change in speed of a particle. Based on the Fig. 5.14 provided, we observe that the particle moves from point P_1 to P_2 under the action of a net force $\vec{F}$

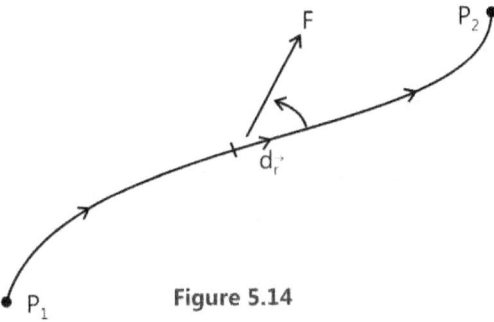

Figure 5.14

$$W = \int_{P_1}^{P_2} \vec{F}.\vec{dr} \; ; \; \vec{F} = F_x \hat{i} + F_y \hat{j} + F_z \hat{k} \; ; \; \vec{dr} = dx\,\hat{i} + d_y\,\hat{j} + dz\,\hat{k}$$

$$W = \int_{P_2}^{P_2} \left(F_x dx + F_y d_y + F_z d_z \right)$$

It is very clear for us now that a particle moves along a curved path from point P_1 to P_2, only when acted upon by a force F that varies in both magnitude and direction. $F_x = ma_x = \dfrac{m\,dv_x}{dt} \; ; \; \int_{P_1}^{P_2} F_x dx = \int_{P_1}^{P_2} m \dfrac{dV_x}{dt} dx$

Treating now v_x as a function of position, we obtain:

$$\frac{dv_x}{dt} = \frac{dv_x}{dx}\left(\frac{dx}{dt}\right) = \frac{dv_x}{dx}.v_x = v_x \frac{dv_x}{dx} \; ; \; \therefore \int_{P_1}^{P_2} F_x d_x = \int_{P_1}^{P_2} m \frac{dv_x}{dt} dx = \int_{P_1}^{P_2} m v_x \frac{dv_x}{dx} dx = \int_{P_1}^{P_2} m v_x dv_x = \frac{1}{2}mv_x^2 \Big|_{v_{x1}}^{v_{x2}} = \frac{1}{2}m\left(v_{x2}^2 - v_{x1}^2\right)$$

v_{x1} = velocity in x-direction at P_1; v_{x2} = velocity in x-direction at P_2.

We now apply the same principle for terms in y and z.

$$W = \frac{1}{2}M\left[v_{x2}^2 + v_{y2}^2 + v_{z2}^2 - \left(v_{x1}^2 + v_{y1}^2 + v_{z1}^2\right) \right] = \frac{1}{2}M\left(v_2^2 - v_1^2\right); \quad W = \frac{1}{2}mv_2^2 - \frac{1}{2}mv_1^2$$

Define: $K = \dfrac{1}{2}mv^2$ = Kinetic energy of particle

KE: Potential of a particle to do work by virtue of its velocity.

We know that the work done on the particle by the net force equals the change in KE of the particle.

$W = K_2 - K_1$ or $\Rightarrow W = \Delta K$ Work–Energy Theorem.

For a particle $\vec{P} = M\vec{v}$ (linear momentum); $\therefore K = \dfrac{1}{2m}P^2$

Regarding KE, the following two points are very significant.

(a) Since, both m and v^2 are always positive, KE is always positive and hence does not depend on the directional parameter of motion of the body.

(b) KE depends on the frame of reference. For example, the KE of a person of mass m in a train moving with speed v is zero in the frame of train, whereas in the frame of earth the KE is $\dfrac{1}{2}mv^2$ for the same person.

NOMORECLASS CONCEPTS

Energy can never be negative.

No! Only kinetic energy can't be negative. If anyone generally speaks about energy, it means the sum of potential and kinetic energies. However, we can always choose such a reference in which this sum is negative. Hence, total energy can be negative.

Illustration 11: A uniform chain of length ℓ and mass m overhangs a smooth table with its two-third parts lying on the table. Find the kinetic energy of the chain as it completely slips off the table.　　　　　　　　　**(JEE MAIN)**

Sol: The initial kinetic energy of the chain is zero. When chain start slipping off table the loss in its potential energy is equal to the gain in its kinetic energy.

Let us take the potential energy at the table as zero. Now, consider a part dx of the chain at a depth x below the surface of the table. The mass of this part is $dm = \dfrac{m}{\ell}dx$ and hence its potential energy is $-(m/\ell\ dx)gx$.

The potential energy of the one-third of the chain that overhangs is given by $U_1 = \displaystyle\int_0^{\ell/3} -\dfrac{m}{\ell}gx\ dx$

$$= -\left[\dfrac{m}{\ell}g\left(\dfrac{x^2}{2}\right)\right]_0^{\ell/3} = -\dfrac{1}{18}mg\ell$$

However, this is also the potential energy of the full chain in the initial position; this is because the part lying on the table has zero potential energy. Now, we can calculate the potential energy of the chain when it completely slips off the table as

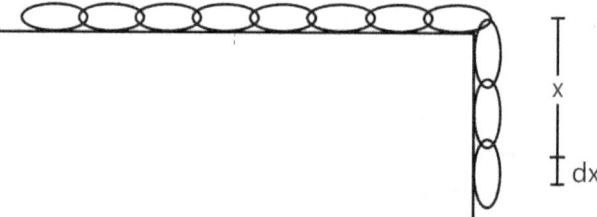

Figure 5.15

$U_2 = \displaystyle\int_0^{\ell} -\dfrac{m}{\ell}gx\ dx = -\dfrac{1}{2}mg\ell$ The loss in potential energy is $= \left(-\dfrac{1}{18}mg\ell\right) - \left(-\dfrac{1}{2}mg\ell\right) = \dfrac{4}{9}mg\ell$.

Basically, this should be equal to the gain in the KE in this case. However, the initial KE is zero. Hence, the KE of the chain as it completely slips off the table is $\dfrac{4}{9}mg\ell$.

Illustration 12: A block of mass m is pushed against a spring of spring constant k fixed at one end to a wall. The block can slide on a frictionless table as shown in the Fig. 5.16. The natural length of the spring is taken as L_0 and it is compressed to half its natural length when the block is released. Now, based on the above find the velocity of the block as a function of its distance x from the wall.　　　　　　　　**(JEE ADVANCED)**

Sol: The block will move under action of restoring force of spring when spring is released. The block will have constant kinetic energy when it looses contact with the spring. In this process the energy of system will be conserved as there are no external forces acting on the system. (Spring + block system)

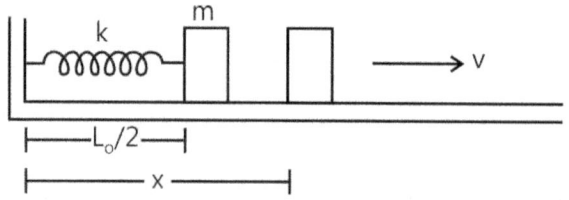

Figure 5.16

When the block is released, naturally the spring pushes it toward right. The velocity of the block keep on inreasing till the block loses contact with the spring and thereafter moves with constant velocity.

Initially, the compression of the spring is $L_0/2$. But when the distance of the block from the wall becomes x, where $x < L_0$, the compression is $(L_0 - x)$. Applying the principle of conservation of energy

$\frac{1}{2}k\left(\frac{L_0}{2}\right)^2 = \frac{1}{2}k(L_0 - x)^2 + \frac{1}{2}mv^2$. Solving this, $v = \sqrt{\frac{k}{m}\left[\frac{L_0^2}{4} - (L_0 - x)^2\right]^{1/2}}$

Thus, when the spring acquires its natural length, then $x = L_0$ and $v = \sqrt{\frac{k}{m}}\frac{L_0}{2}$. Thereafter, the velocity of the block remains constant.

11. MOTION IN A VERTICAL CIRCLE

Let us consider a particle of mass m attached to one end of a string and rotated in a vertical circle of radius r with centre O. The speed of the particle will decrease as the particle travels from the lowest point to the highest point but increases in the reverse direction due to acceleration due to gravity.

Thus, if the particle is moving with velocity v at any instant at A, (where the string is subtending an angle θ with the vertical), then the forces acting on the particle are tension T in the string directed toward AO and weight mg acting downward.

Further, the net force $T - mg\cos\theta$ is directed toward the cenetr and hence provides the centripetal force

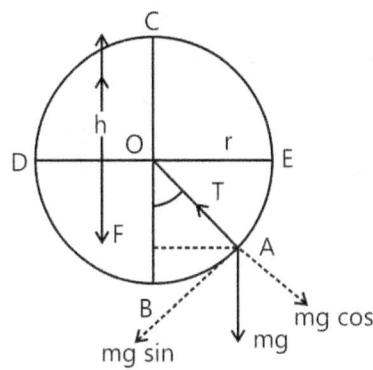

Figure 5.17: Motion in vertical circle

$T - mg\cos\theta = \frac{mv^2}{r}; T = m\left(g\cos\theta + \frac{v^2}{r}\right)$

If v_0 is the speed of the particle at the highest point, then the velocity increases as the particle falls through any height h. However, if it falls from C to A, then the vertical distance h is given by

$h = CF = CO + OF = CO + OA\cos\theta = r + r\cos\theta; h = r(1 + \cos\theta)$

$v^2 = v_0^2 + 2gh = v_0^2 + 2gr(1 + \cos\theta)$ (Because there is no actual work done due to the influence of tension)

(i) At the highest point C, $\theta = 180^0$

Tension at $C = T_c = m\left[\frac{v_0^2}{r} + g\cos(180)\right] = m\left[\frac{v_0^2}{r} - g\right]$ ... (i)

The particle will now fall because the string will slacken if T_c is negative. Therefore, the minimum velocity at the highest point is corresponding to the situation where T_c is just zero, i.e., when $m\left[\frac{v_0^2}{r} - g\right] = 0$, or $v_0 = \sqrt{rg}$

(ii) At the lowest point B, $\theta = 0$, tension T_B is given by $T_B = m\left[\frac{v_B^2}{r} + g\right]$

where v_B is velocity at B. $v_B^2 = v_0^2 + 4rg = rg + 4rg = 5rg$; $\left(\text{using } v^2 = u^2 + 2gh\right); v_B = \sqrt{5rg}$ ... (ii)

Minimum tension at B when the particle completes the circle is given by $T_B = m\left[\frac{5rg}{r} + g\right] = 6mg$

At the point E, when $\theta = 90^0$, $T_E = \frac{mv_E^2}{r}$

Where velocity at E is given by $V_E = V_c^2 + 2rg = rg + 2rg = 3rg$; $V_E = \sqrt{3rg}$

Tension at E corresponding to speed V_E is $T_E = m\left(\dfrac{3rg}{r}\right) = 3mg$

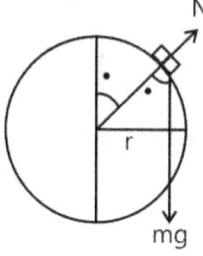

Figure 5.18

(iii) In another case the particle of mass m is not tied to the string but is moving along a circular track of radius r and has normal reaction N. However, it is moving with a velocity v and its radius vector is subtending an angle θ with the vertical, then $mg\cos\theta - N = \dfrac{mv^2}{r}$.

At the highest point, $mg - N = \dfrac{mv^2}{r}$; when ... (iii)

$N = 0$, $V = \sqrt{rg}$ Therefore, $V = \sqrt{rg}$ is the minimum speed with which the particle can move at the highest point without losing contact.

Condition of Looping the Loop ($u \geq \sqrt{5gR}$)

The particle will complete the circle only if the string does not slack even at the highest point ($\theta = \pi$). Thus, tension in the string should be obviously greater than or equal to zero ($T \geq 0$) at $\theta = \pi$. In the critical case, however, by substituting $T = 0$ and $\theta = \pi$ in Eq. (iii), we obtain

$mg = \dfrac{mv_{min}^2}{R}$ or $v_{min}^2 = gR$ or $V_{min} = \sqrt{gR}$ (at the highest point)

Further, by substituting $\theta = \pi$ in Eq. (i), $h = 2R$

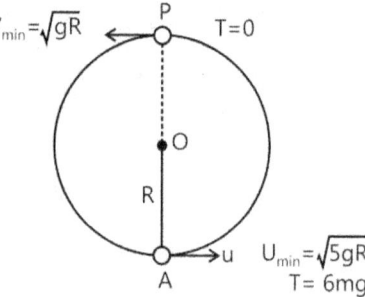

Figure 5.19

Therefore, from Eq. (ii) $u_{min}^2 = v_{min}^2 + 2gh$ or $u_{min}^2 = gR + 2g(2R)$ or $u_{min} = \sqrt{5gR}$

Thus, if $u \geq \sqrt{5gR}$, then the particle will complete the circle.

At $u = \sqrt{5gR}$, the velocity at the highest point is $v = \sqrt{gR}$ and the tension in the string is zero.

By substituting $\theta = 0°$ and $v = \sqrt{5gR}$ in Eq. (iii), we get $T = 6mg$ or in the critical condition tension in the string at the lowest position is 6mg as shown in the Fig. 5.19. If $u < \sqrt{5gR}$, then the following two cases are possible.

Condition of Leaving the Circle ($\sqrt{2gR} < u < \sqrt{5gR}$)

If $u < \sqrt{5gR}$, then the tension in the string will be zero before reaching the highest point. From Eq. (iii), tension in the string is zero ($T=0$) where, $\cos\theta = \dfrac{-v^2}{Rg}$ or $\cos\theta = \dfrac{2gh - u^2}{Rg}$

Now, by substituting, this value of $\cos\theta$ in Eq. (i), we obtain $\dfrac{2gh - u^2}{Rg} = 1 - \dfrac{h}{R}$ or $h = \dfrac{u^2 + Rg}{3g} = h_1$ (say) ... (iv)

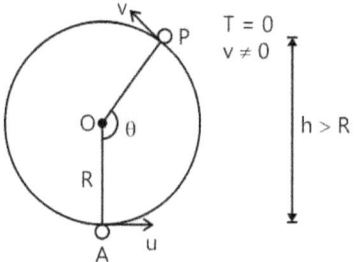

Figure 5.20

Or, in other words, we can say that at height h_1 tension in the string becomes zero. Further, if $u < \sqrt{5gR}$, then the velocity of the particle becomes zero when $0 = u^2 - 2gh$ or $h = \dfrac{u^2}{2g} = h_2$ (say)   ... (v)

i.e., at height h_2 velocity of the particle becomes zero. Now, the particle will move out from the circle if tension alone in the string becomes zero but not the velocity or T=0 but $v \ne 0$. This is possible only when $h_1 < h_2$ or

$\dfrac{u^2 + Rg}{3g} < \dfrac{u^2}{2g}$ or $2u^2 + 2Rg < 3u^2$ or $u^2 > 2Rg$ or $u > \sqrt{2Rg}$.

Therefore, if $\sqrt{2gR} < u < \sqrt{5gR}$, the particle moves out from the circle.

From Eq.(iv), we observe that $h > R$ if $u^2 > 2Rg$. Thus, the particle, will move out of the circle when $h > R$ or $90° < \theta < 180°$. This situation is shown in the Fig. 4.75.

$\sqrt{2gR} < u < \sqrt{5gR}$ or $90° < \theta < 180°$

Note, however, that after leaving the circle, the particle will follow a parabolic path.

Condition of Oscillation $(0 < u < \sqrt{2gR})$

The particle will oscillate, however, only if velocity of the particle becomes zero but not tension in the string. Or, in other words, $v = 0$, but $T \ne 0$. This is possible only when $h_2 < h_1$.

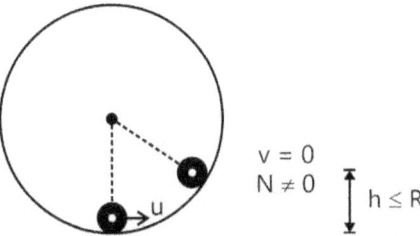

$v = 0$
$N \ne 0$ $h \le R$

Figure 5.21

Or $\dfrac{u^2}{2g} < \dfrac{u^2 + Rg}{3g}$ or $3u^2 < 2u^2 + 2Rg$ or $u^2 < 2Rg$ or $u < \sqrt{2Rg}$

Moreover, if $h_1 = h_2$, $u = \sqrt{2Rg}$ then both tension and velocity becomes zero simultaneously.

Further, from Eq (iv), we observe that $h \le R$ if $u \le \sqrt{2Rg}$. Thus, for $0 < u \le \sqrt{2gR}$, the particle oscillates in the lower half of the circle $(0° < \theta \le 90°)$. This situation is shown in the Fig. 5.21. $(0 < u < \sqrt{2gR})$ or $(0° < \theta \le 90°)$

Note: The above three conditions have been derived for a particle that is moving only in a vertical circle and attached to a string. The same conditions apply, however, if a particle moves inside a smooth spherical shell also of radius R. The only difference here is that the tension is replaced by the normal reaction N.

Condition of Looping the Loop is $(u \ge \sqrt{5gR})$	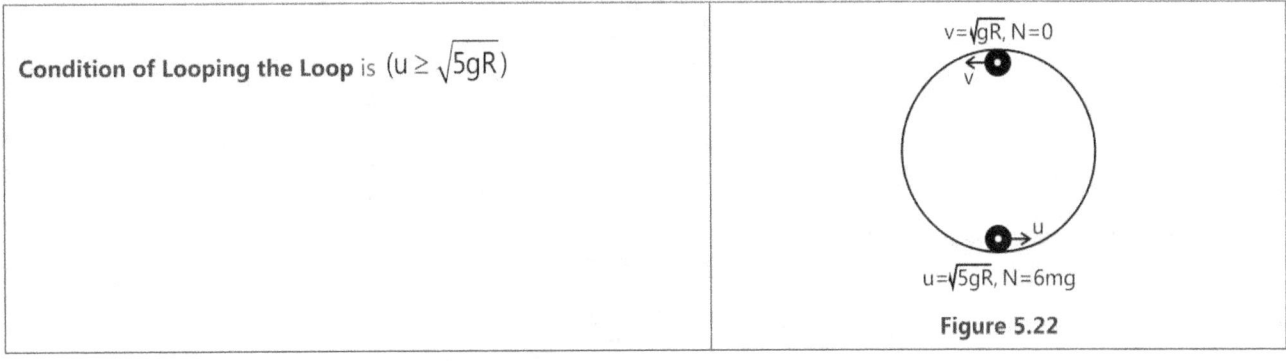 $v = \sqrt{gR}, N = 0$ v $u = \sqrt{5gR}, N = 6mg$ **Figure 5.22**

Condition of Leaving the Circle $(\sqrt{2gR} < u < \sqrt{5gR})$	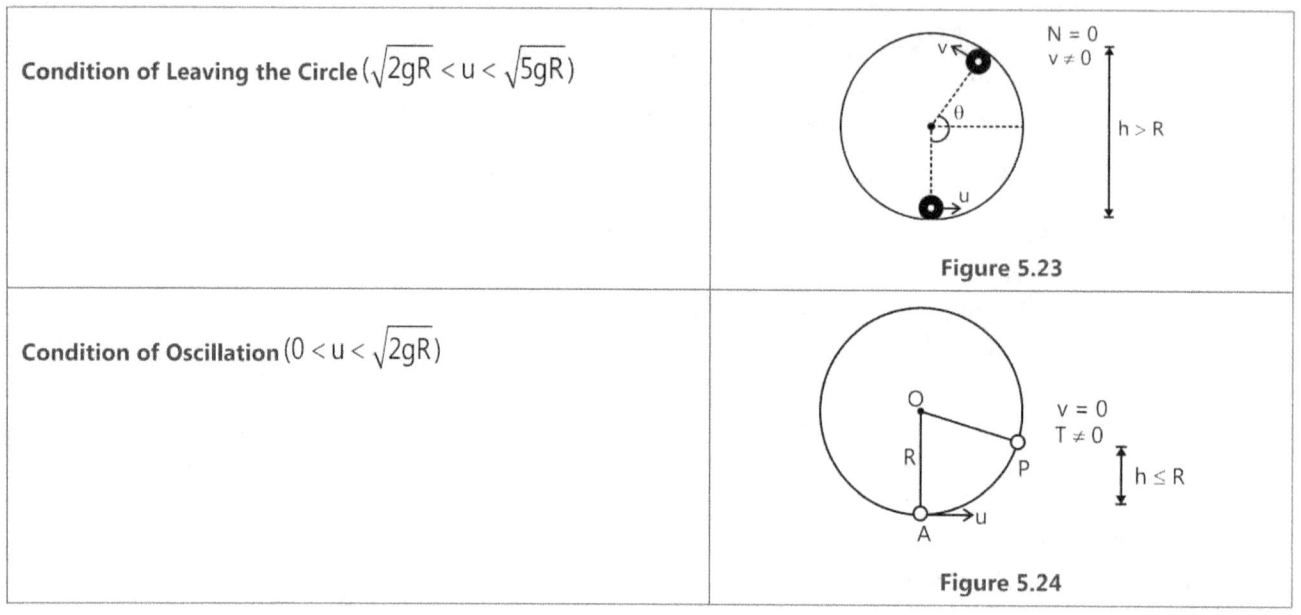 Figure 5.23
Condition of Oscillation $(0 < u < \sqrt{2gR})$	Figure 5.24

Illustration 31: A heavy particle hanging from a fixed point by a light inextensible string of length l is projected horizontally with speed $\sqrt{gl}$. Now, find the speed of the particle and the inclination of the string to the vertical at the instant of the motion when the tension in the string is equal to the weight of the particle. **(JEE ADVANCED)**

Sol: Loss in the kinetic energy of the particle is equal to the gain in the potential energy. Apply Newton's second law along the direction of the string.

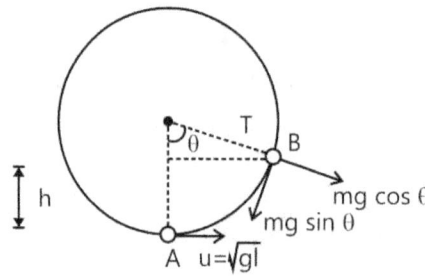

Figure 5.25

Let T = mg at angle θ as shown in the Fig. 5.25.

H = l (1–cos θ) ... (i)

Applying the principle of conservation of mechanical energy between points A and B, we obtain $\frac{1}{2}m(u^2 - v^2) = mgh$

Here, $u^2 = gl$ ... (ii)

and v = speed of particle in position B ∴ $v^2 = u^2 - 2gh$ ... (iii)

Further, $T - mg\cos\theta = \frac{mv^2}{l}$ or $mg - mg\cos\theta = \frac{mv^2}{l}$ (T = mg)

Or $v^2 = gl(1 - \cos\theta)$ ... (iv)

Now, by substituting the values of v^2, u^2 and h from Eqs. (iv), (ii) and (i) in Eq. (iii), we obtain

$gl(1 - \cos\theta) = gl - 2gl(1 - \cos\theta)$ or $\cos\theta = \frac{2}{3}$ or $\theta = \cos^{-1}\left(\frac{2}{3}\right)$

Further, by substituting $\cos\theta = \frac{2}{3}$ in Eq. (iv), we obtain $v = \sqrt{\frac{gl}{3}}$

If a particle of mass m is connected to a light rod and whirled in a vertical circle of radius R, then to complete the circle, the minimum velocity of the particle at the bottommost point is not $\sqrt{5gR}$. Because, in this case, velocity of the particle at the topmost point can be zero also. Using conservation of mechanical energy between points A and B as shown in Fig. 5.26(a) we get

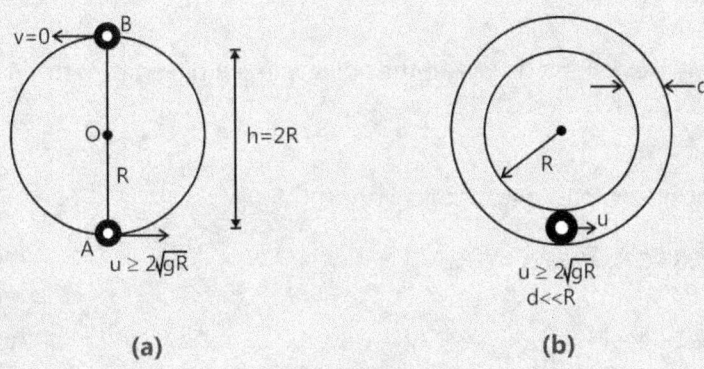

(a)   (b)

Figure 5.26

$$\frac{1}{2}m(u^2 - v^2) = mgh \quad \text{or} \quad \frac{1}{2}mu^2 = mg(2R) \ (\text{as } v = 0) \qquad \therefore \ u = 2\sqrt{gR}$$

Therefore, the minimum value of u in the case is $2\sqrt{gR}$.

Same is the case when a particle is compelled to move inside a smooth vertical tube as shown in Fig 5.26(b).

12. A BODY MOVING INSIDE A HOLLOW TUBE

Our discussion above holds good in this case too, but instead of tension in the string we have the normal reaction of the surface. If we take N is the normal reaction at the lowest point, then $N - mg = \frac{mv_1^2}{r}$; $N = m\left(\frac{v_1^2}{r} + g\right)$ However, at the highest point of the circle, $N + mg = \frac{mv_2^2}{r}$

$N = m\left(\frac{v_2^2}{r} - g\right)$; $N \geq 0 \Rightarrow$ Implies the condition $V_1 \geq \sqrt{5rg}$

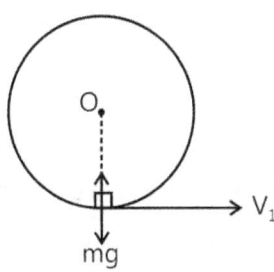

Figure 5.27: Block moving inside hollow sphere

In the same way as shown above, all the other equations similarly can be obtained by just replacing tension T by reaction N.

13. BODY MOVING ON A SPHERICAL SURFACE

Consider that the small body of mass m is placed on top of a smooth sphere whose radius is r.

Now, if the body slides down the surface, at what point does it fly off the surface?

Consider the point C where the mass is, at a certain instant. Now, the acting forces are the normal reaction R and the weight mg. Further, the radial component of the weight is mg cosθ acting toward the center. The centripetal force in this case is taken as $mg\cos\theta - R = \dfrac{mv^2}{r}$ where v is the velocity of the body at O.

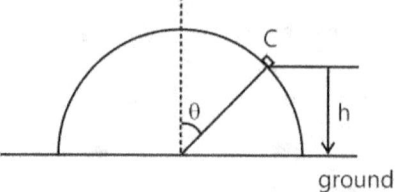

Figure 5.28: Motion of body on spherical surface

$$R = m\left(g\cos\theta - \frac{v^2}{r}\right) \qquad \text{... (i)}$$

Now, it is clear that the body flies off the surface at the point where R becomes zero.

i.e., $g\cos\phi - R = \dfrac{mv^2}{r}$ $\qquad$ ... (ii)

To find v, we apply the principle of conservation of energy

i.e., $\dfrac{1}{2}mv^2 = mg(BN) = mg(OB - ON) = mgr(1 - \cos\phi)$

$v^2 = 2rg(1 - \cos\phi); \qquad 2(1 - \cos\phi) = \dfrac{v^2}{rg}$ $\qquad$ ... (iii)

From equations (ii) and (iii), we obtain

$\cos\phi = 2 - 2\cos\phi; \qquad 3\cos\phi = 2$

$\cos\phi = \dfrac{2}{3}; \qquad \phi = \cos^{-1}\left(\dfrac{2}{3}\right)$ $\qquad$ ... (iv)

This exactly denotes the angle at which the body goes off the surface. The height from the ground of that point is

$= AN = r(1 + \cos\phi) = r\left(1 + \dfrac{2}{3}\right) = \dfrac{5}{3}r$

Illustration 32: A point mass m starts from rest and slides down the surface of a frinctionless solid sphere of radius R as shown in the Fig. 5.29 provided. At what angle will this body break off the surface of the sphere? Also, find the velocity with which it will break off. **(JEE MAIN)**

Sol: As the block slides down, the loss in potential energy is equal to gain in kinetic energy and at time of break off, the normal reaction from the sphere on block is zero.

Applying princliple of conservation of energy (COE), at the points A and B

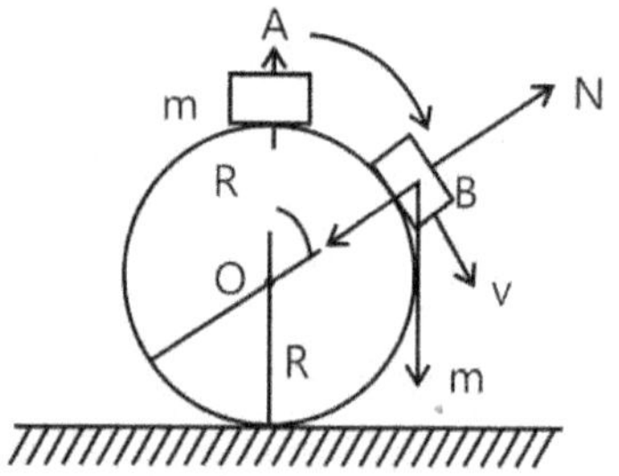

$mgR(1 - \cos\theta) = \dfrac{1}{2}mv^2$ $\qquad$ ... (i)

Force equation in this equation is $mg\cos\theta - N = mv^2/R$ $\qquad$ ... (ii)

N = 0 for break off.

Figure 5.29

$\therefore v = \sqrt{gR\cos\theta}$ $\qquad$ ... (iii)

Replacing this value in (i)

We get $\cos\theta = 2/3$ $\qquad$ Putting this in (iii) we get $v = \sqrt{\dfrac{2}{3}gR}$.

Illustration 33: A heavy particle is suspended by a string of length ℓ. The horizontal velocity of the particle is v_0. However, the string becomes slack at some angle and the particle proceeds on a parabolic path. Find the value of v_0 if the particle passes through the point of suspension. **(JEE ADVANCED)**

Sol: While particle moves in vertical circle, the tension in the string provides the necessary centripetal force. The loss in kinetic energy is equal to the gain in potential energy. At point the string become slack the tension in the string is zero.

Let us suppose the string becomes slack when the particle reaches the point P. We now assume that the string OP makes an angle θ with the upward vertical. Further, the only force acting on the particle at the point P is its weight mg. Further, the radial component of the force is mg cos θ. Now, as the particle moves along the circle upto P,

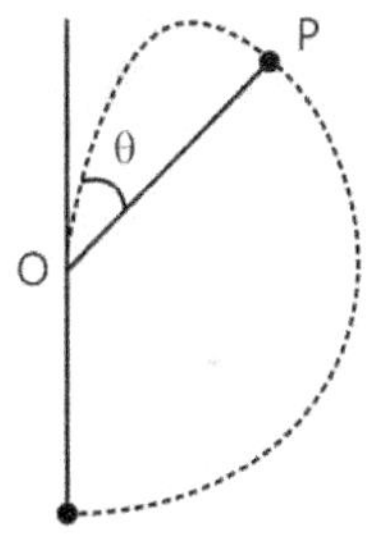

$$mg\cos\theta = m\left(\frac{v^2}{\ell}\right) \Rightarrow v^2 = g\ell\cos\theta \qquad \text{... (i)}$$

Figure 5.30

where v is its speed at the point P. Now, applying the principle of conservation of energy

$$\frac{1}{2}mv_0^2 = \frac{1}{2}mv^2 + mg\ell(1+\cos\theta) \text{ or } v^2 = v_0^2 - 2g\ell(1+\cos\theta) \qquad \text{... (ii)}$$

From (i) and (ii), $v_0^2 = 2g\ell(1+\cos\theta) = g\ell\cos\theta$ or $v_0^2 = g\ell(2+3\cos\theta)$ ... (iii)

From hereon, the particle follows a parabolic path due to acceleration due to gravity. Then as it passes through the point of suspension O, the equations for horizontal and vertical motion give

$$\ell\sin\theta = (v\cos\theta)t \quad \text{and} \quad -\ell\cos\theta = (v\sin\theta)t - \frac{1}{2}gt^2$$

$$\Rightarrow \quad -\cos\theta = (v\sin\theta)\left(\frac{\ell\sin\theta}{v\cos\theta}\right) - \frac{1}{2}g\left(\frac{\ell\sin\theta}{v\cos\theta}\right)^2$$

or, $\quad -\cos^2\theta = \sin^2\theta - \frac{1}{2}g\dfrac{\ell\sin^2\theta}{v^2\cos\theta}$

or, $\quad -\cos^2\theta = 1-\cos^2\theta - \frac{1}{2}\dfrac{g\ell\sin^2\theta}{g\ell\cos^2\theta}\left[\text{From(i)}\right]$

or, $\quad 1 = \frac{1}{2}\tan^2\theta$ or, $\tan\theta = \sqrt{2}$

From (iii), $v_0 = \left[g\ell\left(2+\sqrt{3}\right)\right]^{1/2}$

14. VARIOUS FORMS OF ENERGY: THE LAW OF CONSERVATION OF ENERGY

Conservation of Energy

We observe that in many processes the sum of both the kinetic and potential energies does not remain a constant. This may be due to the influence of dissipative forces such as friction.

(a) The more general form of law of conservation of energy was established by taking into account other forms of energy such as thermal, electrical, chemical, nuclear, etc.

(b) The charges in all forms of energy is given by: $\Delta KE + \Delta U + \Delta$ (all other forms of energy) $\equiv 0$

This is what we mean by the law of conservation of energy and it is one of the most important principles of physics.

"The total energy is neither increased nor decreased in any process. Energy can be transformed from one form to another, and transferred from one body to another, but the total amount remains constant."

PROBLEM-SOLVING TACTICS

(a) One should always isolate the known and unknown quantities and write equations and solve them.

(b) The next step would be to find out a way from unknown to known quantities and write equations and solve them.

(c) One should always be very careful in doing so to avoid silly mistakes such as unit change of parameter.

(d) Energy is scalar in nature. However, get a clear idea of what is being gained or lost by which entity.

(e) Physical visualization of any problem will always help in increaseaing confidence in solving equations pertaining to the same.

(f) Further, problems involving integration would be easy to understand if you go event by event and then solve.

(g) Special cases and boundary conditions of circular motion are definitely recommended to be mastered because many problems break down to these special cases just after few manipulations.

FORMULAE SHEET

S. NO.	DESCRIPTION	FORMULA
1	Kinetic energy of the particle	$K(v) = \dfrac{1}{2}mv^2 = \dfrac{1}{2}m\vec{v}.\vec{v}$
2	Work done by force F	$W = \vec{F}.\vec{r}$ (here $\vec{r}$ is total displacement)
3	Work done by variable force	$w = \int \vec{F}.\vec{dr}$
4	Power generated by force F acting on body	$P = \dfrac{dW}{dt} = \vec{F}.\dfrac{\vec{dr}}{dt} = \vec{F}.\vec{v}$
5	Increase in Kinetic Energy = Decrease in Potential Energy	$KE = -\Delta U$
6	Energy conservation principle	$\Delta K + \Delta U = 0; \ \dfrac{1}{2}mv^2 = mgh$ or, $v = \sqrt{2gh}$
7	For a Spring work done W	$W = \int_{x_1}^{x_2} -kx \ dx = \dfrac{1}{2}k\left(x_1^2 - x_2^2\right)$
8	Work-Energy principle	$W_{net} = \Delta KE = K_f - K_i$
9	Work done by variable forces in short range	For $\vec{F} = \vec{F_1} + \vec{F_2} + \ldots\ldots$ $W = \int \vec{F}.\vec{dr} = \int \left(\vec{F_1} + \vec{F_2} + \ldots..\right).\vec{dr}$
10	For conservative forces, change in potential energy	$U_f - U_i = -\int_{r_i}^{r_f} \vec{F}.\vec{dr}$
11	Elastic Potential Energy	$U = \dfrac{1}{2}kx^2$

Solved Examples

JEE Main/Boards

Example 1: An object of mass 5 kg falls from rest through a vertical distance of 20 m and attains a velocity of 10 m/s. How much work is done by the resistance of the air on the object? $\left(g = 10\,m/s^2\right)$

Sol: According to work energy theorem, the total work done by force of gravity and force of air resistance on object is equal to the change in kinetic energy.

Work done by all forces = Change in KE

$$W_{air} + W_{gravity} = \Delta K.E.; \quad W_{air} + mgh = \frac{1}{2}mv^2$$

$$W_{air} = \frac{1}{2}mv^2 - mgh; \quad = \frac{1}{2} \times 5 \times 10 \times 10 - 5 \times 10 \times 20$$

$$W_{air} = -750\,J$$

Example 2: A block is projected horizontally on a rough horizontal floor. The coefficient of friction between the block and the floor is μ. The block strikes a light spring of stiffness k with a velocity v_0. Find the maximum compression of the spring.

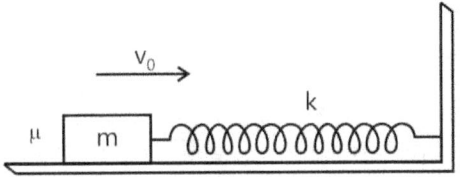

Sol: At the instant of maximum compression the block will come to rest momentarily. By applying work energy theorem the sum of work done by the force of friction, and spring force will be equal to change in kinetic energy. Work done by normal reaction and gravitational force will be zero.

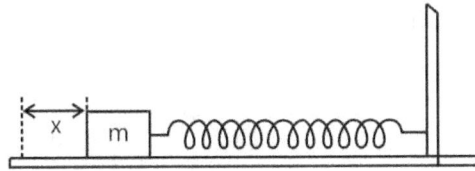

Since the block slides and the spring is compressed through a distance x the net retarding force acting on it

$$= F = -\left(kx + \mu N\right) = -\left(\mu mg + kx\right)$$

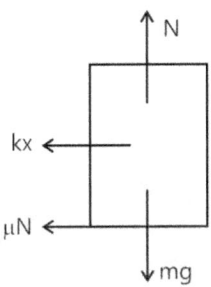

$\Rightarrow$ work done by net force for the displacement

$$\Rightarrow W = \int_0^x F\,dx; \quad \Rightarrow \quad \Delta KE = -\int_0^x \left(\mu mg + kx\right)dx$$

$$\Rightarrow \left(0 - \frac{1}{2}mv_0^2\right) = -\left(\mu mgx + \frac{kx^2}{2}\right)$$

Example 3: Two smooth balls of mass m_1 and m_2 connected by a light inextensible string are at the opposite points of horizontal diameter of a smooth semicylindrical surface of radius R. if m_1 is released, find its speed at any angular distance θ moved by m_2.

Sol: Loss in potential energy of system comprising masses m_1 and m_2, is equal to gain in kinetic energy of the system.

Let the ball m_2 moves through an angle θ, the mass m will fall through a distance $h_1 = R\theta$.

The ball m_2 rises through a height h_2 as,

$h_2 = R\sin\theta$

The change in gravitational potential energy of m1 is

$$\Delta PE_1 = -m_1 gh_1 = m_1 gR\theta$$

(Since m1 loses its potential energy as it falls down). The change in gravitional potential energy of m2 is

$$\Delta PE_2 = -m_2 gh_2 = m_2 gR\sin\theta$$

(Since m_2 gains potential energy as it rises up)

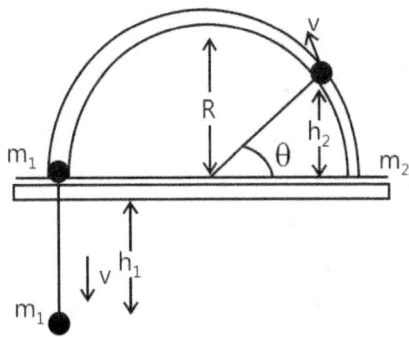

$\Rightarrow$ The total change in gravitational potential energy

$= \Delta PE = -m_1 gR\theta + m_2 gR\sin\theta$

$= gR(m_2 \sin\theta - m_1\theta)$ ... (i)

$= \Delta KE = \dfrac{1}{2}m_1 v^2 + \dfrac{1}{2}m_2 v^2 = \dfrac{(m_1 + m_2)v^2}{2}$ ... (ii)

where v = speed of m1 and m2 at the position as shown in the Fig.5.26 provided. From the principle of conservation of energy, we obtain

$= \Delta KE + \Delta PE = 0$ ... (iii)

Using (i)–(iii), we obtain

$\dfrac{1}{2}(m_1 + m_2)v^2 - gR(m_1\theta - m_2\sin\theta) = 0$

$\Rightarrow v = \sqrt{\dfrac{2gR(m_1\theta - m_2\sin\theta)}{(m_1 + m_2)}}$

Example 4: A locomotive of mass m starts moving so that its velocity varies according to the law $v = \alpha\sqrt{s}$, where α is a constant and s is the distance covered. Find the total work done by all the forces acting on the locomotive during the first second after the beginning of motion.

Sol: Velocity is given as the function of distance covered so we can find the acceleration and by second law of motion we can find the force. As force comes out to be constant the work done by force is product of force and displacement.

Given $v = \alpha\sqrt{s}$

Differentiating w.r.t. 't', we get

$\dfrac{dv}{dt} = \dfrac{1}{2}\alpha s^{-1/2}\dfrac{ds}{dt} = \dfrac{\alpha}{2\sqrt{s}}v = \dfrac{\alpha}{2\sqrt{s}} \times \alpha\sqrt{s} = \dfrac{\alpha^2}{2}$

$\therefore$ Acceleration $a = \dfrac{\alpha^2}{2}$

Now, force acting on the locomotive is

$F = ma = m\dfrac{\alpha^2}{2}$; Here, $u = 0$

Now, using $s = ut + \dfrac{1}{2}at^2$, we have

$s = 0 + \dfrac{1}{2}\dfrac{\alpha^2}{2}t^2 = \dfrac{\alpha^2 t^2}{4}$

Thus total work done on locomotive is when t=1 s is

$W = Fs = \dfrac{m\alpha^2}{2} \times \dfrac{\alpha^2 t^2}{4} = \dfrac{m\alpha^4}{8}$ J

Example 5: A 0.5 kg block slides from the point A on a horizontal track with an initial speed of 3 m/s toward a weightless horizontal spring of length 1 m and force constant 2 N/m. The part AB of the track is frictionless and the part BC has the coefficients of static and kinetic friction as 0.22 and 0.2, respectively. If the distances AB and BD are 2 m and 2.14 m, respectively find the total distance through which the block moves before it comes to rest completely (Take $g = 10\,m/s^2$).

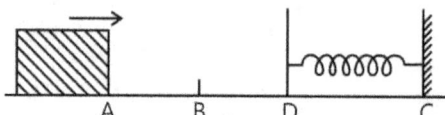

Sol: The sum of work done by force of friction and spring force is equal to change in kinetic energy of the block.

Suppose the block comes to rest at the point E, i.e., let DE = x. The kinetic energy of the block is spent in overcoming friction and compressing the spring through a distance DE = x.

Kinetic energy of the block

$= \dfrac{1}{2}mv^2 = \dfrac{1}{2} \times 0.5 \times 3^2 = 2.25\,J$ ... (i)

As the part AB of the track is frictionless, work done in moving from A to B is zero.

Let normal reaction of the block = mg.

C oefficient of friction = μ

Force due to friction along the track

$BC = \mu mg = 0.2 \times 0.5 \times 10 = 1N$

Distance through which the block moves against the frictional force = 2.14 + x m

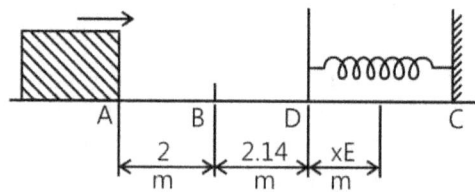

Work done by block against friction before it comes to rest

$= \mu\,mg\,(2.14 + x)$; $= (2.14 + x)\,J$ ... (ii)

Let the spring constant = k

$\therefore$ Work done by the block in compressing the spring through distance X

$= \dfrac{1}{2}kx^2$; $= \dfrac{1}{2}2x^2 = x^2 J$ ... (iii)

Adding (ii) and (iii) and equating it to (i), we get

$2.14 + x + x^2 = 2.25$; or $x^2 + x - 0.11 = 0$

or $100x^2 + 100x - 11 = 0$

or $(10x + 11)(10x - 1) = 0$

$\therefore x = -\dfrac{11}{10}$ or $x = \dfrac{1}{10}$; Since $x \neq -\dfrac{1}{10}$.

$\therefore x = \dfrac{1}{10} = 0.1m$

Restoring force of the spring

$= kx = 2 \times 0.1 = 0.2N$ ... (iv)

Static frictional force of the block

$\mu_{static} mg = 0.22 \times 0.5 \times 10 = 1.1N$ ... (v)

From (iv) and (v) it is clear that the static frictional force is greater than the restoring force of the spring. Therefore, the block will not move in the backward direction. Hence the total distance through which the block moves before it comes to rest completely is

$2.00 + 2.14 + 0.10 = 4.24$ m

Example 6: In a spring gun having spring constant 100 N/m, a small ball of mass 0.1 kg is put in its barrel by compressing the spring through 0.05 m as shown in the Fig. 5.29. Find,

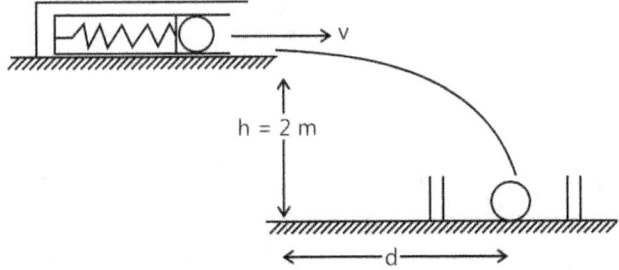

(a) The velocity of the ball when the spring is released.

(b) Where a box should be placed on the ground so that ball falls in it, if the ball leaves the gun horizontally at a height of 2 m above the ground? $(g = 10m/s^2)$.

Sol: As the spring expands the potential energy stored in the spring is converted in the kinetic energy of ball. The horizontal distance moved by the ball will depend on time taken by ball to fall vertical height h

(a) When the spring is released its elastic potential energy is converted into kinectic energy

$\Rightarrow \dfrac{1}{2}mv^2 = \dfrac{1}{2}kx^2$; $\Rightarrow v = \sqrt{\dfrac{5}{2}}m/s$

(b) As vertical component of velocity of the ball is zero, the time taken by the ball to reach the ground is

$h = \dfrac{1}{2}gt^2$; $t = \sqrt{\dfrac{2h}{g}} = \sqrt{\dfrac{2}{5}}$ seconds

Therefore, the horizontal distance traveled by the ball in this time is

$d = v.t = \sqrt{\dfrac{5}{2}} \times \sqrt{\dfrac{2}{5}} = 1m$

Example 7: A block of mass 2 kg is pulled up on a smooth incline of angle 30° with horizontal. If the block moves with an acceleration of 1m/s², find the power delivered by the pulling force at a time t = 4 s after motion starts. What is the average power delivered during these four seconds after the motion starts?

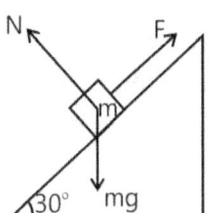

Sol: Apply newton's second law of motion along the direction of incline to find the applied force. As acceleration is constant and initial velocity is zero, the final velocity at time t is v=at along incline and power is $P = F \bullet v$.

The forces acting on the block are shown in the Fig. 5.30 provided.

Resolving forces parallel to incline

$F - mg \sin\theta = ma$; $\Rightarrow F = mg\sin\theta + ma$

$= 2 \times 9.8 \times \sin 30^0 + 2 \times 1 = 11.8N$

The velocity after t = 4 s is v=u+at

$= 0 + 1 \times 4 = 4m/s$

Power delivered by force at t = 4 s

$= \text{Force} \times \text{Velocity} = 11.8 \times 4 = 47.2W$

The displacement during t = 4 s is given by the formula

$v^2 = u^2 + 2as$; $v^2 = 0 + 2 \times 1 \times S$

$\therefore S = 8m$

Work done in t = 4 s is W = Force × distance

$= 11.8 \times 8 = 94.4 J$

$\therefore$ average power delivered

$= \dfrac{\text{workdone}}{\text{time}} = \dfrac{94.4}{4} = 23.6W$

Example 8: A chain of mass m = 0.80 kg and length l = 1.5 m rests on a rough-surfaced table so that one of its ends hangs over the edge. The chain starts sliding off the table all by itself provided the overhanging part equals n = 1/3 of the chain length. What will be the total work performed by the friction forces acting on the chain by the moment it slides completely off the table?

Sol: When chain starts sliding off the table, the friction is limiting. So from this we can find the coefficient of friction. As the chain slides off, the force of friction will be variable in nature as length of chain on table is decreasing. So the work done by the force of friction for infinitesimal displacement ds is $dW = \vec{F}.d\vec{s}$

Let μ be the coefficient of friction between chain and table.

Weight of hanging part = μ

(weight of horizontal part) $nmg = \mu (l-n)mg$

$$\mu = \frac{n}{l-n}$$

Let x be the length of the hanging part at some time instant.

Frictional force f (x) = N (normal reaction)

$-\dfrac{\mu(\ell-x)mg}{\ell}$ The work done by the frictional force if the hanging part increases to (x + dx) is

$dW = f(x)dx$

$$W = \int dW = -\int_\ell^1 \frac{\mu(\ell-x)mg}{\ell}; w = -\frac{\mu mg}{\ell}\left[\ell x - \frac{x^2}{2}\right]_\ell^i$$

$$W = -\mu mg\left[\ell(1-n) - \frac{\ell}{2}(1-n^2)\right]:$$

Substituting the value of μ from (I), we get

$$w = -\frac{n(1-n)mg\ell}{2} = \frac{1}{3}\times\frac{2}{3}\times\frac{mgl}{2} = \frac{mgl}{9}$$

JEE Advanced/Boards

Example 1: In the Fig. 5.31 shown a massless spring of stiffiness k and natural length l_0 is rigidly attached to a block of mass m and is in vertical position. A wooden ball of mass m is released from rest to fall under gravity. Having fallen a height h the ball strikes the spring and gets stuck up in the spring at the top. What should be the minimum value of h so that the lower block will just loose contact with the ground later on? Find also the correspoinding maximum compression in the spring.

Assume that $l_0 >> \dfrac{4mg}{k}$. Neglect any loss of energy.

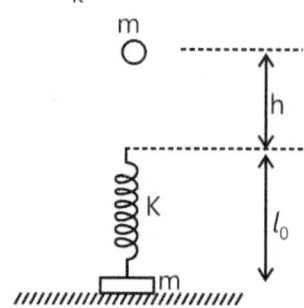

Sol: When ball falls from height h, the loss in its potential energy is equal to the gain in its kinetic energy. At the point of the maximum compression of the spring the ball comes to rest momentarily. After this the ball will again start moving up till the spring is again elongated to the point where the lower block looses contact with ground. For minimum value of h the ball will again come to rest at this point. So the total loss in gravitational potential energy will be equal to gain in the elastic potential energy.

The minimum force needed to lift the lower block is equal to its weight. During upward motion the spring will get elongated. If elongation in the spring for just lifting the block is x_0 then.

$kx_0 = mg;$ $\Rightarrow x_0 = \dfrac{mg}{k}$ ...... (i)

From COE

$$mg(l_0 + h) = mg(l_0 + x_0) + \frac{1}{2}kx_0^2;$$

$$\Rightarrow mgh = mgx_0 + \frac{1}{2}kx_0^2$$

$$\Rightarrow mgh = \frac{(mg)^2}{k} + \frac{1}{2}\frac{m^2g^2}{k}; \quad \Rightarrow h = \frac{3mg}{2k}$$

During downward motion, suppose maximum compression in the spring is x. From COE

$$mg(l_0 + h) = mg(l_0 - x) + \frac{1}{2}kx^2$$

$$\Rightarrow mgh = -mgx + \frac{1}{2}kx^2$$

$$\Rightarrow mg\frac{3mg}{2k} = -mgx + \frac{1}{2}kx^2$$

$$\Rightarrow 3(mg)^2 = -2mg\,kx + k^2x^2$$

$$\Rightarrow k^2x^2 - 2mgkx - 3(mg)^2 = 0$$

$$\Rightarrow x = \frac{2mgk \pm \sqrt{4(mgk)^2 + 12k^2(mg)^2}}{2k^2}$$

$$= \frac{2mgk \pm 4mgk}{2k^2} \Rightarrow x = \frac{3mg}{k}$$

Example 2: A smooth, light horizontal rod AB can rotate about a vertical axis passing through its end A. The rod is fitted with a small sleeve of mass m attached to the end A by a weightless spring of length ℓ_0 and stiffness k. What work must be performed to slowly get this system going and reach the angular velocity ω?

Sol: When system starts moving about a point A, the spring force provides the necessary centripetal force to the sleeve of mass m to move with angular speed ω. The work done by external agent will be equal to the kinetic energy of the spring and elastic potential energy of the spring.

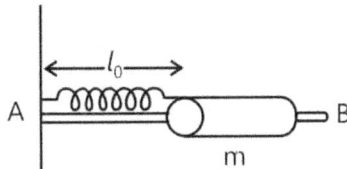

The mass m rotates in a circle of radius ℓ, which is the extended length of the spring. Centripetal force on $m = k(\ell - \ell_0) = m\omega^2\ell$

or, $\ell = \dfrac{\ell_0}{1-n}$ where $n = \dfrac{m\omega^2}{k}$

W = Change in KE of m + energy stored in the spring

$$= \frac{1}{2}m\omega^2 l^2 + \frac{1}{2}k(\ell - \ell_0)^2 = \frac{1}{2}m\frac{l_0^2\,\omega^2}{(1-n)^2} + \frac{1}{2}k\left[\frac{l_0}{1-n} - l_0\right]^2$$

$$W = \frac{1}{2}\frac{k l_0^2}{(1-n)^2}\left[\frac{m\omega^2}{k} + n^2\right]$$

Example 3: A small block is projected with a speed V_0 on a horizontal track which turns into a semicircle (vertical) of radius R. Find the minimum value of v_0 so that the body will hit the point A after leaving the track at its highest point. The arrangement is shown in the figure, given that the straight part is rough and the curved part is smooth. The coefficient of friction is μ.

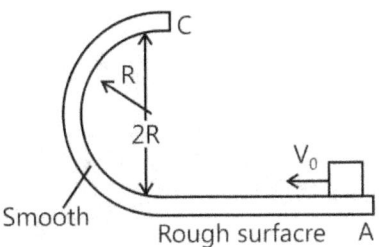

Sol: While block travels on the frictional surface AB, the work done by the frictional force is equal to the change in kinetic energy of the block. The horizontal distance moved by the block after leaving track at point C, will depend on time taken by disc to fall vertical height 2R. At point C, for minimum velocity, normal force on the block is zero.

Let the block escape the point at C with a velocity V horizontally. Since it hits the initial spot A after falling thorugh a height 2R we can write $(2R) = (1/2)gt^2$

where t = time of its fall

$\Rightarrow t = 2\sqrt{R/g}$

$\therefore$ the distance $AB = 2v\sqrt{R/g}$

$\Rightarrow d = 2v\sqrt{R/g}$ ... (i)

Work–energy theorem applied to the motion of the body from A to B leads

$\Delta KE = W_F$

$\Rightarrow \dfrac{1}{2}mv_0^2 - \dfrac{1}{2}mv_1^2 = \mu mgd$

$\Rightarrow v_0 = \sqrt{v_1^2 + 2\mu gd}$ ... (ii)

Energy conservation between B and C yields

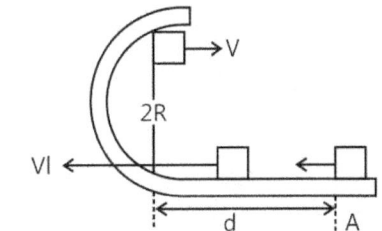

$\Rightarrow \dfrac{1}{2}mv_0^2 - \dfrac{1}{2}mv_1^2 = mg(2R)$

$\Rightarrow v_1 = \sqrt{v^2 + 4gR}$ ... (iii)

When the disc escapes C, its minimum speed v can be given as

$\dfrac{mv^2}{R} = mg$ ($\because$ the normal contact force = 0)

$\Rightarrow v = \sqrt{gR}$ ... (iv)

By using (iii) and (iv), we obtain

$v_1 = \sqrt{5gR}$ ... (v)

Using (i) and (iv), we obtain $d = (\sqrt{gR})2\left(\sqrt{\dfrac{R}{g}}\right) = 2R$ ... (vi)

Putting the values of v_1 and d in (ii), we obtain

$v_0 = \sqrt{5gR + 2\mu g(2R)}$

$\Rightarrow v_0 = \sqrt{(5 + 4\mu)gR}$

Example 4: Two bodies A and B connected by a light rigid bar of 10 m long move in two frictionless guides as shown in the Figure. If B starts from rest when it is vertically below A, find the velocity of B when X = 6 m.

Sol: As the blocks A and C fall vertically downwards, the loss in its potential energy is equal to gain in kinetic energy of blocks A, B and C.

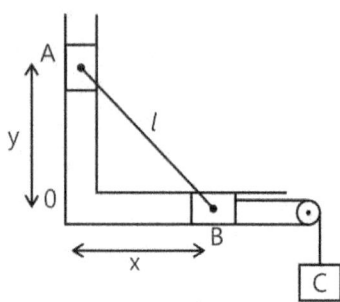

Assume

$m_A = m_B = 200$kg

and $m_C = 100$kg

At the instant, when the bar is as shown in the Figure

$$x^2 + y^2 = l^2; \therefore 2x\frac{dx}{dt} + 2y\frac{dy}{dt} = 0 \qquad \dots (i)$$

$$\therefore x\frac{dx}{dt} = -y\frac{dy}{dt} \qquad \dots (ii)$$

where $\frac{dx}{dt}$ = velocity of B and $\frac{dy}{dt}$ = velocity of A

Applying the law of conservation of energy, loss of potential energy of A, if it is going down when the rod is vertical to the position as shown in the Fig.

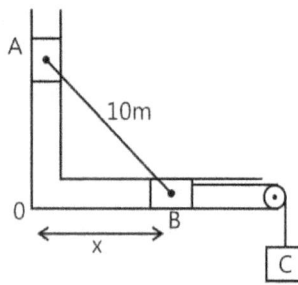

$= m_A g(10-8) = 2 \times 200 \times 9.8$

C moves down 6 m since B moves 6 m along x-axis.

Total loss of potential energy

$= 200 \times 9.8 \times 2 - 100 \times 9.8 \times 6 = 100 \times 9.8 \times 10 = 9800\,J.$

This must be equal to kinetic energy gained

Kinetic energy gained

$$= \frac{1}{2}m_A(v_A)^2 + \frac{1}{2}m_B(v_B)^2 + \frac{1}{2}m_C(v_C)^2$$

$$= \frac{1}{2} \times 200\left(\frac{dy}{dt}\right)^2 + \frac{1}{2} \times 200\left(\frac{dx}{dt}\right)^2 + \frac{1}{2} \times 100\left(\frac{dx}{dt}\right)^2$$

$$= 100\left(\frac{dy}{dt}\right)^2 + 150\left(\frac{dx}{dt}\right)^2$$

$$= 100\left[\frac{x\,dx}{y\,dt}\right]^2 + 150\left(\frac{dx}{dt}\right)^2 \quad \text{from (ii)}$$

$$= 100\left[\frac{6\,dx}{8\,dt}\right]^2 + 150\left(\frac{dx}{dt}\right)^2$$

$$= \left[100 \times \frac{9}{16} + 150\right]\left(\frac{dx}{dt}\right)^2 = \frac{3300}{16}v_B^2$$

$$\therefore \frac{3300}{16}v_B^2 = 9800$$

$$\therefore v_B = \sqrt{\frac{98 \times 16}{33}} = 7 \times 4\sqrt{\frac{2}{33}} = 6.9\,ms^{-1}$$

$\therefore$ Velocity of B at the required moment is $= 6.9\,ms^{-1}$

Example 5: A particle is suspended by a string of length 'l'. It is projected with such a velocity v along the horizontal such that after the string becomes slack it flies through its initial position. Find v.

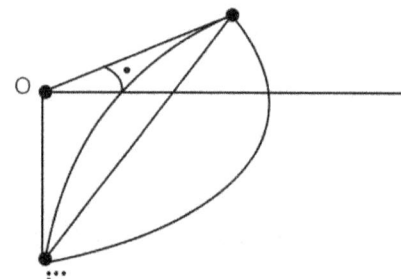

Sol: As the string becomes slack, the tension in the string becomes zero. Apply the Newton's second law of motion along the direction of string at the instant of slacking. The loss in kinetic energy is equal to gain in potential energy as the particle moves in vertical plane.

Let the velocity be v' at B where the string become slack and the string makes angle θ with horizontal by the law of conservation of energy.

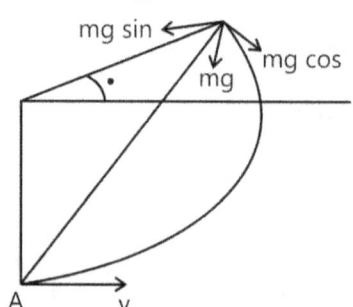

1.26

$$\frac{1}{2}mv^2 = \frac{1}{2}mv'^2 + mg\ell(1+\sin\theta) \qquad \ldots \text{(i)}$$

$$\text{or } v'^2 = v^2 - 2g\ell(1+\sin\theta) \qquad \ldots \text{(ii)}$$

By the dynamics of circular motion

$$mg\sin\theta = \frac{mv'^2}{\ell}; \Rightarrow v'^2 = g\ell\sin\theta \qquad \ldots \text{(iii)}$$

From equations (ii) and (iii), we get

$$\therefore g\ell\sin\theta = v^2 - 2g\ell(1+\sin\theta) \qquad \ldots \text{(iv)}$$

At B the particle becomes a projectile of velocity v' at $90-\theta$ with the horizontal.

Here $u_x = v'\sin\theta \, \& \, u_y = v'\cos\theta$

$a_x = 0 \, \& \, a_y = -g$

$$\therefore \ell\cos\theta - v'\sin\theta t \qquad \ldots \text{(v)}$$

$$\therefore t = \frac{\ell\cos\theta}{v'\sin\theta} \, \& - \ell(1+\sin\theta) = v'\cos\theta$$

$$\frac{\ell\cos\theta}{v'\sin\theta} - \frac{1}{2}g\frac{\ell^2\cos^2\theta}{v'^2\sin^2\theta}; \quad \Rightarrow 2\sin^3\theta + 3\sin^2\theta - 1 = 0$$

$$\therefore \sin\theta = \frac{1}{2} \text{ is the acceptable solution}$$

$$\therefore v^2 = 2g\ell + 3g\ell \times \frac{1}{2} = \frac{7g\ell}{2} \Rightarrow v = \sqrt{\frac{7g\ell}{2}}$$

[from equation (iv)]

Example 6: A motorcar of mass 1000kg attains a speed of 64 km/hr when running down an inclination of 1 in 20 with the engine shut off. It can attain a speed of 48 km/hr up the same incline when the engine is switched on. Assuming that the resistance varies as the square of the velocity, find the power developed by engine.

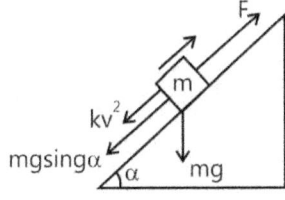

Sol: While moving down the incline plane car attains constant speed of 64 km/hr and while moving up the incline plane it attains the constant speed of 48 km/hr.While moving down the plane force of gravity is balanced by force of friction. While moving up on incline plane, the force developed by the engine of the car is balanced by the frictional force and force of gravity. The power developed by the engine is given by $P = \vec{F} \cdot \vec{V}$

When the motor car is moving down the plane there is force Mg sin α down the plane. This is opposed by the resistance, which is proportional to square of the velocity. That is $Mg\sin\alpha \propto V^2$

$$Mg \times \frac{1}{20} = kV^2 \text{ where k is a constant.}$$

$$\therefore \frac{1000 \times g}{20} = k\left(64 \times \frac{5}{18}\right)^2$$

$$k = \frac{1000 \times g}{20} \times \left(\frac{18}{64 \times 5}\right)^2 \qquad \ldots \text{(i)}$$

When the engine is on, let the tractive force (force exerted by engine) be F. This is used to overcome the force due to incline and the resistance offered.

$$\therefore F = k\left(48 \times 5/18\right)^2 + \frac{Mg}{20};$$

$$F = k\left(48 \times 5/18\right)^2 + \frac{1000 \times g}{20}$$

$$F = \frac{1000 \times g}{20} \times \left(\frac{18}{64 \times 5}\right)^2 \times \left(48 \times \frac{5}{18}\right)^2 + \frac{1000 \times g}{20}$$

$$= \frac{1000 \times 9.8}{20}\left[\frac{9}{16} - 1\right]$$

$$= \frac{50 \times 9.8 \times 2.5}{16} = 765.6 \text{N}$$

Power developed = Force x Velocity

$$= 765.6 \times 48 \times 5/18 = 10208 \text{W} = 10.2 \text{W}$$

JEE Main/Boards

Exercise 1

Q.1 What is meant by zero work? State the conditions under which a force does no work. Give any one example.

Q.2 Two bodies of unequal masses have same linear momentum. Which one has greater K.E.?

Q.3 Two bodies of unequal masses have same K.E. Which one has greater linear momentum?

Q.4 How do potential energy and K.E. of a spring vary with displacement? Is this variation different from variation in potential energy and K.E. of a body in free fall?

Q.5 Explain what is meant by work. Obtain an expression for work done by a constant force.

Q.6 Discuss the absolute and gravitational units of work on m.k.s. and c.g.s systems.

Q.7 What is meant by positive work, negative work and zero work? Illustrate your answer with two example of each type.

Q.8 Obtain graphically and mathemativally work done by a variable force.

Q.9 What are conservative and non-conservative forces, explain with examples. Mention some of their properties.

Q.10 What is meant by power and energy? Give their units.

Q.11 Explain the meaning of K.E. with examples. Obtain an expression for K.E. of a body moving uniformly?

Q.12 State and explain work energy principle.

Q.13 What do you mean by potential energy? Give any two examples of potential energy other than that of the gravitational potential energy.

Q.14 Obtain an expression for gravitational potential energy of a body.

Q.15 Explain what is meant by potential energy of a spring? Obtain an expression for it and discuss the nature of its variation.

Q.16 Mention some of the different forms of energy and discuss them briefly.

Q.17 A particle moves along the x-axis from $x=0$ to $x=5m$ under the influence of a force given by $f = 7 - 2x + 3x2$. Calculate the work done.

Q.18 The relation between the displacement x and the time t for a body of mass 2kg moving under the action of a force is given by $x = t^3/3$, where x is the metre and t is in second. Calculate work done by the body in first 2 seconds.

Q.19 A woman pushes a trunk on a railway platform which has a rough surface. She applies a force of 100N over a distance of 10m. Thereafter, she gets progressively tired and her applied force reduces linearly with distance to 50N. The total distance through which trunk has been moved is 20m. Plot the force applied by the woman and frictional force which is 50N against the distance. Calculate the work done by the two forces over 20m.

Q.20 A body of mass 50kg has a momentum of 1000kg ms^1. Calculate its K.E.

Q.21 A bullet of mass 50g moving with a velocity of 400ms-1 strikes a wall and goes out from the other side with a velocity of $100ms^{-1}$. Calculate work done in passing through the wall.

Q.22 A body dropped from a height H reaches the ground with a speed of $1.2\sqrt{gH}$. Calculate the work done by air-friction.

Q.23 A bullet weighting 10g is fired with a velocity of 800ms^1. After passing thorugh a mud wall 1m thick, its velocity decreases to 100 m/s. Find the average resistance offered by the mud wall.

Q.24 A particle originally at rest at the highest point of a smooth vertical circle of radius R, is slightly displaced. Find the vertical distance below the highest point where the particle will leave the circle.

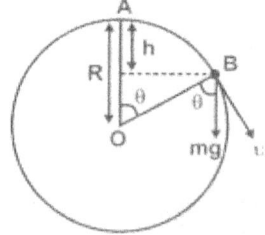

Exercise 2

Single Correct Choice Type

Q.1 A forceF $\vec{F} = -k\left(y\,\hat{i} + x\,\hat{J}\right)$, where k is a positive constant, acts on a particle moving in the xy plane.

Starting from the origin, the particle is taken along the positive x-axis to the point (a, 0), and the parallel to the y-axis to the point (a, a). The total work done by the force on the particle is:

(A) $2ka^2$ (B) $2ka^2$ (C) $-ka^2$ (D) ka^2

Q.2 Supposing that the earth of mass m moves around the sun in a circular orbit of radius 'R', the work done in half revolution is:

(A) $\dfrac{mv^2}{R} \times \pi R$ (B) $\dfrac{mv^2}{R} \times 2R$

(C) Zero   (D) None of these

Q.3 A string of mass 'm' and length 'l' rests over a frictionless table with 1/4th of its length hanging from a side. The work done in bringing the hanging part back on the table is:

(A) $mgl / 4$ (B) $mgl / 32$

(C) $mgl / 16$ (D) None of these

Q.4 A weight mg is suspended from a spring. If the elongation in the spring is x_o, the elastic energy stored in it is:

(A) $\dfrac{1}{2}mgx_o$ (B) $2mgx_0$ (C) mgx_0 (D) $\dfrac{1}{4}mgx_0$

Q.5 A ball is thrown up with a certain velocity at angle θ to the horizontal. The kinetic energy KE of the horizontal. The kinetic energy KE of the ball varies with horizontal displacement x as:

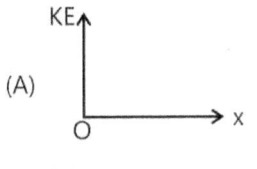
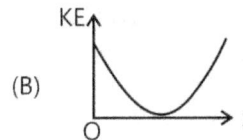
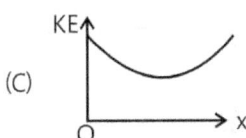
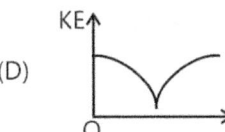

Q.6 A body m_1 is projected upwards with velocity v_1 another body m_2 of same mass is projected at an angle of 45°. Both reach the same height. What is the ratio of their kinetic energies at the point of projection:

(A) 1   (B) 1/2   (C) 1/3   (D) 1/4

Q.7 A uniform chain of length L and mass M is lying on a smooth table and one third of its length is hanging

vertically down over the edge of the table. If g is acceleration due to gravity, then the work required to pull the hanging part onto the table is:

(A) MgL   (B) $\dfrac{MgL}{3}$ (C) $\dfrac{4MgL}{9}$ (D) $\dfrac{MgL}{18}$

Q.8 A body is moved along a straight line by a machine delivering constant power. The distance moved by the body in time t is proportional to:

(A) $t^{1/2}$ (B) $t^{3/4}$ (C) $t^{3/2}$ (D) t^2

Q.9 An alpha particle of energy 4 MeV is scattered through 180° by a fixed uranium nucleus. The distance of the closest approach is of the order of

(A) $1 \overset{\circ}{A}$ (B) 10^{-10} cm

(C) 10^{-12} cm   (D) 10^{-15} cm

Q.10 A simple pendulum has a string of length and bob of mass m. When the bob is at its lowest position, it is given the minimum horizontal speed necessary for it to move in a circular path about the point of suspension. The tension in the string at the lowest position of the bob is:

(A) 3mg   (B) 4mg   (C) 5mg   (D) 6mg

Q.11 A horse pulls a wagon with a force of 360N at an angle of 60^0 with the horizontal at a speed of 10Km/hr. The power of the horse is:

(A) 1000 W   (B) 2000 W

(C) 500 W   (D) 750 W

Q.12 A man pulls a bucket of water from a well of depth H. If the mass of the rope and that of the bucket full of water are m and M respectively, then the work done by the man is:

(A) $(m+M)gh$ (B) $\left(\dfrac{m}{2}+M\right)gh$

(C) $\left(\dfrac{m+M}{2}\right)gh$ (D) $\left(m+\dfrac{M}{2}\right)gh$

Q13 A small block of mass m is kept on a rough inclined surface of inclination θ fixed in a elevator. The elevator goes up with a uniform velocity v and the block does not slide on the wedge. The work done by the force of friction on block in time t will be-

(A) Zero   (B) mgvtcosθ

(C) mgvtsinθ   (D) mgvtsin2θ

Q.14 Two equal masses are attached to the two ends of a spring of spring constant k. the masses are pulled out symmetrically k. the masses are pulled out symmetrically to stretch the spring by a length x over its natural length. The work done by the spring on each mass is-

(A) $\frac{1}{2}kx^2$ (B) $-\frac{1}{2}kx^2$ (C) $\frac{1}{4}kx^2$ (D) $-\frac{1}{4}kx^2$

Q.15 A particle is acted by a force F-kx, where k is a +ve constant. Its potential energy at x-0 is zero. Which curve correctly represents the variation of potential energy of the block with respect to x?

(A)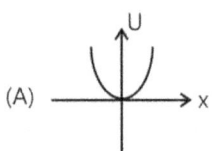

(B)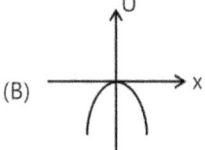

(C)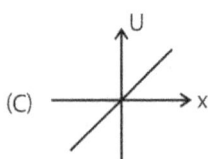

(D)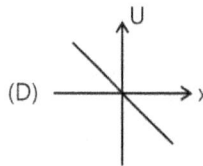

Q.16 If W_1, W_2 and W_3 represent the work done in moving a particle from A to B along there different paths 1, 2 and 3 respectively (as shown) in the gravitational field of a point mass m, find the correct relation between W_1, W_2 and W_3.

(A) $W_1 > W_2 > W_3$ (B) $W_1 = W_2 = W_3$

(C) $W_1 < W_2 < W_3$ (D) $W_2 > W_1 > W_3$

Q.17 An ideal spring with spring-constant k is hung from the ceiling and a block of mass M is attached to its lower end. The mass is released with the spring initially unstretched. Then the maximum extension in the spring is,

(A) $k = \frac{(2+\sqrt{3})mg}{\sqrt{3}R}$ (B) $\frac{2Mg}{k}$

(C) $\frac{Mg}{k}$ (D) $\frac{4Mg}{2k}$

Q.18 The total work done on a particle is equal to the change in its kinetic energy:

(A) Always

(B) Only if the forces acting on it are conservative

(C) Only if gravitational force alone acts on it

(D) Only if elastic force alone acts on it

Previous Years Questions

Q.1 Two masses of 1g and 4g are moving with equal kinetic energies. The ratio of the magnitudes of their momenta is: *(1980)*

(A) 4:1   (B) $\sqrt{2}$:1   (C) 1:2   (D) 1:16

Q.2 A stone tied to a string of length L is whirled in a vertical circle with the other end of the string at the centre. At a certain instant of time, the stone is at its lowest position, and has a speed μ. The magnitude of the change in its velocity as it reaches a position, where the string is horizontal, is *(1998)*

(A) $\sqrt{u^2 - 2gL}$ (B) $\sqrt{2gL}$

(C) $\sqrt{u^2 - gL}$ (D) $\sqrt{2(u^2 - gL)}$

Q.3 A wind-powered generator converts wind energy into electric energy. Assume that the generator converts a fixed fraction of the wind energy intercepted by its blades into electrical energy. For wind speed v, the electrical power output will be proportional to *(2000)*

(A) v   (B) v^2 (C) v^3 (D) v^4

Q.4 An ideal spring with spring constant k is hung from the ceiling and a block of mass M is attached to its lower end. The mass is released with the spring initially unstretched. Then the maximum extension in the spring is *(2002)*

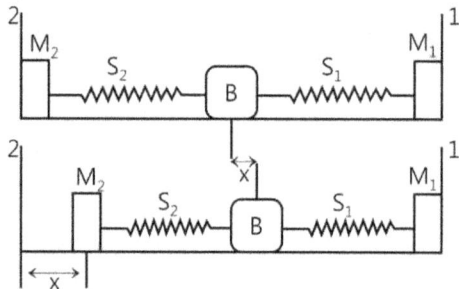

(A) $\frac{4Mg}{k}$ (B) $\frac{2Mg}{k}$ (C) $\frac{Mg}{k}$ (D) $\frac{Mg}{2k}$

Q.5 A block (B) is attached to two unstretched springs S_1 and S_2 with spring constants k and 4k, respectively. The other ends are attached to supports M_1 and M_2 not attached to the walls. The springs and supports have negligible mass. There is no friction anywhere. The block B is displaced towards wall I by a small distance x and released. The block returns and moves a maximum distance y towards wall 2. Displacement x and y are measured with respect to the equilibrium position of the block B.

The ratio $\dfrac{y}{x}$ is *(2008)*

(A) 4   (B) 2   (C) $\dfrac{1}{2}$ (D) $\dfrac{1}{4}$

Q.6 This question has Statement-I and Statement-II. Of the four choices given after the statements, choose the one that best describes the two statements.

If two springs S_1 and S_2 of force constants k_1 and k_2, respectively, are stretched by the same force, it is found that more work is done on spring S_1 than on spring S_2. *(2012)*

Statement-I: If stretched by the same amount, work done on S_1, will be more than that on S_2

Statement-II: $k_1 < k_2$

(A) Statement-I is false, Statement-II is true

(B) Statement-I is true, Statement-II is false

(C) Statement-I is true, Statement-II is the correct explanation for Statement-I

(D) Statement-I is true, Statement-II is true, and Statement-II is not the correct explanation for Statement-I.

Q.7 A person trying to lose weight by burning fat lifts a mass of 10 kg upto a height of 1 m 1000 times. Assume that the potential energy lost each time he lowers the mass is dissipated. How much fat will he use up considering the work done only when the weight is lifted up? Fat supplies 3.8×10^7 J of energy per kg which is converted to mechanical energy with a 20 % efficiency rate. Take g = 9.8 ms^{-2}: *(2016)*

(A) 6.45×10^{-3} kg   (B) 9.89×10^{-3} kg

(C) 12.89×10^{-3} kg   (D) 2.45×10^{-3} kg

JEE Advanced/Boards

Exercise 1

Q.1 A small disc of mass m slides down a smooth hill of height h without initial velocity and gets onto a plank of mass M lying on the horizontal plane at the base of hill as shown in the Figure. Due to friction between the disc and the plank, disc slows down and beginning with a certain moment, moves in one piece with the plank. Find out total work performed by the frictional forces in this process.

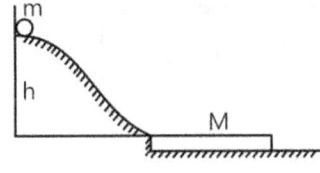

Q.2 A block of mass m starts from rest to slide along a smooth frictionless track of the shape shown in the Figure. What should be height h so that when the mass reaches point A on the track, it pushes the track with a force equal to thrice it weight?

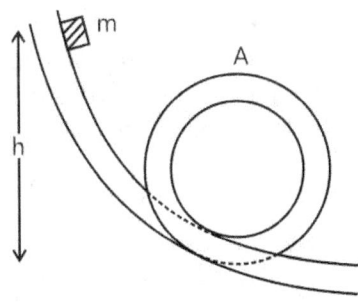

Q.3 A 0.5kg block slides from point A on a horizontal track with a initial speed of 3m/s towards a weightless spring of length l m and having a force constant 2 N/m. The part AB of the track is frictionless and the part BC has coefficent of static and kinetic friction as 0.22 and 0.20 respectively. If the distacnces AB and BD are 2m and 2.14m respectively, find the total distance through which the block moves before it comes to rest completely. $\left(g = 10 m / s^2\right)$

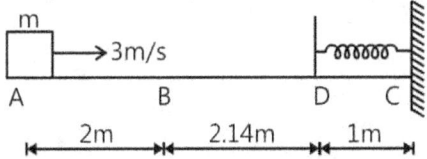

Q.4 A particle is suspended from a fixed point by a string of length 5m. It is projected horizontally from the equilibrium position with such a speed that the string slackens after the particle has reached a height of 8m above the lowest point. Find the speed of the particle just before the string slackens and the height to which the particle will rise further.

Q.5 Two blocks of masses $m_1 = 2$kg and $m_2 = 5$kg are moving in the same direction along a fricionless surface with speeds 10m/s and 3m/s respectively, m_2 being ahead of m_1. An ideal spring with k=1120N/m is attached to the back side of m_2. Find the maximum compression of the spring when the blocks collide.

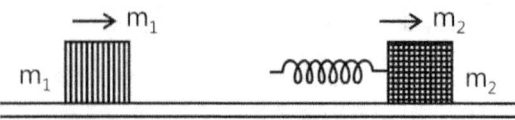

Q.6 An automobile of mass m accelerates, starting from rest, while the engine supplies constant power P; show that:

(a) The velocity is given as a function of time by

$v = \left(2Pt / m\right)^{1/2}$

(b) The position is given as a function of time by

$s = \left(8P / 9m\right)^{1/2} t^{3/2}$

Q.7 One end of a light spring of natural length d and spring constant k is fixed on a rigid wall and the other end is fixed to a smooth ring of mass m which can slide without friction on a vertical rod fixed at a distance d from the wall. Initially the spring makes an angle of 37° with the horizontal as shown in the Figure. If the system is released from rest, find the speed of the ring when the spring becomes horizontal $\left[\sin 37^0 = 3/5\right]$

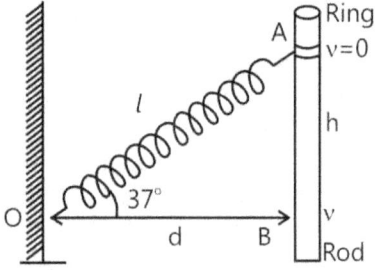

Q.8 A ring of mass m=0.3kg slides over a smooth vertical rod A. Attached to the ring is a light stirng passing over a smooth fixed pulley at a distance of 0.8m from the rod as shown in the figure. At the other end of the string there is a mass M=0.5kg. The ring is held in level with the pully and then released.

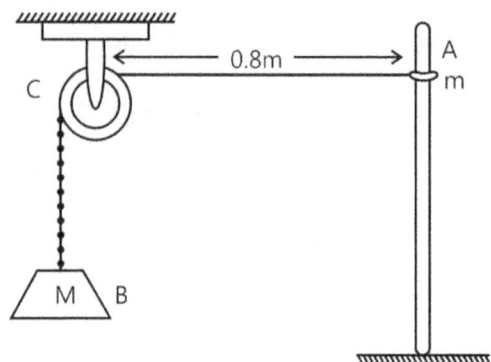

(a) Determine the distance by which the mass m moves

down before coming to rest for the first time.

(b) How far below the initial position of m is the equilibrium position of m located?

Q.9 A string, with one end fixed on a rigid wall, passing over a fixed frictionless pulley at a distance of 2m from the wall, has a point mass M=2kg attached to it at a distance of 1m from the wall. A mass m=0.5kg attached at the free end is held at rest so that the string is horizontal between the wall and the pulley and vertical beyond the pulley. Find the speed with which the mass M will hit the wall when the mass m is released?

Q.10 A massless platform is kept on a light elastic spring, as shown in the Figure. When a sand particle of mass 0.1kg is dropped on the pan from a height of 0.2m, the particle strikes the pan and sticks of 0.2m, the particle strikes the pan and sticks to it while the spring compresses by 0.01m. From what height should be particle be dropped to cause a compression of 0.04m?

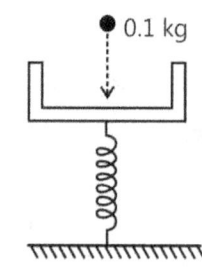

Q.11 Two blocks A and B each having mass of 0.32kg are connected by a light string passing over a smooth pully as shown in the Figure.

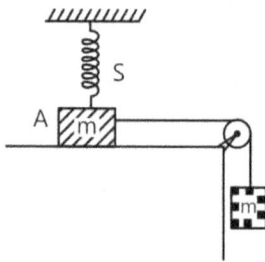

The horizontal surface on which the block A slides is smooth. The block A is attached to a spring of force constant 40N/m whose other end is fixed to a support 0.40m above the horizontal surface. Initially, when the system is releasesd to move, the spring is vertical and unstretched. Find the velocity of the block A at the instant it breaks off the surface below it. $\left[g = 10m/s^2\right]$

Q.12 A block of mass m is held at rest on a smooth horizontal floor. A light frictionless, small pulley is fixed at a height of 6m from the floor. A light inextensible string of length 16m, connected with A passes over the pulley and another identical block B is hung from the string. Initial height of B is 5m from the floor as shown in the figure. When the system is released from rest, B starts to move vertically downwards and A slides on the floor towards right.

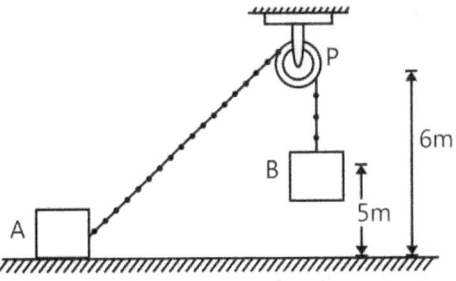

(a) If at an instant the string makes an angle θ with the horizontal, calculate relation between velocity u of A velocity v of B.

(b) Calculate v when B strikes the floor. $\left(g = 10m/s^2\right)$

Q.13 Two blocks are connected by a string as shown in the Figure. They are released from rest. Show that after they have moved a distance L, their common speed is given by $\sqrt{\dfrac{2\left(m_2 - \mu ml\right)gl}{\left(m_1 - m_2\right)}}$, where μ is the coeffiecient of friction between the floor and the blocks.

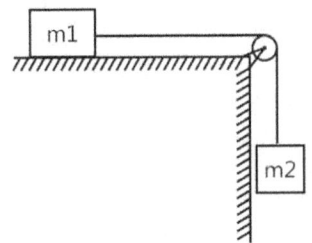

Q.14 A particle of mass m is moving in a circular path of constant radius r such that is centripetal acceleration α_c is varying with time t as $\alpha_c = k^2 rt^2$ when k is a constant. what is the power delivered to the particle by the forces acting on it?

Q.15 A body of mass m was slowly pulled up the hill as shown in the Figure. by a force F which at each point was directed along a tangent to the trajectory.

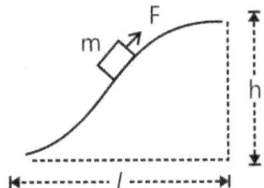

Find the work performed by this force, if the height of the hill is h, the length of its base l, and the co-effiecient of frinction between m and the hill is.

Q.16 A chain A B of length I is loaded in a smooth horizontal tube so that a part of its length h hangs freely and touches the surface of the table with its end B. At a certain moment, the end A of the chain is set free, with what velocity will this end of the chain slip out of the tube?

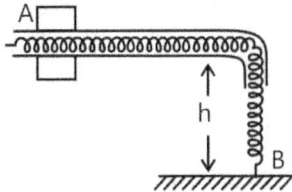

Q.17 A system consists of two identical cubes, each of mass m, linked together by the compressed weightless spring constant k. The cubes are also connected by a thread which is burned through at a certain moment. Find:

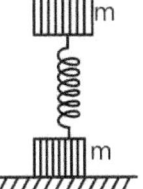

(a) At what values of Δl the initial compression of the spring, the lower cube will bounce up after the thread has been burned through:

(b) To what height h the centre of gravity of this system will rise if the initial compression of the spring Δl=7mg/k

Q.18 A stone with weight w is thrown vertically upward into the air with initial speed v_0. If a constant force f due to air drag acts on the stone throughout its flight:

(a) Show that the maximum height reached by the stone is $h = \dfrac{v_0^2}{2g\left[1 + \left(f/w\right)\right]}$.

(b) Show that the speed of the stone upon impact with the ground is $v = v_0\left(\dfrac{w-f}{w+f}\right)^{1/2}$

Q.19 One end of spring of natural length h is fixed at the ground and the other end is fitted with a smooth ring of mass m which is allowed to slide on a horizontal rod fixed at a height h as shown in Figure. Initially, the spring makes an angle of 37° with the vertical when the system is released from rest. Find the speed of the ring when the spring becomes vertical.

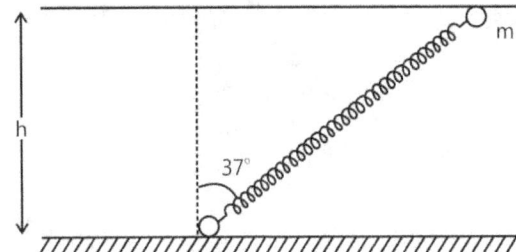

Q.20 A nail is located at a certain distance verticaly below the point of suspension of a simple pendulum.

1.33

The pendulum bob is released from the position where the string makes an angle of 60° with the downward vertical. Find the distance of the nail from the point of suspension such that the bob will just perform a complete revolution with the nail as centre. The length of the pendulum is 1m.

Q.21 A partical is suspended vertically from a point O by an inextensible massless string of length L.A vertical line A B is at a distance of L/8 form O as shown in the Figure. The particle is given a horizontal velocity u. At some point, its motion cases to be circular and eventually the object passes through the line A B. At the instant of crossing A B, its velocity is horizontal. Find u.

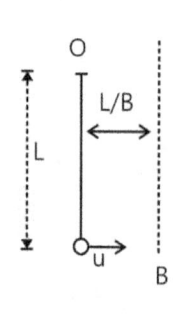

Q.22 A spherical ball of mass m is kept at the highest point in the space between two fixed, concentric spheres A and B. The smaller sphere A has a radius R and the space between the two spheres has a width d. the ball has a diameter very slightly less that d. All surfaces are frictionless. The ball is given a gental push (towards the right in the Figure). The angle made by the radius vector of the ball with the upwards vertical is denoted by (see Figure).

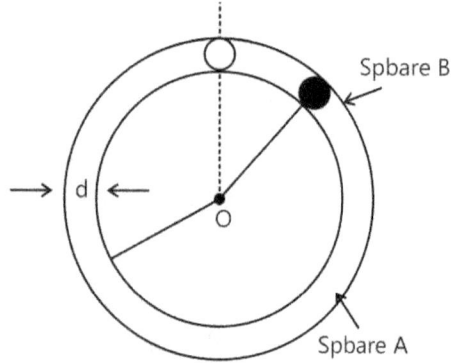

(a) Express the total normal reaction force exerted by the spheres on the ball as a function of angle θ

(b) Let N_A and N_B denote the magnitudes of the normal reaction forces on the ball exerted by the spheres A and B, respectively. Sketch the variations of N_A and N_B as fuctions of $\cos \theta$ in the range $0 \leq \theta \leq \pi$ by drawing two separate graphs in your answer book, taking $\cos\theta$ on the horizontal axes.

Q.23 Figure shows a smooth track, a part of which is a circle of radius R. A block of mass m is pushed against a spring of spring constant k fixed at the left end and is then released. Find the initial compression of the spring so that the block presses the track with a force mg when it reaches the point P, where the radius of the track is horizontal.

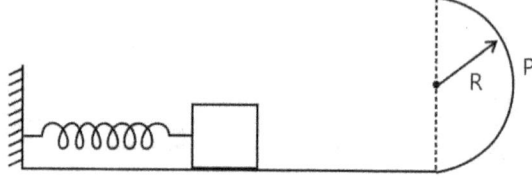

Q.24 A particle of mass 100g is suspended from one end of a weightless string of length 100cm and is allowed to swing in a vertical plane. The speed of the mass is 200cm/s when the string makes an angle of 60° with the vertical.

Determine

(a) The tension in the string at 60° and

(b) The speed of the particle when it is at the lowest position. (Take g=980 cm/s²)

Q.25 A smooth horizontal rod AB can rotate about a vertical axis passing thorugh its end A. The rod is fitted with a small sleeve of mass m attached ot the end A by a weightless spring of length l and spring constant k. what work must be performed to slowly get this system going and reaching to slowly get this system going and reaching the angular velocity ω

Q.26 Figure shows a smooth track which consists of a straight inclined part of length L joining smoothly with the circular part. A particle of mass m is projected up the inclined from its bottom.

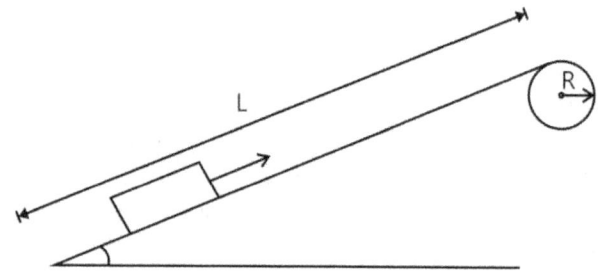

(a) Find the minimum projection speed u_0 for which the particle reaches the top of track.

(b) Assuming that the projection speed is 2 u_0 and that the block does't lose contact with the track before reaching its top. Find the force acting on it when it reaches the top.

(c) Assuming that the projection speed is only slightly grater than v_0, where will the block lose contact with the track?

Q.27 A small block of mass m slides along a smooth frictional track as shown in the Figure.

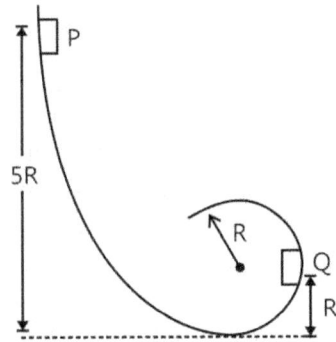

(a) If it starts from rest at P, what is the resultant force acting on it at Q?

(b) At what height above the bottom of the loop should the block be released so that the force it exerts against the track at the top of the loop equals its weight?

Q.28 A particle slides along a track with elevated ends and a flat central part as shown in the Figure. The flat central part as shown in the Figure. The flat part has a length l=3.0m. The curved portions of the track are frictionless. For the flat part the coefficient of kinectic friction is $\mu_k = 0.20$, the partical is released at point A which is at a height h=1.5m above the flat part of the track. Where does the particle finally come to rest?

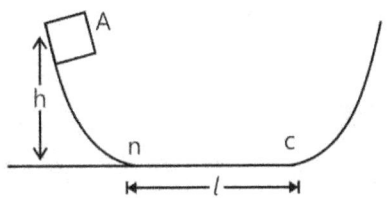

Q.29 The system of mass A and B shown in the Figure is released from rest with x=0, determine

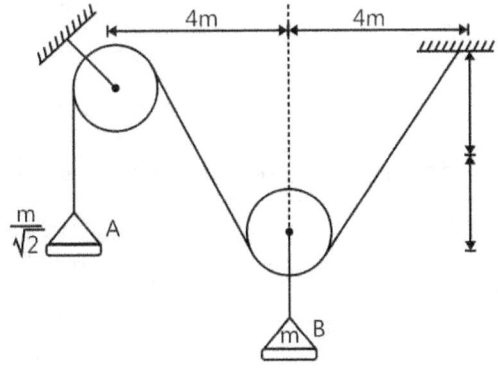

(a) The velocity of mass B when x=3m.

(b) The maximum displacement of mass B.

Exercise 2

Single Correct Choice Type

Q.1 When water falls from the top of a water fall 100m high:

(A) It freezes

(B) It warms up slightly

(C) It evaporates

(D) There is no change in temperature.

Q.2 A 2kg block is dropped from a height of 0.4m on a spring of force constant 2000N/m. the maximum compression of the spring is:

(A) 0.1m   (B) 0.2m   (C) 0.01m   (D) 0.02m

Q.3 A partical of mass M is moving in a horizontal circle of radius 'R' under the centripetal force equal to K/R^2, where K is constant. The potential energy of the particle is

(A) K/2R   (B) –K/2R   (C) K/R   (D) –K/R

Q.4 A linear harmonic oscillator of force constant 2×10^6N and amplitude 0.01 m has a total mechanical energy of 160 J. Its

(A) Maximum potential energy is 100 J

(B) Maximum kinetic energy is 100 J

(C) Maximum potential energy is 160 J

(D) Maximum potential energy is zero.

Q.5 The potential energy of particle varies with position x according to the relation

$U(x) = 2x_4 - 27x$ the po int x = 3 / 2 is point of

(A) Unstable equilibrium   (B) Stable equilibrium

(C) Neutral equilibrium   (D) None of these

Q.6 A particle of mass m is fixed to one end of a light rigid rod of length l and rotated in a vertical circular path about is other end. The minimum speed of the particle at its hightest point must be

(A) Zero   (B) $\sqrt{g\ell}$ (C) $\sqrt{1.5g\ell}$ (D) $\sqrt{2g\ell}$

Q.7 A particle of mass m is fixed to one end of a light spring of force constant k and unstretched length l. The system is rotated about the other end of the spring with an angular velocity ω, in gravity free space. The increase in length of the spring will be:

(A) $\sqrt{1.5g\ell}$ (B) $\sqrt{g\ell}$ (C) $\sqrt{2g\ell}$ (D) None

Q.8 A particle of mass m is moving in a circular path of constant radius r such that its centripetal acceleration a_c is varying with time as $a_c = k^2 r t^2$ where k is a constant. The power delivered to the particle by the forces acting on it is:

(A) $2\pi m k^2 r^2 t$

(B) $m k^2 r^2 t$

(C) $\frac{1}{3} m k^4 r^2 t^5$

(D) 0

Q.9 A simple pendulum having a bob of mass m is suspended from the ceiling of a car used in a stunt film shooting. The car moves up along an inclined cliff at a speed v and makes a jump to leave the cliff and lands at some distance. Let R be the maximum height of the car from the top of the cliff. The tension in the string when the car is in air is

(A) mg
(B) $mg\dfrac{mv^2}{R}$
(C) $mg\dfrac{mv^2}{R}$
(D) Zero

Q.10 A particle, which is constrained to move along the x-axis, is subjected to a force in the same direction which varies with the distance x of the particle from the origin as $F(x) = -kx + ax^3$. Here k and a are positive constants. For $x \geq 0$, the functional form of the potential energy $U(x)$ of particle is

(A)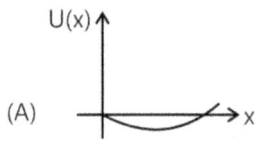

(B)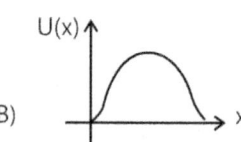

(C)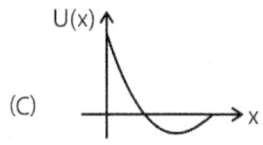

(D)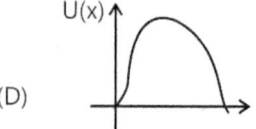

Q.11 A wind-powered generator converts wind energy into electric energy. Assume that the generator converts a fixed fraction of the wind energy intercepted by its blades into electric energy. For wind speed V, the electrical power output will be proportional to

(A) V
(B) v^2
(C) v^3
(D) v^4

Q.12 A block of mass M is hanging over a smooth and light pulley thorugh a light string. The other end of the string is pulled by a constant force F. The string energy of the block increases by 20 J in 1s.

(A) The tension in the string is Mg.

(B) The tension in the string is F.

(C) The work done by the tension on the block is 20J in

the above 1s.

(D) The work done by the force of gravity is -20J in the above 1s.

Q.13 Consider two observers moving with respect to each other at a speed v along a straight line. They observe a block of mass m moving a distance ℓ on a rough surface. The following quantities will be same as observed by the two observers.

(A) Kinetic energy of the block at time t

(B) Work done by friction.

(C) Total work done on the block

(D) Acceleration of the block.

Multiple Correct Choice Type

Q.14 A particle of mass m is attached to a light string of length ℓ, the other end of which is fixed. Initially the string is kept horizontal and the particle is given an upward velocity v, the particle is just able to complete a circle.

(A) The string becomes slack when the particle reaches its highest point

(B) The velocity of the particle becomes zero

(C) The kinetic energy of the ball in initial position was $\frac{1}{2} mv^2 = mg\ell$.

(D) The particle again passes through the initial position.

Q.15 The kinetic energy of a particle continuously increases with time.

(A) The resultant force on the particle must be parallel to the velocity at all instants

(B) The resultant force on the particle must be at an angle less than 90^0 all the time

(C) Its height above the ground level must continuously decreases.

(D) The magnitude of its linear momentum is increasing continuously.

Q.16 One end of a light spring of constant k is fixed to a wall and the other end is tied to block placed on a smooth horizontal surface. In a displacement, the work done by the spring is $\frac{1}{2} kx^2$. The possible cases are.

(A) The spring was initially compressed by a distance x and was finally in its natural length.

(B) It was initially stretched by a distance x and finally was in its natural length.

(C) It was initially in its natural length and finally in a compressed position.

(D) It was initially in its natural length and finally in a stretched position.

Q.17 No work is done by a force on an object if,

(A) The force is always perpendicular to its velocity

(B) The force is always perpendicular to its acceleration

(C) The object is stationary but the point of application of the force moves on the object

(D) The object moves in such a way that the point of application of the force remains fixed

Q.18 A particle is acted upon by a force of constant magnitude which is always perpendicular to the velocity of the particle. The motion of the particle takes place in a plane. It follows that,

(A) Its velocity is constant

(B) It acceleration is constant

(C) Its kinetic energy is constant

(D) It moves in a circular path

Q.19 A heavy stone is thrown from a cliff of height h in a given direction. The speed with which it hits the ground:

(A) Must depend on the speed of projection

(B) Must be larger then the speed of projection

(C) Must be independent of the speed of projection

(D) May be smaller than the speed of projection

Q.20 You lift a suitcase from the floor and keep it on a table. The work done by you on the suitcase does not depend on:

(A) The path taken by the suitcase

(B) The time taken by you in doing so

(C) The weight of the suitcase

(D) Your weight

Assertion Reasoning Type

(A) Both assertion and reason are true and reason is the correct explanation of Assertion.

(B) Both assertion and reason are true and reason is not

the correct explanation of assertion.

(C) Assertion is true but reason is false

(D) Assertion is false but reason is true.

Q.21 Assertion: For stable equilibrium force has to be zero and potential energy should be minimum.

Reason: For equilibrium, it is not necessary that the force is not zero.

Q.22 Assertion: The work done in pushing a block is more than the work done in pulling the block is more than the work done in pulling the block on a rough surface.

Reason: In the pushing condition normal reaction is more

Q.23 Assertion: Potential energy is defined for only conservation forces $\text{Reason} \, \bar{F} = -\dfrac{d\hat{U}n}{dr}$

Q.24 Assertion: An object of mass m is initially at rest. A constant force F acts on it. Then the velocity gained by the object in a fixed displacement is proportional to $\dfrac{1}{\sqrt{m}}$

Reason: For a given force and displacement velocity is always inversely proportional to root of mass.

Comprehension Type

Paragraph 1

The work done by external forces on a body is equal to change of kinetic energy of the body. This is true for both constant and variable force (variable in both magnitude and direction). For a particle $W = \Delta k$. For a system,

$W_{net} = W_{cal} + W_{pseudo} = \Delta Kcm$ or

$W_{ext} + W_{nonconservative} = \Delta K + \Delta U$.

In the absence of external and non conservative forces, total mechanical energy of the system remain conserved.

Q.25 I-work done in raising a box onto a platform depends on how fast it it raised II-work is an inter convertible form of energy.

(A) 1-False II-true   (B) 1-False II-False

(C) 1-True II-False   (D) 1-True II-true

Q.26 Consider a case of rigid body rolling without sliding over a rough horizontal surface

(A) There will be a non-zero conservative force acting on the body and work done by non-conservative force will be positive.

(B) There will be non-zero non-conservative force acting on body and work done by non-conservative force will be negative.

(C) There will be no non-conservative force acting on the body but totoal mechanical energy will not be conserved.

(D) There will be no non-conservative force acting on the body and total mechanical energy will be conserved.

Q.27 Now consider a case of rigid body rolling with sliding along rough horizontal plane and Vcm is linear. Velocity by $\omega = V_{cm} / 2R$, R is radius of body at (t=0)

(A) There is no non-conservative force acting on body.

(B) There is a non-conservative force acting on body and direction of force is opposite to direction of velocity.

(C) There is a non-conservative force acting on body and direction of the force along the direction of velocity.

(D) None of these.

Q.28 In the above problem if W=3Vcm/R where Vcm velocity of centre of mass at t=0

(A) There is non-conservative force acting on body.

(B) There is non-conservative force acting on body the direction of velocity of centre of mass.

(C) There is a non-conservative force acting on body opposite to the direction of velocity

(D) None of these

Paragraph 2

Two idedtical beads are attached to free ends of two identical springs of spring constant $k = \dfrac{(2+\sqrt{3})mg}{\sqrt{3}R}$. Initially both springs make an angle of 60^0 at the fixed point normal length of each spring is 2R. Where R is the radius of smooth ring over which bead is sliding. Ring is placed on vertical plane and beads are at symmetry with respect to vertical line as diameter.

Q.29 Normal reaction on one of the bead at initial moment due to ring is

(A) mg / 2   (B) $\sqrt{3}$mg / 2

(C) mg   (D) Insufficient data

Q.30 Relative acceleration between two beads at the initial moment:

(A) g/2 vertically away from each other

(B) g/2 horizontally towards each other

(C) $2g / \sqrt{3}$ Vertically away from each other

(D) $2g / \sqrt{3}$ Horizontally towards each other

Q.31 The speed of bead when spring is at normal length

(A) $\sqrt{\dfrac{(2-\sqrt{3})gR}{\sqrt{3}}}$ (B) $\sqrt{\dfrac{(2+\sqrt{3})gR}{\sqrt{3}}}$

(C) $\sqrt{\dfrac{2gR}{\sqrt{3}}}$ (D) $\sqrt{3gR}$

Q.32 Choose the correct statement

(A) Maximum angle made by spring after collision is same as that at initial moment.

(B) If the collision is perfectly inelastic, the total energy is conserved.

(C) If the collision is perfectly elastic, each bead undergoes SHM.

(D) Both linear momentum and angular momentum with respect to centre of smooth ring are conserved only at the instant of collision.

Match the Columns

Q.33 A single conservative force acts on a body of mass 1kg that moves along the x-axis. The potential energy U(x) is given by $U(x) = 20 + (x - 2)^2$ where x is the meters. At x=5.0m the particle has a kinetic energy of 20 J then:

Column-I		Column-II	
(A)	Minimum value of x in meters	(p)	29
(B)	Maximum value of x in meters	(q)	7.38
(C)	Maximum potential energy in joules	(r)	49
(V)	Maximum kinetic Energy in joules	(s)	-3.38

Q.34 A body of mass 75 kg is lifted by 15m with an acceleration of g/10 by an ideal string. If work done by tension in string is W_1, magnitude of work done by gravitational force is W_2, kinetic energy when it has lifted is K and speed is W_1, magnitude of work done by gravitational force is W_2, kinetic energy when it has lifted is K and speed of mass when it has lifted is v then: (data in column is given in SI units) $(g=10 \text{ m/s}^2)$

Column I		Column II	
(A)	W_1	(p)	10800
(B)	W_2	(q)	1080
(C)	K	(r)	11880
(D)	v	(s)	5.47

Previous Years' Questions

Q.1 The displacement x of a particle moving in one dimension, under the action of a constant force is related to the time t by the equation $t = \sqrt{x} + 3$. Where x is in metre and t in second, Find: (a) The displacement of the particle when its velocity is zero, and (b) The work done by the force in the first 6s. **(1980)**

Q.2 A body of mass 2kg is being dragged with a uniform velocity of 2m/s on a rough horizontal plane. The coefficient of friction between the body and the surface is 0.20, J=4.2 J/cal and g=9.8m/s². Calculate the amount of heat generated in 5s. **(1980)**

Q.3 Two blocks A and B are connected to each other by a string and a spring; the string passes over a frictionless pulley as shown in the Figure. Black B slides over the horizontal top surface of a stationary block C and the block A slides along the vertical side of C, both with the same uniform speed. The coefficient of friction between the surfaces of block is 0.2. Force constant of the spring is 1960 N/m. If mass of block A is 2kg. Calculate the mass of block B and the energy stored in the spring. **(1982)**

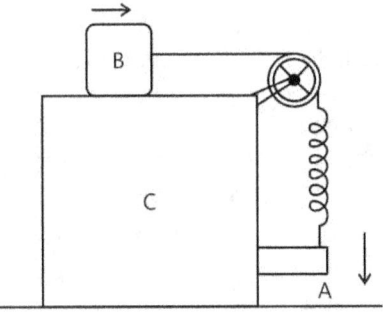

Q.4 A string, with one end fixed on a rigid wall, passing over a fixed frictionless pulley at a distance of 2m from the wall has a point mass M=2kg attached to it at a distance of 1m from the wall. A mass m=0.5kg attached at the free end is held at rest so that the string is horizontal between the wall the pulley and vertical

beyond the pulley. What will be the speed with which the mass M will hit the wall when the mass the m is released? (Take g=9.8 m/s²) **(1985)**

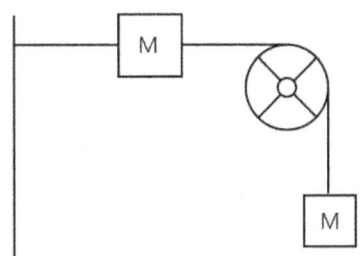

Q.5 A bullet of mass M is fired with a velocity 50m/s at an angle θ with the horizontal. At the highest point of its trajectory, it collides head-on with a bob of mass 3M suspended by a massless string of length 10/3m gets embedded in the bob. After the collision the string moves through an angle of 120°. Find (a) The angle θ (b) The vertical and horizontal coordinates of the initial position of the bob with respect to the point of firing of the bullet (take g = 10 m/s²) **(1988)**

Q.6 A particle is suspended vertically from a point O by an inextensible massless string of length L. A vertical line AB is at a distance L/8 from O as shown in Figure. The object is given a horizontal velocity u.

At some point, its motion ceases to be circular and eventually the object passes through the line AB. At the instant of crossing AB, its velocity is horizontal. Find u. **(1999)**

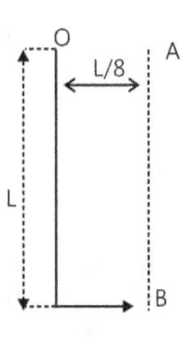

Q.7 A light inextensible string that goes over a smooth fixed pulley as shown in the Figure connects two blocks of masses 0.36 kg and 0.72kg. taking $g = 10 \text{ ms}^{-2}$, find the work done (in Joule) by string on the block of mass 0.36kg during the first second after the system is released from rest **(2009)**

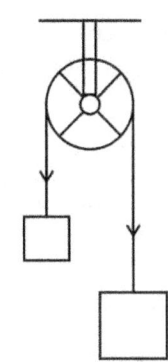

Q.8 A block of mass 0.18kg is attached to a spring of force constant 2N/m. The coefficient of friction between the block and the floor is 0.1. Initially the block is at rest and the spring is unstretched. An impulse is given to the block as shown in the Figure.

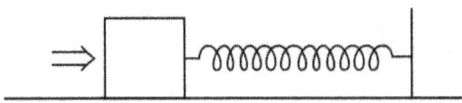

The block slides a distance of 0.06m and comes to rest for the first time. The initial velocity of the block in m/s is $v = \dfrac{N}{10}$. Then N is. **(2011)**

Q.9 The work done on a particle of mass m by a force

$$K\left[\frac{x}{\left(x^2+y^2\right)^{3/2}}\hat{i} + \frac{y}{\left(x^2+y^2\right)^{3/2}}\hat{j}\right]$$ (K being a constant of

appropriate dimensions, when the particle is taken from the point (a, 0) to the point (0, a) along a circular path of radius a about the origin in the x-y plane is **(2013)**

(A) $\dfrac{2K\pi}{a}$ (B) $\dfrac{K\pi}{a}$ (C) $\dfrac{K\pi}{2a}$ (D) 0

Q.10 A bob of mass m, suspended by a string of length l1 is given a minimum velocity required to complete a full circle in the vertical plane. At the highest point, it collides elastically with another bob of mass m suspended by a string of length l2, which is initially at rest. Both the strings are mass-less and inextensible.

If the second bob, after collision acquires the minimum speed required to complete a full circle in the vertical plane, the ratio l_1/l_2 is **(2013)**

Q.11 A particle of mass 0.2 kg is moving in one dimension under a force that delivers a constant power 0.5 W to the particle. If the initial speed (in m/s) of the particle is zero, the speed (in m/s) after 5 s is **(2013)**

Q.12 A tennis ball is dropped on a horizontal smooth surface. It bounces back to its original position after hitting the surface. The force on the ball during the collision is proportional to the length of compression of the ball. Which one of the following sketches describes the variation of its kinetic energy K with time t most appropriately? The figures are only illustrative and not to the scale. **(2014)**

Q.13 A block of mass m_1 = 1 kg another mass m_2 = 2kg, are placed together (see figure) on an inclined plane with angle of inclination θ. Various values of θ are given in List I.

(A)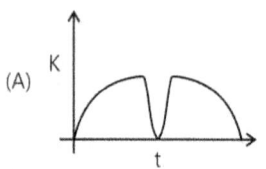

(B)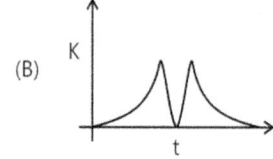

(C)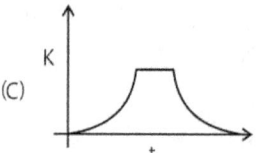

(D)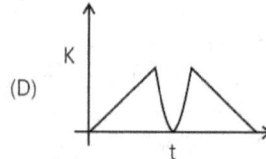

The coefficient of friction between the block m_1 and the plane is always zero. The coefficient of static and dynamic friction between the block m_2 and the plane are equal to μ = 0.3. In List II expressions for the friction on the block m_2 are given. Match the correct expression of the friction in List II with the angles given in List I, and choose the correct option.

The acceleration due to gravity is denoted by g.

[Useful information: tan (5.5°) ≈ 0.1; tan (11.5°) ≈ 0.2; tan (16.5°) ≈ 0.3] **(2014)**

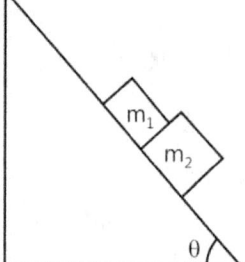

	List I		List II
I	$\theta = 5°$	p.	$m_2g \sin\theta$
II	$\theta = 10°$	q.	$(m_1+m_2) g \sin\theta$
III	$\theta = 15°$	r.	$mm_2g \cos\theta$
IV	$\theta = 20°$	s.	$\mu(m_1 + m_2)g \cos\theta$

Code:

(A) P-1, Q-1, R-1, S-3

(B) P-2, Q-2, R-2, S-3

(C) P-2, Q-2, R-2, S-4

(D) P-2, Q-2, R-3, S-3

Q.14 Consider two different metallic strips (1 and 2) of same dimensions (lengths ℓ, width w and thickness d) with carrier densities n_1 and n_2, respectively. Strip 1 is placed in magnetic field B_1 and strip 2 is placed in magnetic field B_2, both along positive y-directions. Then V_1 and V_2 are the potential differences developed between K and M in strips 1 and 2, respectively. Assuming that the current I is the same for both the strips, the correct option(s) is(are) **(2015)**

(A) If $B_1 = B_2$ and $n_1 = 2n_2$, then $V_2 = 2V_1$

(B) If $B_1 = B_2$ and $n_1 = 2n_2$, then $V_2 = V_1$

(C) If $B_1 = 2B_2$ and $n_1 = n_2$, then $V_2 = 0.5V_1$

(D) If $B_1 = 2B_2$ and $n_1 = n_2$, then $V_2 = V_1$

Important Questions

JEE Main/Boards

Exercise 1

Q. 18   Q.19   Q.22   Q.23

Exercise 2

Q.1   Q.5   Q.10   Q.13

Q.15   Q.16   Q.17

Previous Years' Questions

Q.3   Q.4   Q.8

JEE Advanced/ Boards

Exercise 1

Q.3   Q.5   Q.7   Q.11

Q.13   Q.19   Q.21   Q.30

Exercise 2

Q.4   Q.10   Q.12   Q.33

Q.34

Previous Years' Questions

Q.3   Q.4   Q.6   Q.8

Answer Key

JEE Main/Boards

Exercise 1

Q.1 Zero

Q.18 16 J

Q.19 1750 J; -1000 J

Q.20 10^4 J

Q.21 3.75×10^3 J

Q.22 -0.28 mgH

Q.23 3150N

Exercise 2

Single Correct Choice Type

Q.1 C	**Q.2** C	**Q.3** B	**Q.4** A	**Q.5** C
Q.6 B	**Q.7** C	**Q.8** C	**Q.9** C	**Q.10** D
Q.11 C	**Q.12** B	**Q.13** C	**Q.14** D	**Q.15** B
Q.16 B	**Q.17** B	**Q.18** A		

Previous Years' Questions

Q.1 C	**Q.2** D	**Q.3** C	**Q.4** B	**Q.5** C
Q.6 A	**Q.7** C			

JEE Advanced/Boards

Exercise 1

Single Correct option

Q.1 $mgh\left(\dfrac{M}{M+m}\right)$

Q.2 $4\,R$

Q.3 $4.24\,M$

Q.4 $5.42M/S^{1/2}, 0.97M$

Q.5 $0.25\,M$

Q.6 $s=\left(\dfrac{8P}{9m}\right)^{1/2}t^{3/2}$

Q.7 $d\sqrt{\dfrac{3g}{2d}+\dfrac{k}{16m}}$

Q.8 (a) $1.5m$ (b) $0.6\,m$

Q.9 3.3 m/s

Q.10 3.32 m

Q.11 154 m/s

Q.12 $\dfrac{40}{\sqrt{41}}$ m/s

Q.14 mk^2r^2t

Q.15 $mgh+\mu mgl$

Q.16 $\sqrt{2hg\,\ell n\left(\dfrac{\ell}{h}\right)}$

Q.17 (a) $\Delta\ell>\dfrac{3mg}{k}$ (b) $h=\dfrac{8mg}{k}$

Q.18 $v_0\left(\dfrac{w-f}{w+f}\right)^{1/2}$

Q.19 $\dfrac{h}{4}\sqrt{\dfrac{k}{m}}$

Q.20 $0.8m$

Q.21 $2.14\sqrt{gL}$

Q.22 (a) $N=mg(\cos\theta-2)$; (b) for $\theta\le\cos^{-1}(2/3), N_B=0, N_A=mg(3\cos\theta-2)$

and for $\theta\ge\cos^{-1}(2/3)N_A=0, N_B=mg(2-3\cos\theta)$

Q.23 $\sqrt{\dfrac{3mgR}{k}}$

Q.24 2.12 m/s

Q.25 $\dfrac{k\ell^2 x'(1+x')}{2(1-x')^2}$

Q.26 (a) $\sqrt{2g\left[R(1-\cos\theta)+L\sin\theta\right]}$; (b) $6mg\left(1-\cos\theta+\dfrac{L}{R}\sin\theta\right)$

(c) The radius through the particle makes an angle $\cos^{-1}\left(\dfrac{2}{3}\right)$ with the vertical.

Q.27 $3\,R$

Q.28 Third trip

Q.29 $8\sqrt{2}\,m$

Exercise 2

Single Correct Choice Type

Q.1 B	**Q.2** A	**Q.3** C	**Q.4** C	**Q.5** B
Q.6 A	**Q.7** B	**Q.8** B	**Q. 9** D	**Q.10** D
Q.11 C	**Q.12** B	**Q.13** D		

Multiple Correct Choice Type

Q.14 A, D	**Q.15** B, D	**Q.16** A, B	**Q.17** A, C, D	**Q.18** C, D
Q.19 A, B	**Q.20** A, B, D			

Assertion Reasoning Type

Q.21 B   Q.22 A   Q.23 A   Q.24 B

Comprehension Type

Paragraph 1

Q.25 A   Q.26 D   Q.27 B   Q.28 B

Paragraph 2

Q.29 C   Q.30 D   Q.31 C   Q.32 D

Match the Columns

Q.33 A → s; B → q; C → r; D → p   Q.34 A → r; B → p; C → q; D → s

Previous Years' Questions

Q.1 0,0

Q.2 9.33 cal

Q.3 0.098J

Q.4 3.29m/s

Q.5 30^0, (108.25m, 31.25m)

Q.6 $u = \sqrt{gL\left(2 + \dfrac{3\sqrt{3}}{2}\right)}$

Q.7 8 J

Q.8 D

Q.9 D

Q.10 5

Q.11 5

Q.12 B

Q.13 D

Q.14 A, C

Solutions

JEE Main/Boards

Exercise 1

Sol 1: If the vector product $\vec{F}.\vec{s} = 0$, then work done is zero when

$\vec{F}$ = force acting on the body

$\vec{s}$ = displacement of the body

Force does no work under given condition

(i) Direction of force and displacement is perpendicular

(ii) Displacement is zero

Example:

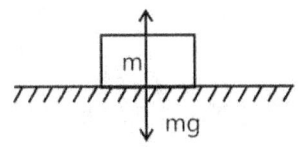

Work done by gravity and normal force is zero.

Sol 2: Two bodies have same linear momentum so

$m_1v_1 = m_2v_2$

m_1 = mass of first object

v_1 = velocity of first object

m_2 = mass of second object

v_2 = velocity of second object

Case – I : $m_1 > m_2$

$m_1v_1 = m_2v_2 \Rightarrow v_1 < v_2$

kinetic energy KE. = $\dfrac{1}{2}\, m_1v^2$

$KE_1 = \dfrac{1}{2}\, m_1v_1{}^2 ; K_2 = \dfrac{1}{2}\, m_2v_2{}^2 \Rightarrow K_2 > K_1$

Case – II : $m_2 > m_1$

$m_1v_1 = m_2v_2$

$\Rightarrow v_2 < v_1 \Rightarrow KE_1 > KE_2$

Sol 3: K.E. of object 1 is $KE_1 = \dfrac{1}{2} m_1 u_1^2$

K.E. of object 2 is $KE_2 = \dfrac{1}{2} m_2 u_2^2$

$KE_1 = KE_2$

$\Rightarrow \dfrac{1}{2} m_1 u_1^2 = \dfrac{1}{2} m_2 u_2^2$

$\Rightarrow m_1 u_1^2 = m_2 u_2^2 \; ; \; \dfrac{(m_1 u_1)^2}{m_1} = \dfrac{(m_2 u_2)^2}{m_2}$

Suppose $m_1 > m_2 \Rightarrow m_1 u_1 > m_2 u_2$

$\Rightarrow$ Linear momentum of 1st object is more than 2nd object.

Sol 4: Springs are generally taken as massless and spring transfers its potential energy to the kinetic and energy of the body to which it is attached whereas during a free fall, PE of a body is converted into its KE

In other words, for a spring all the energy is stored as potential energy whereas for body, potential energy is converted into kinetic energy.

Sol 5: work is the vector product $\vec{F}.\vec{s}$

$w = \vec{F}.\vec{s}$

$\vec{F}$ = force acting on the object

$\vec{s}$ = displacement of the object

If a constant force F displaces a body through displacements then the work done, w is given by

$w = Fs \cos \theta$

s = net displacement

θ = angle between force and displacement

Sol 6: Absolute unit of work on m.k.s is Joule (J)

Absolute unit of work on c.g.s. is erg

Gravitational unit of work on m.k.s. is kg-m

Gravitational unit of work on c.g.s is gm-cm

Sol 7: Work done $w = \vec{F}.\vec{s}$

$w = |\vec{F}| \, |\vec{s}| \cos \theta$

θ = angle between force and displacement when $0 < \theta, \pi/2$

$w = |F||s| \cos \theta$ is positive.

when $\theta = \pi/2 \Rightarrow W = 0$

when $\pi/2 < \theta < \pi$

$W = |F||s| \cos \theta$ is negative

Examples:

(1) Positive work

Work done by F is positive

Free falling object (Positive work)

Work done by gravity is negative

(2) Negative work

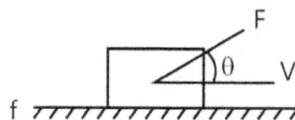

Work done by friction is negative

Zero work

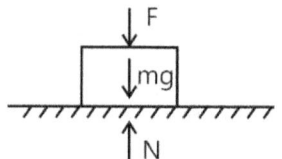

Work done by F is zero

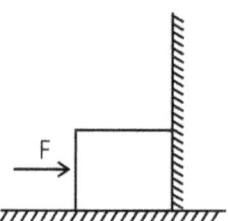

Work done by F is zero

Sol 8: Graphically work is the area under the force displacement graph

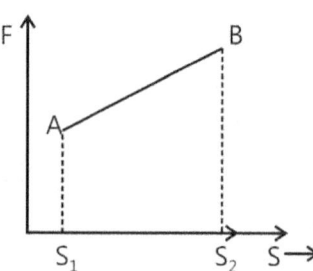

$$W_{A \to B} = \int_{s_1}^{s_2} \vec{F}.\overline{ds}$$

Mathematically, work is the integral dot product of force vector and infinitesimal displacement vector

$$W = \int_{A}^{B} \vec{F}.\overline{ds}$$

1.44

Sol 9: Conservative force: It is a force with the property that the work done in moving a particle between two points is independent of the taken path. Example gravity

Non-conservative force – it is a force with the property that the work done in moving a particle between two points is dependent on the path taken. Example friction

Properties of conservative force

(i) Work from one point to another point on any path is same

(ii) $\dfrac{dF_x}{dy} = \dfrac{dF_y}{dx}$; $\dfrac{dF_x}{dz} = \dfrac{dF_2}{dx}$; $\dfrac{dF_y}{dz} = \dfrac{dF_2}{dy}$

Where $\vec{F} = F_x\hat{i} + F_y\hat{j} + F_2\hat{k}$

Properties of non-conservative force

(i) Work done from one point to another is dependent on path taken.

(ii) Work cannot be recovered back

Sol 10: Power: It is defined as the rate at which the work is done SI unit of power is (J/s)

Energy: Energy of a body is the capacity of the body to do work. SI unit of energy is J.

Sol 11: It is the energy possessed by a body by virtue of its motion. A body of mass m moving with a velocity v has a kinetic energy.

$E_k = \dfrac{1}{2}\,mv^2$

$KE = w = \int \vec{F}.d\vec{s} = \int m\vec{a}ds = m\int v\dfrac{dv}{ds}\,ds = \dfrac{1}{2}\,mv^2$

Sol 12: When a particle is acted upon by various forces and undergoes a displacement, then its kinetic energy changes. By an amount equal to the total work done w_{net} on the particle by all the forces

$w_{net} = sk$

$W_{net} = w_c + w_{Nc} + w_{oth}$

w_c = work done by conservative force

w_{Nc} = work done by non-conservative force

w_{oth} = work done by other forces which are not included in above category

Sol 13: Potential energy – It is the energy of a body possessed by virtue of its position on the energy possessed by the body due to its state

Example:

(i) Energy stored in the compressed spring

(ii) When a rubber band is stretched potential energy is stored in it.

Sol 14:

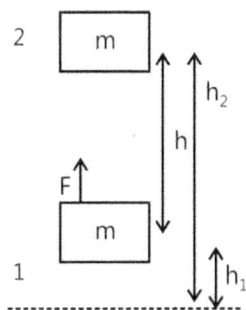

Suppose a mass m is raised from point 1 to point 2 and suppose change in kinetic energy is negligible. Potential energy is negative of the work done

$\Delta U = -W$

$W = \int_{h_1}^{h_2} mg\,ds$

$W = -mg\,(h_2 - h_1)$

$W = -mgh \Rightarrow \Delta U = mgh$

Sol 15: Potential energy of spring is the energy stored in the spring when compressed or stretched relative to its natural length.

Suppose a force F is acting on the spring of spring constant k

To keep the spring compressed in this position, the applied force should be same as kx

i.e. $F_s = kx$

Work done in compressing by a distance x is given by

$w = \int_0^x kx'dx' \Rightarrow w = \dfrac{1}{2}\,kx^2$

Work done is equal to change in potential energy of a spring

$w = U$

$U = \dfrac{1}{2}\,kx^2$

Potential energy is zero at spring's natural length and is proportional to the square of the distance from mean position

Sol 16: Different forms of energy are

(i) Kinetic Energy – It is the energy possessed by a body

by virtue of its motion.

(ii) Potential energy – It is the energy of a object on a system due to the position of the body and the arrangement of the particles of the system.

Example: Gravitational potential energy –

(iii) Mechanical energy – it is the sum of potential energy and kinetic energy

Sol 17: Force $f = 7 - 2x + 3x^2$

Work done

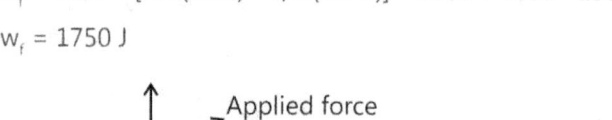

$$w = \int F.dx = \int_0^5 (7 - 2x + 3x^2)\,dx$$

$$= \left[7x - x^2 + x^3\right]_0^5 = 7(5 - 0) - [5^2 - 0^2] + [5^3 - 0^3]$$

$$W = 35 - 25 + 125 = 135 \text{ N-m}$$

So, work done is 135 N-m

Sol 18: Mass = 2 kg

$x = t^3/3;\ dx = t^2\,dt$

We need to find the force by first finding the acceleration of the body

$$v = \frac{dx}{dt} = \frac{d(t^2/3)}{dt} = t^2$$

$$a = \frac{dv}{dt} = 2t$$

$$F = ma = 2 \times 2t = 4t$$

Work done

$$w = \int F.dx = \int_0^2 4t.t^2 dt = \int_0^2 4t^3 dt = \left[t^4\right]_0^2 = 2^4 = 16 \text{ J}$$

$$w = 16 \text{ J}$$

Sol 19:

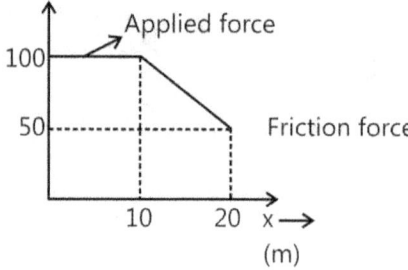

Work done by friction $w_f = \int F dx$

Since friction is constant all over the motion so displacement = 20 m

$$w_f = + f\,\Delta x = -50 \times 20$$

$$w_f = -1000 \text{ J}$$

[∵ Direction of friction force is opposite to displacement.

So force is – 50]

Work done by force f is

$$w_f = w_{f1\to2} + w_{f2\to3} = \int_0^{10} F_1 dx + \int_0^{10} F_2 dx$$

$$= 100 \times 10 + \int_0^{10} (100 - 5x)\,dx = 1000 + \left[100x - \frac{5x^2}{2}\right]_0^{10}$$

$$w_f = 1000 + [100(10-0) - 5/2\,(10^2 - 0)] = 1000 + 1000 - 250$$

$$w_f = 1750 \text{ J}$$

Sol 20: Mass = 50 kg = m

Momentum = 1000 kg m/s = mv

$$\text{K.E.} = \frac{1}{2} mv^2 = \frac{1}{2} \frac{(mv)^2}{m} = \frac{1}{2} \frac{(1000)^2}{50} = 10000 \text{ J}$$

Sol 21: Mass = 50 g = 0.05 kg

Initial velocity before collision = v_i = 400 m/Δ

Final velocity after passing through the wall = v_f = 100 m/s

By work energy theorem, Work done by the bullet is equal to the negative change in kinetic energy of the bullet

$$w = \Delta KE = -KE_f + KE_i = \frac{-1}{2} \times 0.05\,[(100)^2 - (400)^2]$$

$$w = -\frac{5}{2}\,[1 - 16] \times 100 = 3.75 \times 10^3 \text{ J}$$

Sol 22: By work energy theorem

$$w_{net} = \Delta KE$$

$$w_{gravity} + w_{friction} = \Delta KE$$

$$mgH + w_{friction} = \frac{1}{2} m\,(1.44\,gH - 0)$$

$$mgH + w_{friction} = 0.72\,mgH$$

$$w_{friction} = (0.72 - 1)\,mgH = -0.28\,mgH$$

Sol 23: Mass of bullet = 10 g = 0.01 kg

Initial velocity of bullet = 800 m/s

Thickness of mud wall = 1 m

Final velocity of bullet = 100 m/s

By work energy theorem

$$W_{net} = \Delta KE = \frac{1}{2} m [v_f^2 - v_i^2]F$$

$$w_r = \frac{1}{2} (0.01) [(100)^2 - (800)^2] = \frac{100}{2} [1^2 - 8^2]$$

$$w = -50 \times 63 = -3150 \, N$$

So average resistance offered is 3150 N

Sol 24:

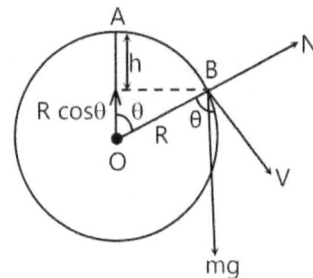

This is a very standard problem for a JEE aspirant.

Let us say at point B, the particle loses its contact. At point B say the particle has velocity v.

$$mg \cos\theta = N + \frac{mv^2}{R}$$

$$N = mg \cos\theta - \frac{mv^2}{R} \qquad \dots (i)$$

Now when the particle is about to lose contact, the normal reaction between the particle and surface becomes zero.

$$\therefore N = 0$$

$$\Rightarrow mg \cos\theta = \frac{mv^2}{R} \qquad \dots (ii)$$

Now energy at point A, taking O as reference;

$$E_A = 0 + mg R$$

$$E_B = \frac{1}{2} mv^2 + mg R \cos\theta$$

Using Energy conservation

$$E_A = E_B$$

$$\Rightarrow Mg R = \frac{1}{2} mv^2 + mg R \cos\theta$$

$$2 mg R (1 - \cos\theta) = mv^2$$

$$2mg (1 - \cos\theta) = \frac{mv^2}{R} \qquad \dots (iii)$$

Putting this value of $\frac{mv^2}{R}$ in eq^n $\qquad \dots (iv)$

$$mg \cos\theta = 2mg (1 - \cos\theta)$$

$$3 \cos\theta = 2$$

$$\cos\theta = \frac{2}{3} \Rightarrow \theta = \cos^{-1}\left(\frac{2}{3}\right)$$

And now $h = R (1 - \cos\theta) = R\left(1 - \frac{2}{3}\right)$

$$h = \frac{R}{3} \, m.$$

Exercise 2

Single Correct Choice Type

Sol 1: (C) Work done $= \int F_x \cdot dx + \int F_y \cdot dy$

$$F = F_x \, \hat{i} + F_y \, \hat{j}$$

$$F_x = -ky; \quad F_y = -kx$$

$$w = w_{1-2} + w_{2-3}$$

$$\int_0^a F_y dy + \int_0^a F_x dx$$

$$-\int_0^a K \times 0 \times dy + \int_0^a -ka\,dx = 0 + -ka (a - 0) = -ka^2$$

Sol 2: (C) Work done is zero as force and displacement are perpendicular to each other so $\vec{F}.\vec{ds} = 0$

Sol 3: (B) By work energy theorem

$$w_{net} = \Delta KE[\text{assuming negligible K.E.}]$$

$$W_{gravity} + W_{force} = 0$$

$$W_{force} = -\left(-\frac{m}{4} g\left(\frac{1}{8}\right)\right) = \frac{mgl}{32}$$

Sol 4: (A) by force equilibrium

$$kx_0 = \frac{mg}{x_0}$$

Elastic energy $= \frac{1}{2} kx_0^2 = \frac{1}{2} \frac{mg}{x_0} x_0^2 = \frac{mgx_0}{2}$

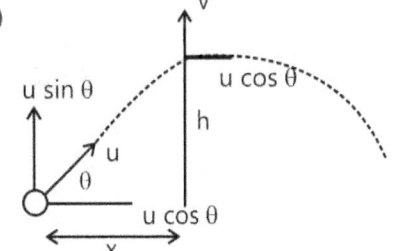

Sol 5: (C)

Vertical velocity at height h is v_y

$v_y = u \sin \theta - gt$

$t = \dfrac{x}{u\cos\theta}$

$v_y = u \sin \theta - \dfrac{gx}{u\cos\theta}$

$KE = \dfrac{1}{2} m[v_y^2 + v_x^2] = \dfrac{1}{2} m\left[u^2 + \dfrac{g^2x^2}{u^2\cos^2\theta} - 2\tan\theta gx\right]$

so graph (c) is correct

Sol 6: (B)

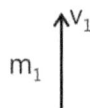
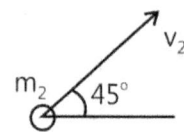

Maximum height $H = \dfrac{u^2 \sin^2\theta}{2g}$

$H_1 = \dfrac{v_1^2}{2g}$; $H_2 = \dfrac{v_2^2\left(\dfrac{1}{\sqrt{2}}\right)^2}{2g} = \dfrac{v_2^2}{ug}$

$H_1 = H_2$

$\Rightarrow \dfrac{v_1^2}{2} = \dfrac{v_2^2}{4}$

$v_2^2 = 2v_1^2$

$\dfrac{KE_1}{KE_2} = \dfrac{1/2m_1v_1^2}{1/2m_2v_2^2} = \dfrac{v_1^2}{v_2^2} = \dfrac{1}{2}$

Sol 7: (C) By work energy theorem

$w_{net} = \Delta KE$ [Assuming negligible K.E.]

$w_{gravity} + w_{force} = 0$

$W_{net} = \left(-\dfrac{1}{18}mg\ell\right) - \left(-\dfrac{1}{2}mg\ell\right) = \dfrac{4}{9}mg\ell$

Sol 8: (C) $P = F.V$

$P = ma.V$

$P = \left[\dfrac{mdV}{dt}V\right]$

$\displaystyle\int_0^t \dfrac{Pdt}{m} = \int_0^v vdv \Rightarrow \dfrac{p}{m} = \dfrac{v^2}{2}$

$v = \sqrt{\dfrac{2Pt}{m}} \Rightarrow x \propto t^{\frac{1}{2}+1}$

$x \propto t^{3/2}$

Sol 9: (C) Mechanical energy conservation

$KE_{alpa} + PE_i + KE = PE_{electrostatic}$

$4 \times 10^6 \times 10^{-19} = \dfrac{9 \times 10^{10} \times 10^{-19} \times 100 \times 10^{-9}}{r}$

$r \sim 10^{-14}$ m

$r \sim 10^{-12}$ cm

Sol 10: (D) Minimum velocity required is $v = \sqrt{5gR}$

by Newton's second law

$T - mg = \dfrac{mv^2}{R}$

$T = mg + \dfrac{m5gR}{R} = 6mg$

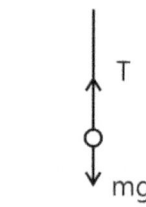

Sol 11: (C) $F = 360$ N

Work done is 1 hr

is $w = 360 \cos 60° \, 10 \times 100$

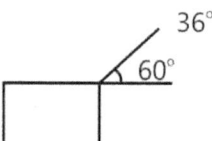

Work done per sec $= \dfrac{3600 \times 10^3 \cos 60°}{60 \times 60}$

$= 1000\,\dfrac{1}{2} = 500$ w

Sol 12: (B) By work energy theorem

$w_{net} = \Delta KE = 0$

$w_{man} + w_{gravity} = 0$

$w_{man} - MgH - mg\dfrac{H}{2} = 0$

$w_{man} = \left(\dfrac{m}{2} + M\right)gH$

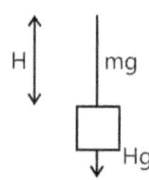

Sol 13: (C)

By force equilibrium

$f = mg \sin \theta$

Work done by friction f is equal to $f \sin \theta \, vt$

$w = mg \, vt \sin \theta$

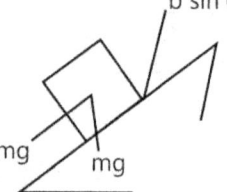

Sol 14: (D)

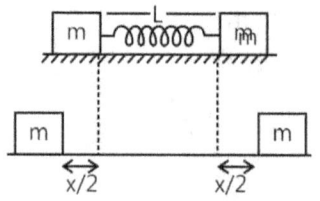

Potential energy of spring is ½ kx²

By work energy theorem

$W_{net} = \Delta KE$

$W_{force} + W_{spring} = 0$

$W_{spring} = \dfrac{-1}{2} kx^2$

Since displacements of both the masses are same so work done by spring on both masses is same.

So work done on each mass = $\dfrac{-1}{4} kx^2$

Sol 15: (B) F = kx

$U = -\int Fdx = -\int kxdx \Rightarrow U = -\dfrac{kx^2}{2} + c$

$U(x = 0) = 0 \Rightarrow 0 = 0 + c \Rightarrow c = 0$

$U = \dfrac{-kx^2}{2}$

Sol 16: (B) $w_1 = w_2 = w_3$ as gravitational force is conservative and work done by conservative forces is independent of path taken.

Sol 17: (B) By work energy theorem

$w_{net} = \Delta K$

h is the maximum extension of the spring

$W_{gravity} + W_{spring} = 0$

$+mgh - \dfrac{1}{2} kh^2 = 0$

$h = \dfrac{2mg}{k}$

Sol 18: (A) Work energy theorem includes all the forces. Conservation as well as non-conservative. This theorem is always true.

Previous Years' Questions

Sol 1: (C) $p = \sqrt{2Km}$

or $p \propto \sqrt{m}$, $\dfrac{m_1}{m_2} = \dfrac{1}{4}$

$\therefore \dfrac{p_1}{p_2} = \dfrac{1}{2}$

Sol 2: (D)

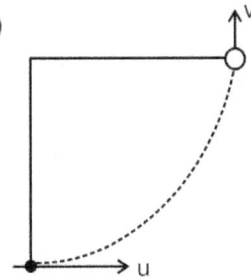

From energy conservation $v^2 = u^2 - 2gL$

Now, since the rwo velocity vectors shown in figure are mutually perpendicular, hence the magnitude of change of velocity will be given by

$|\Delta \vec{v}| = \sqrt{u^2 + v^2}$

Substituting value of v^2 from eq. (i)

$|\Delta \vec{v}| = \sqrt{u^2 + u^2 - 2gL} = \sqrt{2(u^2 - gL)}$

Sol 3: (C) Power = $\vec{F} . \vec{v}$ = Fv

$F = v\left(\dfrac{dm}{dt}\right) = v\left\{\dfrac{d(\rho \times volume)}{dt}\right\}$

$= \rho v\left\{\dfrac{d(volume)}{dt}\right\} \rho v(Av) = \rho Av^2$

$\therefore$ power P = ρAv^3 or $P \propto v^3$

Sol 4: (B) Let x be the maximum extension of the spring. From conservation of mechanical energy

Decrease in gravitational potential energy = increase in elastic potential energy

$\therefore$ Mg x = $\dfrac{1}{2} kx^2$

or x = $\dfrac{2Mg}{k}$

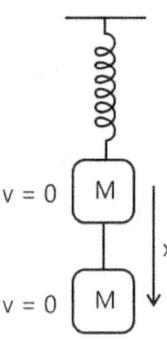

Sol 5: (C) From energy conservation,

$\dfrac{1}{2} kx^2 = \dfrac{1}{2} (4k)y^2$

$\dfrac{y}{x} = \dfrac{1}{2}$

Sol 6: (A) F = $k_1 S_1 = k_2 S_2$

$W_1 = FS_1, W_2 = FS_2$

$k_1 S_1^2 > k_2 S_2^2$

$S_1 > S_2$

$k_1 < k_2$

$W \propto k$

$W_1 < W_2$

Sol 7: (C) Let fat used be 'x' kg

$\Rightarrow$ Mechanical energy available $= x \times 3.8 \times 10^7 \times \dfrac{20}{100}$

Work done in lifting up $= 10 \times 9.8 \times 1000$

$\Rightarrow x \times 3.8 \times 10^7 \times \dfrac{20}{100} = 9.8 \times 10^4$

$\Rightarrow x \approx 12.89 \times 10^3$ kg.

JEE Advanced/Boards

Exercise 1

Sol 1: By work energy theorem for point A to point B

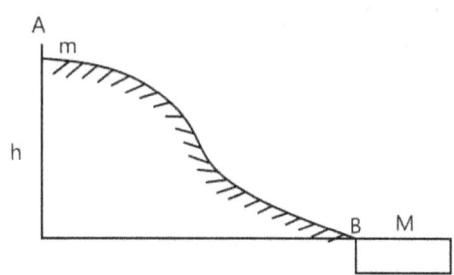

$w_{net} = \Delta KE$

$w_g = KE_f - KE_i$

$mgh = \dfrac{1}{2} mv^2$

$v = \sqrt{2gh}$

Now m, M both move together so by conservation of linear momentum

$mv = (M + m) v'$

$v' = \dfrac{m\sqrt{2gh}}{M+m}$

v' is the combined velocity of (m + M) system.

Applying work energy theorem for the whole process

$w_{net} = \Delta KE$

$w_{gravity} + w_{friction} = kE_f - kE_i \ (kE_i = 0)$

$mgh + w_{friction} = \dfrac{1}{2} (M + m) \left(\dfrac{m\sqrt{2gh}}{M+m} \right)^2$

$mgh + w_{friction} = \dfrac{m^2 gh}{M+m}$

$w_{friction} = \dfrac{m^2 gh}{M+m} - mgh$

$w_{friction} = - \dfrac{Mmgh}{(M+m)}$

Sol 2:

Drawing FBD at point A

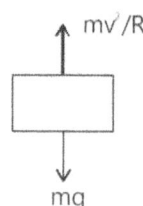

So net upward force exerted by the mass is $\dfrac{mv^2}{R} - mg$

Which is equal to 3 mg $\Rightarrow \dfrac{mv^2}{R} - mg = 3 mg$

$\dfrac{mv^2}{R} = 4 mg$

$v = \sqrt{4gR} = 2\sqrt{gR}$

Now applying work-energy theorem

$w_{net} = \Delta KE$

$w_{gravity} = K_f$

$mg(h-2R) = \dfrac{1}{2} m (4gR)$

$h - 2R = 2R$

$h = 4R$

Sol 3: Mass $= \dfrac{1}{2}$ kg

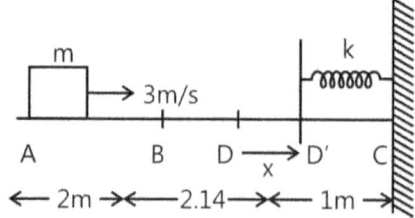

Let us assume that block stops at point D' which is at distance x m from D.

1.50

By applying work energy theorem

$w_{net} = \Delta KE$

$w_{friction} + w_{spring} = K_f - K_i$ ... (i)

$(K_f = 0)$

$w_{friction} = -\mu\, mg\, (BD')$

$w_{friction} = -(0.20)\, mg\, (2.14 + x)$ ... (ii)

$w_{spring} = -\dfrac{1}{2} kx^2$ ... (iii)

[∵ block is in motion from point B to D' so we will take kinetic friction]

Substituting (i), (ii) and (iii) we get

$(0.2)\, mg\, (-2.14 - x) - \dfrac{1}{2} kx^2 = -\dfrac{1}{2} m\, (g)$

Substituting value of m we get

$2.14 + x + \dfrac{1}{2} \times 2x^2 = +\dfrac{9}{4}$

$x^2 + x + 2.14 - 2.25 = 0$

$x^2 + x - 0.11 = 0$

$(x - 0.1)\,(x + 1.1) = 0$

$x = 0.1$ m

So, total distance thought which block moves is 2 + 2.14 + 0.1 m $\Rightarrow$ 4.24 m

Sol 4:

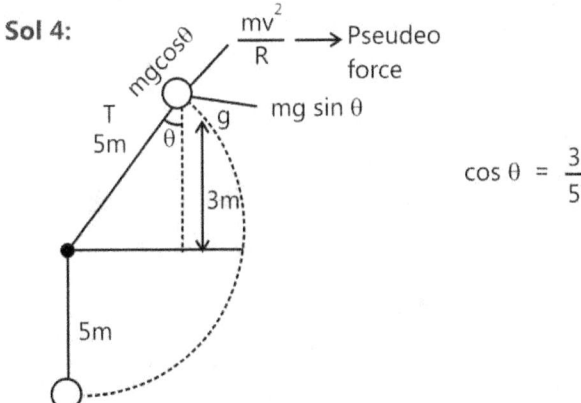

$\cos \theta = \dfrac{3}{5}$

When particle is at 8m height from lowest point tension is just zero. So balancing force in the direction of string

We get $mg \cos \theta - \dfrac{mv^2}{R} = 0$

$gR \cos \theta = v^2$

$v = \sqrt{gR \cos \theta}$

Substituting values we get

$v = \sqrt{9.8 \times 5 \times 0.6} \Rightarrow 5.42$ m/s

Maximum height attained $= \dfrac{u^2 \sin^2 \theta}{2g}$

$= \dfrac{9.8 \times 5 \times 0.6 \times (0.8)^2}{2 \times 9.8} = 0.97$ m

Sol 5: At maximum compression, velocity of both blocks will be equal so, let v be the final velocity of both the blocks x be the compression in the spring. Applying moment conservation, we get

$2 \times 10 + 5 \times 3 = (2 + 5)\, V$

$\Rightarrow v = \dfrac{35}{7} = 5$ m/s

Applying work energy theorem

$w_{net} = \Delta KE = KE_f - KE_i$

$-\dfrac{1}{2} kx^2 = \dfrac{1}{2}\,(5 + 2)\,(5)^2 - \left[\dfrac{1}{2}(5)(3)^2 + \dfrac{1}{2}(2)(10)^2 \right]$

$\Rightarrow \dfrac{1}{2} kx^2 = 100 + \dfrac{45}{2} - \dfrac{25 \times 7}{2}$

$x^2 = 0.0625$

$x = 0.25$ m

Sol 6: P = Power = F.V = constant

(a) $F = ma = \dfrac{mdv}{dt}$

$P = \dfrac{mdv}{dt} \cdot v$

$\int v\, dv = \int \dfrac{P}{m}\, dt$

$\int_0^v v\, dv = \dfrac{P}{m} \int_0^t dt$

[∵ p and m are constant]

$\dfrac{v^2}{2} = \dfrac{pt}{m}$

$v = \sqrt{\dfrac{2pt}{m}}$

(b) $F = ma = \dfrac{mvdv}{dx} \Rightarrow P = \dfrac{mvdv}{dx} v$

$\int_0^v v^2\, dv = \int_0^s \dfrac{Pdx}{m}$

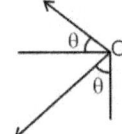

$$\frac{v^3}{3} = \frac{ps}{m}$$

$$s = \frac{mv^3}{3P} = \frac{m}{3P}\left(\frac{2Pt}{m}\right)^{3/2}$$

$$s = \left(\frac{8P}{9m}\right)^{1/2} t^{3/2}$$

Sol 7: Length of the spring at point A = $\dfrac{d}{\cos 37°} = \dfrac{5d}{3}$

By work energy theorem

$$w_{net} = \Delta KE$$

$$\underbrace{mgh}_{w_{gravity}} + \frac{1}{2}k(x^2 - 0^2) = \frac{1}{2}mv^2$$

$$[x = l - d = \frac{d}{4}]$$

$$h = l\sin\theta = \frac{5d}{4} \cdot \frac{3}{5} = \frac{3d}{4}$$

$$v = d\sqrt{\frac{3g}{2d} + \frac{k}{16m}}$$

Sol 8: (a) When mass m comes to rest for the first time kinetic energy of both the masses is zero.

work energy theorem

$$w_{net} = \Delta KE = KE_f - KE_i$$

$$-Mgh_2 + mgh_1 = 0 \Rightarrow Mh_2 = mh_1$$

$$\Rightarrow h_1 = \frac{5}{2}h_2 \qquad \qquad \dots (i)$$

Length of the string is constant, so

$$BC + Ac = A'C + B'C$$

$$BC - B'C = A'C - AC$$

$$h_2 + 0.8 = A'C$$

$$(0.8)^2 + h_1^2$$

$$= (h_2 + 0.8)^2$$

$$h_1^2 = h_2^2 + 1.6\,h_2$$

By (i)

$$\frac{25}{9}h_2^2 = h_2^2 + 1.6\,h_2$$

$$\frac{16h_2^2}{9} = 1.6\,h_2$$

$$h_2 = \frac{9}{10}$$

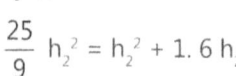

$$h_1 = \frac{9}{10} \times \frac{5}{3} = 1.5\ m$$

(b) Force equilibrium on M

$$T = Mg$$

Force equilibrium on m

$$T\sin\theta = mg$$

$$Mg\sin\theta = mg$$

$$\sin\theta = \frac{M}{m} = \frac{3}{5}$$

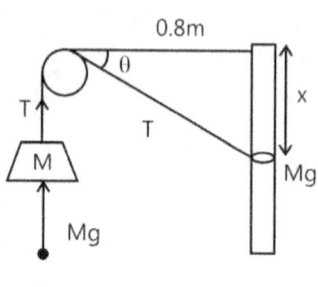

Sol 9: M "falls" and loses potential energy. This loss of potential energy is converted to gain in potential energy of m and gain in kinetic of energy for m and M both.

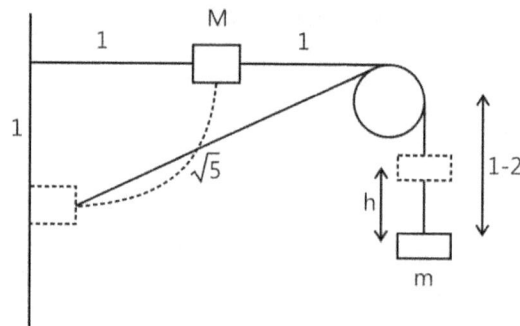

Let the total length of the string be l. So, the length of the hanging part in the beginning = l-2.

Since, total mechanical energy is conserved.

Loss in M.E. = Gain in M.E.

$$Mgl = \frac{1}{2}mv^2 + mgh + \frac{1}{2}MV^2 \dots\dots *$$

h can be obtained from the conservation of the length of the string.

$$h = l - 2 - (l - \sqrt{5} - 1) = \sqrt{5} - 1$$

We want V, v can be obtained in terms of V.

As M "falls", it moves in circular path with its velocity along the tangent. The velocity along the tangent can be resolved into two components, one along the length of the string and the other perpendicular to the length of the string. The component along the length of the string is same as the velocity of m as m always moves along the length of the string.

$$V\cos\theta = v$$

$$\cos\theta = \frac{2}{\sqrt{5}}$$

From,

$*, 2 \times 9.8 \times 1 = \frac{1}{2} \times 0.5 \times V^2 \cos^2 \theta + 0.5 \times 9.8 \times (\sqrt{5}-1) + \frac{1}{2} \times 2V^2$

V can be obtained.

Sol 10: Work energy theorem

$W_{net} = \Delta KE$

Since initial and final velocity of sand particles are zero
so $\Delta KE = 0$

$W_{gravity} = mg (0.2 + 0.01)$

$W_{net} = 0$

$W_{gravity} + W_{spring} = 0$

$(0.1) \times (10) \times (0.21) - \frac{1}{2} k (0.1)^2 = 0$

$k = 0.42 \times 10^4 = 4200$ N/m

Now if compression is 0.04 m

$W_{gravity} + W_{spring} = 0$

$\Rightarrow (0.1) \times (10)(h + 0.04) - \frac{1}{2} \times 4200 (0.04)^2 = 0$

$h + 0.04 = 2100 \times 16 \times 10^{-4}$

$h + 0.04 = 0.21 \times 16$

$h = 3.36 - 0.04 = 3.32$ m

Sol 11: Let the extension kx

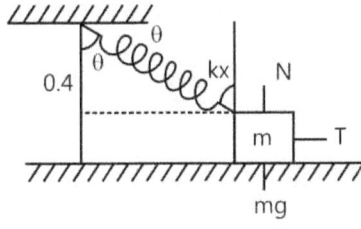

By Newton's second law

$h = 0.3$ m

$N - mg + kx \cos \theta = 0$

$N = 0$

$kx \cos \theta = mg$ ... (i)

By geometry $(x + 0.4) \cos \theta = 0.4$

By (i) $x \cos \theta = \dfrac{0.32 \times 10}{40} = \dfrac{3.2}{40} = 0.08$

$(x + 0.4) \dfrac{0.08}{x} = 0.4$

$x + 0.4 = 5x$

$x = 0.1$ m

By work energy theorem

$w = \dfrac{1}{2} mv^2 + \dfrac{1}{2} mv^2$

$-\dfrac{1}{2} \times 40 (0.1)^2 + 0.32 \times 10 \times 0.3 = 0.32 \, v^2$

$v = 1.54$ m/s

Sol 12:

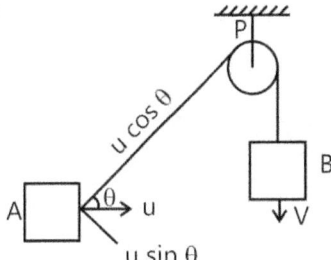

(a) Velocity of A along the string is equal velocity of B along the string

Length of string is constant

$\Rightarrow AP + BP = $ constant

Differentiate w.r.t

$\dfrac{d(AP)}{dt} + \dfrac{d(BP)}{dt} = 0$

$-u \cos \phi + v = 0$

$v = u \cos \phi$

(b)   When B strikes ground length

$BP = 6$ cm

So length of AP = 16 – 6 = 10 m

$\sin \phi = \dfrac{6}{10} = \dfrac{3}{5}$

$\cos \phi = \dfrac{4}{5} \Rightarrow u = \dfrac{v}{4/5} = \dfrac{5v}{4}$

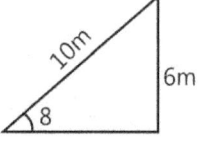

By work energy theorem

$W_{net} = \Delta KE$

$W_{gravity} + W_{string} = \dfrac{1}{2} mv^2 + \dfrac{1}{2} mu^2 - 0$

$mg \times 5 = \dfrac{1}{2} mv^2 + \dfrac{1}{2} m \left(\dfrac{25v^2}{16} \right)$

$10 g = v^2 + \dfrac{25}{16} v^2$

$$v^2 = \frac{10g \times 16}{41}$$

$$v = \frac{\sqrt{10g \times 16}}{41} = \frac{40}{\sqrt{41}} \text{ m/s}$$

Sol 13: By work energy theorem [Velocity of both blocks is same]

$$w_{net} = \Delta KE$$

$$W_{friction} + W_{gravity} = \frac{1}{2} m_1 u^2 + \frac{1}{2} m_2 u^2$$

$$-\mu m_1 g L + m_2 gL = \frac{1}{2}(m_1 + m_2) u^2$$

$$u = \sqrt{\frac{2(m_2 - \mu m_1)gL}{m_1 + m_2}}$$

Sol 14: Since $\alpha_c = k^2 r t^2$

$$\frac{v^2}{r} = k^2 r t^2 \Rightarrow v^2 = k^2 r^2 t^2$$

$$\Rightarrow v = krt \Rightarrow F = m\frac{dv}{dt} = mkr$$

Power = F.v = (mkr)×(krt) = $mk^2 r^2 t$

Sol 15:

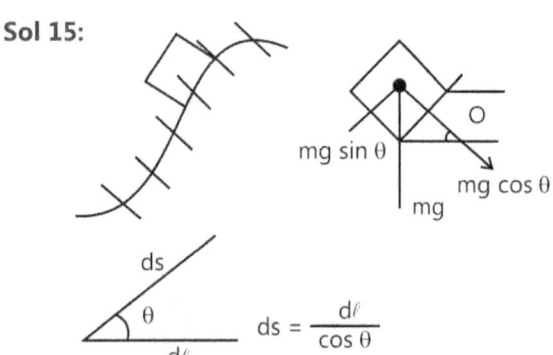

At any point on the path frictional force f = μ mg cos θ

θ = angle between path and the horizontal surface at some point

$$W_{friction} = -\int(\mu mg\cos\theta)\, ds$$

$$= -\int \mu mg\cos\theta\, \frac{d\ell}{\cos\theta} = -\int_0^\ell \mu mg\, d\ell = -\mu mg\ell$$

$$W_{gravity} = -mgh$$

By work energy theorem

$$w_{net} = \Delta KE$$

$$w_f + w_{gravity} + w_{frictoin} = 0$$

$$w_f - \mu mgl - mgh = 0$$

$$w_f = \mu\, mgl + mgh$$

Sol 16: At any point of time, let the length of chain remaining in tube be x

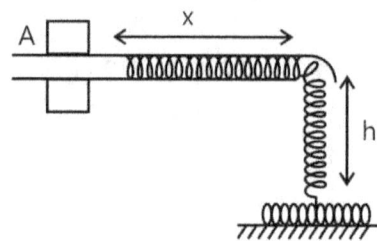

m′ = mass of chain above ground = $\frac{m}{\ell}(h + x)$

Now by Newton's second law on chain of length (x + h). As length of chain which has fallen on ground has no effect on the upward chain.

$$F = m'a$$

$$\left(\frac{m}{\ell}h\right)g = \frac{m(h+x)}{\ell}a$$

$$\frac{mhg}{\ell} = \frac{m(h+x)}{\ell}\left(-\frac{vdv}{dx}\right)$$

[∵ x is decreasing with increase in length]

$$\int_{\ell - h}^{0} -\frac{hg}{h+x}\, dx = \int_0^{v'} vdv$$

[v′ is the velocity of the end]

$$-\Big[hg\log(h+x)\Big]_{\ell-h}^{0} = \frac{v^2}{2}\Big]_0^{v'}$$

$$-hg\log\left(\frac{h}{\ell}\right) = \frac{v'^2}{2}$$

$$v'^2 = 2hg\log\left(\frac{\ell}{h}\right)$$

$$v' = \sqrt{2hg\log\left(\frac{\ell}{h}\right)}$$

Sol 17:

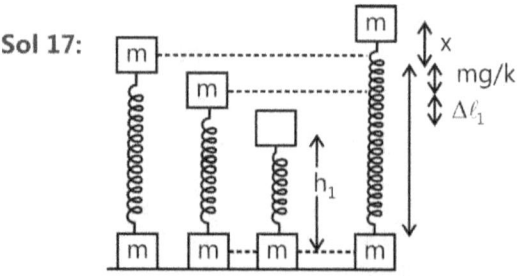

Let the natural length of spring be λ_1

Initially there is same compression x in spring in equilibrium

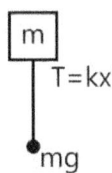

$$T = kx'$$

$$mg$$

$$mg = kx' \Rightarrow x' = \frac{mg}{k}$$

Now it is further compressed by $\Delta\lambda_1$ by thread

Now if thread is burnt it will go at upward extreme which is x distance above natural length of spring. Spring will just lift the lower block so by newton 2nd law; $T = kx = mg$

$$x = \frac{mg}{k}$$

By mechanical energy conservation

$$PE_{spring} + PE_{gravity} = PE'_{gravity} + PE'_{spring}$$

$$\frac{1}{2} k \left(\Delta\ell_1 + \frac{mg}{k} \right)^2 + mgh_1$$

$$= mg \left(\frac{2mg}{k} + \Delta\ell_1 + h_1 \right) + \frac{1}{2} k \left(\frac{mg}{k} \right)^2$$

$$\frac{1}{2} k \left(\Delta\ell_1^2 + \left(\frac{mg}{k} \right)^2 + 2\frac{mg}{k} \Delta\ell_1 \right)$$

$$= \frac{2(mg)^2}{k} + mg\Delta\lambda_1 + \frac{1}{2} k \left(\frac{mg}{k} \right)^2$$

$$\frac{1}{2} k \Delta\lambda_1^2 = \frac{2(mg)^2}{k}$$

$$\Delta\lambda_1 = \frac{2mg}{k}$$

To lift block of mass m

$$\Delta\ell > \Delta\lambda_1 \quad \frac{mg}{k}$$

$$\Delta\ell > \frac{3mg}{k}$$

(b)

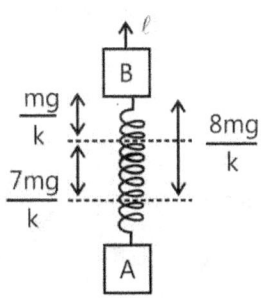

We will find the velocity of block B when block A will just lift upwards

$$\frac{1}{2} k \left(\frac{7mg}{k} \right)^2 - \frac{1}{2} k \left(\frac{mg}{k} \right)^2 - \frac{8mg}{k} mg = \frac{1}{2} mv^2$$

$$v^2 = \frac{32m^2g^2}{k}$$

Now block A and B together form a system with acceleration $-g$, $V_{cm} = \frac{v}{2}$

So,

$$v^2 = u^2 + 2as$$

$$0 = v_{cm}^2 - 2gs$$

$$v^2 = 8gs$$

$$s = \frac{v^2}{8g} = \frac{4mg}{k}$$

Movement of centre of gravity

$$= \left(\frac{8mg}{k} \right) \frac{m}{2m} + \frac{4mg}{k} = \frac{8mg}{k} \text{ upwards}$$

Sol 18: (a)

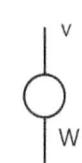

$$v$$

$$W$$

By Newton's second law

$$(w + f) = \left(\frac{w}{g} \right) a$$

$$a = -g \left(1 + \frac{f}{w} \right) = \frac{vdv}{dx}$$

$$\int_{v_0}^{0} vdv = -\int_{0}^{s} g \left(1 + \frac{f}{w} \right) dx$$

$$0 - \frac{v_0^2}{2} = -g \left(1 + \frac{f}{w} \right) s$$

$$s = \frac{v_0^2}{2g \left(1 + \frac{f}{w} \right)}$$

Final velocity = v

(b) By work energy theorem

$$W_{friction} + W_{gravity} = \Delta KE$$

$$-2fs + 0 = \frac{1}{2} m (v^2 - v_0^2); \quad v^2 - v_0^2 = \frac{-4fs}{m}$$

$$v^2 = \frac{-4b}{m} \cdot \frac{v_0^2}{2g\left(1 + \frac{f}{w}\right)} + v_0^2$$

$$\Rightarrow v^2 = v_0^2 \frac{[-4f + 2f + 2w]}{2(w+f)} \quad [mg = w] = v^2 = v_0^2 \frac{[w-f]}{[w+f]}$$

$$v = v_0 \left(\frac{w-f}{w+f}\right)^{1/2}$$

Sol 19: Let the speed of ring is v

Length of spring $= \dfrac{h}{\cos 37°} = \dfrac{5h}{4}$

By mechanical energy conservation

$PE_1 + KE_1 = PE_2 + KE_2$

$$\frac{1}{2} K \left(\frac{5h}{4} - h\right)^2 + 0 = 0 + \frac{1}{2} mv^2$$

$$K \cdot \frac{1}{2} \cdot \frac{h^2}{16} = \frac{1}{2} mv^2$$

$$v = \sqrt{\frac{k}{m}} \frac{h}{4}$$

Sol 20:

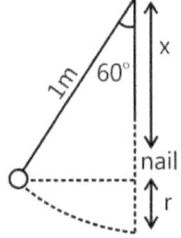

$x + r = 1$

For velocity of pendulum when string becomes vertical is v

Work energy theorem

$W_{gravity} = KE$

$m/g \,(1 - \cos 60°) = \frac{1}{2} mv^2$

$$v = \sqrt{2g \times \frac{1}{2}}$$

$v = \sqrt{g}$ m/s   ...(i)

For circular motion to be just completed

$v = \sqrt{5gr}$ ... (ii)

By (i) and (ii)

$\sqrt{g} = \sqrt{5gr}$

$r = \dfrac{1}{5}$ m

$x = \left(1 - \dfrac{1}{5}\right)$ m = 0.8 m

Sol 21:

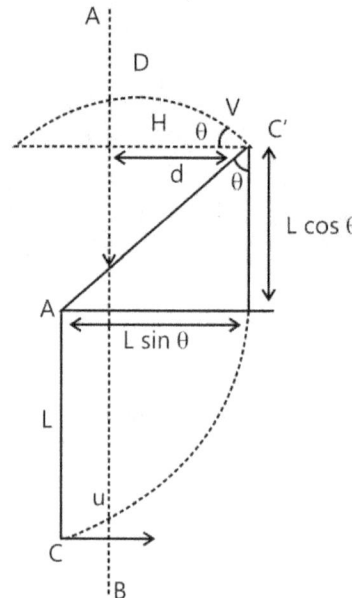

$$H = \frac{v^2 \sin^2 \theta}{2g}$$

By Newton's second law

$$mg \cos \theta = \frac{mv^2}{R}$$

$v^2 = g\ell \cos \theta$ ... (i)

Let us assume that pendulum leaves circular motion at point C with velocity v making an angle θ with horizontal

Applying work energy theorem from point S to C'

$W_{net} = \Delta K$

$$W_{gravity} = \frac{1}{2} m \, (v^2 - a^2)$$

$$-mgr \,(1 + \cos\theta) = \frac{1}{2} m \, (v^2 - u^2)$$

$v^2 = u^2 - 2g\ell \,(1 + \cos \theta)$ ... (ii)

From point C' to D it will follow parabolic path and velocity at line AB is horizontal

$$L \sin \theta - \frac{L}{4} = \frac{v^2 \sin 2\theta}{2g}$$ ... (iii)

By (i) and (ii)

$u^2 = g\ell\,(2 + 3\cos\theta)$

By (ii) and (iii)

$u = 2.14\,\sqrt{g\ell}$

Sol 22:

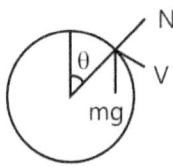

By work energy conservation

$w_1 = \Delta KE$

$mg\,R\,(1 - \cos\theta) = \dfrac{1}{2}\,mv^2$

$2mg\,(1 - \cos\theta) = \dfrac{mv^2}{R}$

By newton's 2nd law

$Mg\cos\theta - N = 2\,mg\,(1 - \cos\theta)$

$N = mg\,(3\cos\theta - 2)$

For $\theta \leq \cos^{-1}(2/3)$; $N_B = 0$,

$N_A = mg\,(3\cos\theta - 2)$

For $\theta \geq \cos^{-1}(2/3)$ $N_A = 0$,

$N_B = mg\,(2 - 3\cos\theta)$

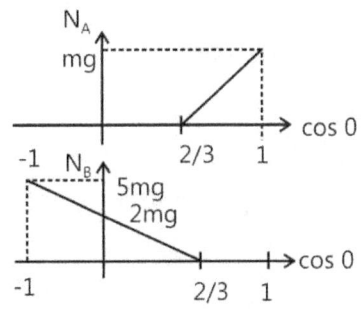

Sol 23: Let the initial compression be x

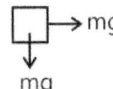

By Newton's second law F = ma

$mg = \dfrac{mv^2}{R}$

By work energy theorem

$w_{net} = \Delta K$

$w_{spring} + w_{gravity} = \dfrac{1}{2}\,mv^2 - 0$

$\dfrac{1}{2}\,kx^2 - mgR = \dfrac{1}{2}\,mgR$

$\dfrac{1}{2}\,kx^2 = \dfrac{3mgR}{2x}$

$x = \sqrt{\dfrac{3mgR}{k}}$

Sol 24:

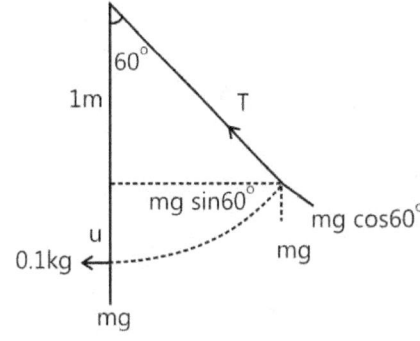

(a) By newton's second law

$T - mg\cos60° = \dfrac{mv^2}{R}$

$T = \dfrac{mg}{2} + \dfrac{m}{R}\,(4) = (4.9 + 4)\,m$

$= 8.9 \times 0.1 = 0.89\,N = 8.9 \times 10^4\,dyne$

(b) By work energy theorem

$w_{net} = \Delta kE$

$mg\ell\,(1 - \cos60°) = u^2 - v^2$

$u^2 = v^2 + mg\ell\,(1 - \cos60°)$

$= 4 + 0.1 \times 9.8 \times 1 \times \dfrac{1}{2} = 4 + 0.49$

$u^2 = 4.49$

$u = 2.12\,m/s$

Sol 25:

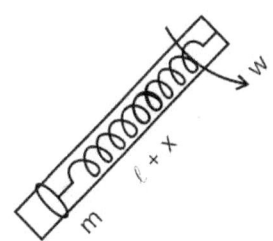

By Newton's second law

$Kx = m\omega^2\,(\ell + x)$

$x = \dfrac{m\omega^2\ell}{k - m\omega^2}$

Let $x' = \dfrac{m\omega^2}{k}$

$x = \dfrac{x'\ell}{1 - x'}$

By work energy theorem

$W_{net} = \Delta KE$

$W_{spring} + W_{force} = \dfrac{1}{2} m \omega^2 (\ell + x)^2$

$\dfrac{-1}{2} kx^2 + W_{force} = \dfrac{1}{2} m\omega^2 (\ell + x)^2$

$W_{force} = \dfrac{1}{2} kx^2 + \dfrac{1}{2} m\omega^2 (\ell + x)^2$

$= \dfrac{1}{2} k \dfrac{x'^2 \ell^2}{(1-x')^2} + \dfrac{1}{2} m\omega^2 \left(\ell + \dfrac{x'\ell}{1-x'} \right)^2$

$= \dfrac{k\ell^2}{2(1-x')^2} \left[x^2 + \left(\dfrac{m\omega^2}{k} \right) \right]$

$W_{force} = \dfrac{k\ell^2}{2(1-x')^2} [x'^2 + x']$

$= \dfrac{k\ell^2 x'(1+x')}{2(1-x')^2}$ where $x' = \dfrac{m\omega^2}{k}$

Sol 26: (a) Minimum speed is required so in the limiting case velocity of block at highest point is zero

By work energy theorem

$W_{net} = \Delta KE$

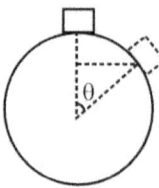

$W_{gravity} = 0 - \dfrac{1}{2} mu_0^2$

$- mg [L \sin \theta + R (1-\cos \theta)] = -\dfrac{1}{2} mu_0^2$

$u_0 = \sqrt{2g[L \sin\theta + R(1-\cos\theta)]}$

(b) Let the final velocity be v at top point

$W_{gravity} = \dfrac{1}{2} mv [v^2 - 4 u_0^2]$

$-mg [L \sin \theta + R (1 - \cos \theta)] = \dfrac{1}{2}m[v^2 - 4u_0^2]$

$v^2 = 3u_0^2$

By Newton's second law

$Force = ma = \dfrac{mv^2}{R} = \dfrac{m(3u_0^2)}{R}$

$= \dfrac{3m}{R} [2G(R(1-\cos\theta) + L \sin \theta]$

$= 6 mg [1 - \cos \theta + \dfrac{L}{R} \sin \theta]$

(c) If the projection speed is slightly greater than u_0, then speed at top most point is just than zero.

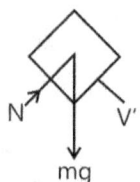

Particle will lose contact when normal just becomes zero.

So by Newton's second law

$mg \cos \theta = \dfrac{mV'^2}{R}$

$v'^2 = g R \cos \theta$

By work energy theorem, $W_{net} = \Delta KE$

$mg R (1-\cos \theta) = \dfrac{1}{2} mv'^2 = \dfrac{1}{2} mg R \cos \theta$

$2(1 - \cos \theta) = \cos \theta$

$2 - \cos \theta = - \cos \theta$

$\cos \theta = \dfrac{2}{3}$

So it will lose contact when particle makes an angle $\cos^{-1}\left(\dfrac{2}{3}\right)$ with vertical.

Sol 27: (a) By work energy theorem

$W_{net} = \Delta KE$

$mg (5R - R) = \dfrac{1}{2} mv^2$

$v^2 = 8gR$

By Newton's second law force exerted in horizontal direction $= \dfrac{mv^2}{R} = \dfrac{8mgR}{R} = 8mg$

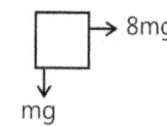

Net force $= \sqrt{8^2 + 1} = \sqrt{65}$ mg

By Newtons second law

$Force = \dfrac{mv^2}{R}$

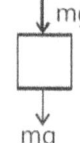

$mg + mg = \dfrac{mv^2}{R} \Rightarrow v^2 = 2gR$

Let the height be h

By work energy theorem

$mg(h - 2R) = \dfrac{1}{2} m\, 2gR$

$h - 2R = R \qquad \Rightarrow h = 3R$

Sol 28: (Coming to a stop) A particle can slide along a track with elevated ends and a flat central part, as shown in the figure below. The flat part has length L = 40 cm. The curved portions of the track are frictionless, but for the flat part the coefficient of kinetic friction is μ_k = 0.2. The particle is released from rest at point A, which is at height h = L/2. How far from the left edge of the flat part does the particle finally stop?

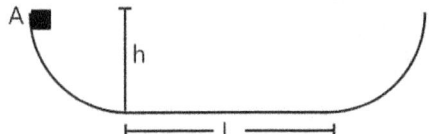

The initial energy is mgh = mgL/2. On the level ground, the particle experiences a constant friction force f = μ_k N = $\mu_k$mg. It will stop once the work W = -fs done by friction has dissipated all the initial energy:

$$mgL/2 = fs = \mu_k mgs \quad \Rightarrow \quad s = \frac{L}{2\mu_k} = 100 \, cm$$

So the particle will make two full passes (one moving right, one moving left) over the flat area, then stop halfway across (20 cm from the left edge) on its third trip.

Sol 29:

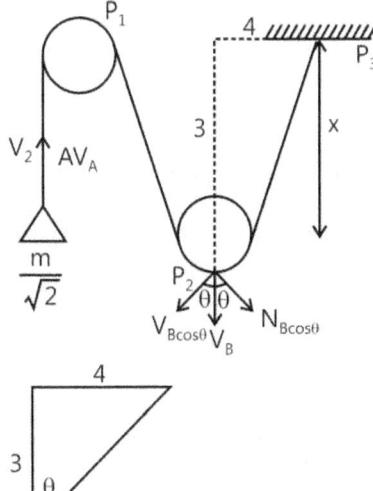

$$\cos\theta = \frac{3}{5}$$

$$\sin\theta = \frac{4}{5}$$

(a) Length of string is constant

$$\Rightarrow AP_1 + P_1P_2 + P_2P_3 = 0$$

$$\frac{d(AP_1)}{dt} + \frac{d(P_1P_2)}{dt} + \frac{d(P_2P_3)}{dt} = 0$$

$$-V_A + V_B + V_B \cos\theta = 0$$

$$V_A = V_B (\cos\theta + \cos\theta)$$

$$V_A = \frac{6}{5} V_B$$

Now, by work energy theorem

$$W_{net} = \Delta KE$$

$$W_{gravityA} + W_{gravityB} = \Delta KE \qquad \qquad .....(i)$$

Initially length of string between P_1 and P_3 is 8m

Finally length of string between P_1 and P_3 is 10 m so A has moved (10 – 8) upward

By (i)

$$-\frac{m}{\sqrt{2}} g \times 2 + mg \times 3 = \frac{1}{2} mv_B^2 + \frac{1}{2} \frac{m}{\sqrt{2}} v_A^2$$

$$mg[3 - \sqrt{2}] = \frac{1}{2} m \left[v_B^2 + \frac{36}{25} \frac{V_B^2}{\sqrt{2}} \right]$$

$$V_B = \sqrt{\frac{(3 - \sqrt{2})g \times 2}{\left(1 + \frac{36}{25\sqrt{2}}\right)}} \approx 4m/s$$

(b) Velocity of A and B is zero at maximum displacement

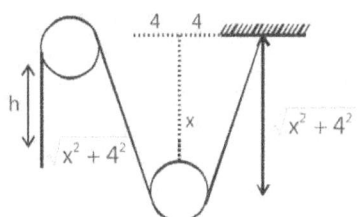

$$h = \sqrt{x^2 + 4^2} - 8$$

By work energy theorem

$$-\frac{m}{\sqrt{2}} gh + mg x = 0$$

$$h = \sqrt{2} x$$

$$2\sqrt{x^2 + h^2} - 8 = \sqrt{2} x$$

$$x = 8\sqrt{2} \, m$$

Exercise 2

Single Correct Choice Type

Sol 1: (B) Potential of water after falling down will convert in heat and sound. So temperature will increase slightly.

Sol 2: (A) By work energy theorem

$w_{net} = \Delta KE$

$w_{gravity} + w_{spring} = 0$

$mg\,(0.4 + x) + -\dfrac{1}{2}\,kx^2 = 0$

$20 \times 0.4 + 20\,x - 1000\,x^2 = 0$

$1000\,x^2 - 20\,x - 8 = 0$

$x = \dfrac{20 \pm \sqrt{400 + 32000}}{2000}$

$x = \dfrac{20 \pm 180}{2000} = \dfrac{200}{2000} = \dfrac{1}{10} = 0.1 \text{ m}$

Sol 3: (C) $F - \dfrac{-dv}{dR}$

$U = -\int F\,dR = -\left[\dfrac{k}{R^2}dR\right] = -\left[\dfrac{-k}{R}\right] = \dfrac{k}{R}$

Sol 4: (C) Mechanical energy is ME = KE + PE

Maximum potential energy is 160 J when Kinetic energy is zero i.e. at end points.

Sol 5: (B) $u = 2x^4 - 27\,x$

$F = \dfrac{-dU}{dx}$

$F = -[8x^3 - 27]$

at $x = 3/2$

Force is zero

$f\left(\dfrac{3^+}{2}\right) = \text{-ve}$

$F\left(\dfrac{3^-}{2}\right) = \text{+ve}$

So this is stable equilibrium

Sol 6: (A) Minimum speed must be zero as it is connected to rod so it will not leave the circular motion at any point in the path

Sol 7: (B)

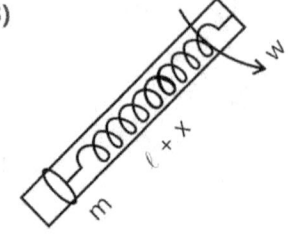

By Newton's second law

$Kx = m\omega^2\,(\ell + x)$

$x = \dfrac{m\omega^2\ell}{k - m\omega^2}$

Let $x' = \dfrac{m\omega^2}{k}$

$x = \dfrac{x'\ell}{1 - x'}$

By work energy theorem

$w_{net} = \Delta KE$

$w_{spring} + w_{force} = \dfrac{1}{2}\,m\,\omega^2\,(\ell + \lambda)^2$

$\dfrac{-1}{2}\,kx^2 + w_{force} = \dfrac{1}{2}\,m\omega^2\,(\ell + x)^2$

$w_{force} = \dfrac{1}{2}\,kx^2 + \dfrac{1}{2}\,m\omega^2\,(\ell + x)^2$

$= \dfrac{1}{2}\,k\,\dfrac{x'^2\,\ell^2}{(1 - x')^2} + \dfrac{1}{2}\,m\omega^2\left(\ell + \dfrac{x'\ell}{1 - x'}\right)^2$

$= \dfrac{k\ell^2}{2(1 - x')^2}\left[x^2 + \left(\dfrac{m\omega^2}{k}\right)\right]$

$w_{force} = \dfrac{k\ell^2}{2(1 - x')^2}\,[x'^2 + x']$

$= \dfrac{k\ell^2 x'(1 + x')}{2(1 - x')^2}$ where $x' = \dfrac{m\omega^2}{k}$

Sol 8: (B) $a_c = k^2\,rt^2 = \dfrac{v^2}{R}$

$v = krt$

$a_t = \dfrac{dv}{dt} = kr$

Power P = F.V= $ma_t.v = mkr.\,Krt = m\,k^2\,r^2\,t$

Sol 9: (D) Tension is zero as can and pendulum are falling freely under gravity

Sol 10: (D)

$f(x) = -kx + ax^2$

$U(x) = \int -f(x)\,dx$

$U(x) = -\left[-\dfrac{kx^2}{2} + \dfrac{ax^4}{4} \right] + c = \dfrac{kx^2}{2} \dfrac{-ax^4}{4} + c$

It corresponds to graph (D) for $c = 0$

Sol 11: (C) Power = F.v

Force = rate of change of linear momentum of wind

$\dfrac{dm}{dt} = \rho AV$ where ρ = density

A = area of blades

V = velocity

$F = \dfrac{d(m.v)}{dt} = v\dfrac{dm}{dt} = v\,\rho AV$

$F = \rho AV^2$

Power = ρAV^3

$P \propto v^2$

Sol 12: (B)

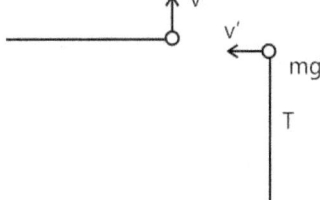

By force equilibrium

$F = T$

By Newton's second law

$T - mg = ma$

$T = m(g + a)$

By work energy theorem

$w_{net} = \Delta KE$

$w_{tension} + w_{gravity} = 20J$

$w_{tension} = 120 - w_{gravity}$

$w_{gravity} = -Mgh = -Mg\,v \times 1$

$= -Mg\sqrt{\dfrac{40}{M}} = -g\sqrt{40M}$

$KE = 20\,J = \dfrac{1}{2}Mv^2$

$v = \sqrt{\dfrac{40}{M}}$

Sol 13: (D)

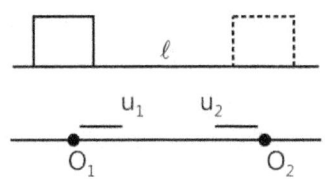

$u_1 + u_2 = V$

Acceleration will be same for both as acceleration of both observers is zero.

Kinetic energy will be different so by work energy theorem work done will also depend on kinetic energy.

Multiple Correct Choice Type

Sol 14: (A, D)

By Newton's second law

$T + mg = \dfrac{mv^2}{\ell}$

$T = 0$ at highest point

So string becomes slack at the highest point.

Sol 15: (B, D)

KE is increasing

$\Rightarrow$ Velocity is increasing

$\Rightarrow$ Resultant force must be at an angle less than 90° so that a component of force in the direction of velocity will increase its velocity

$\Rightarrow$ Linear momentum is increasing

Sol 16: (A, B)

In (C) and (D) work done by spring is $\dfrac{-1}{2}kx^2$ but in (A) and (B) work done is $\dfrac{+1}{2}kx^2$

Sol 17: (A, C, D) Work done = $\vec{F}.\overrightarrow{ds}$

(A) If force is always perpendicular to velocity then,

$\vec{F}.\overrightarrow{ds} = 0$

(B) If there is some initial velocity in the direction of force then, work done can be non-zero

(C), (D) Work done depends only on the displacement of point of application of force.

1.61

Sol 18: (C, D) Work done = 0 so kinetic energy is constant

Since in velocity and acceleration, direction is changing so they are not constant.

Sol 19: (A, B) By work energy theorem

$w_{net} = \Delta KE$

$mgh = \dfrac{1}{2} m (v_b^2 - v_i^2)$

So final velocity is larger than initial and will depend on speed of projection.

Sol 20: (A, B, D) By work energy theorem

$w_{net} = \Delta KE = 0$

$w_{you} + w_{gravity} = 0$

$w_{you} = -w_{gravity} = + mgh$

Assertion Reasoning Type

Sol 21: (B) Force has to be zero

Sol 22: (A)

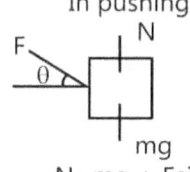   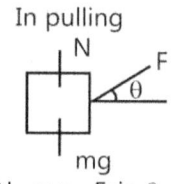

In pushing   In pulling

N = mg + Fsin θ   N = mg - Fsin θ

Sol 23: (A) $U = -\int \vec{F}.\vec{dr}$

Assume a closed loop in space

Since $\vec{F}$ is a conservative force, the line integral is zero. Thus U is a state function.

So potential energy is defined only for conservative force.

Sol 24: (B) By work energy theorem $w_{net} = \Delta KE$

$F.\Delta S = \dfrac{1}{2} mv^2$; $v \propto \dfrac{1}{\sqrt{m}}$

Comprehension Type

Sol 25: (A) Work done in raising box = $-w_{gravity}$

= –(–mgh) = mgh

1 – false

Sol 26: (D) There is no friction and non-conservative force so mechanical energy is conserved.

Sol 27: (B) As there is sliding at t =0 so friction will act opposite to the direction of velocity

Sol 28: (B) $\omega > \dfrac{v_{cm}}{R}$

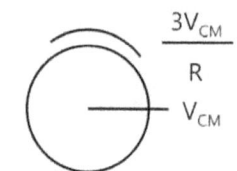

So friction will act in the direction of velocity to increase the velocity and decrease the angular acceleration

Sol 29: (C)

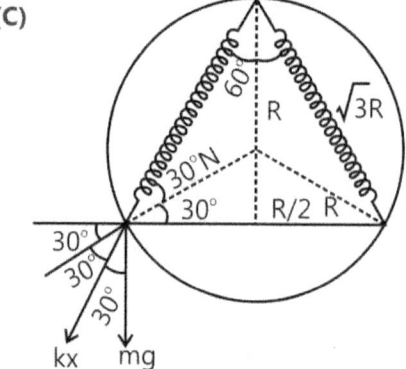

Since acceleration of bead in the normal direction is zero. So by Newton's second law

N – kx cos 30° = mg cos 60°

$N = \dfrac{mg}{2} + \dfrac{kx\sqrt{3}}{2} = \dfrac{mg}{2} + \dfrac{\sqrt{3}}{2} \dfrac{(2+\sqrt{3})mg}{\sqrt{3}R} (2-\sqrt{3}) R$

$= \dfrac{mg}{2} + \dfrac{mg}{2} = mg$

Sol 30: (D) Newton's second law in direction perpendicular to normal

(i) mg cos 30° + kx cos 60° = ma

$\dfrac{mg\sqrt{3}}{2} + \dfrac{(2+\sqrt{3})mg}{\sqrt{3}R} (2-\sqrt{3}) \times \dfrac{1}{2} = ma$

$\dfrac{mg\sqrt{3}}{2} + \dfrac{mg}{2\sqrt{3}} = ma;$

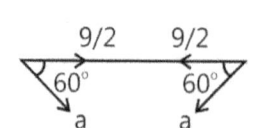

$\dfrac{mg}{2}\left[\sqrt{3} + \dfrac{1}{\sqrt{3}}\right] = ma$

$a = \dfrac{2g}{\sqrt{3}}$

From figure relative acceleration is $\dfrac{a}{2} + \dfrac{a}{2} = a = \dfrac{2g}{\sqrt{3}}$

Sol 31: (C) By mechanical energy conservation

$PE_i + KE_i = PE_f + KE_f$

$$\frac{mgR}{x} + \frac{1}{2}kx^2 + 0 = 0 + \frac{1}{2}mv^2$$

$$\frac{mgR}{2} + \frac{(2+\sqrt{3})mg}{\sqrt{3}R}(2-\sqrt{3})^2 R^2 = mv^2$$

$$v^2 = \frac{gR}{2} + \left(\frac{2}{\sqrt{3}} - 1\right)gR$$

$$v = \sqrt{\frac{2gR}{\sqrt{3}}}$$

Sol 32: (D) (A) Wrong, collision can be inelastic

(B) In perfectly inelastic collision energy is not conserved

(C) For SHM, θ should be small.

(D) At the instant of collision, they are at the bottom

$\Rightarrow \Sigma F = 0$ and $\Sigma M = 0$

$\Rightarrow$ Momentum conserved

Match the Columns

Sol 33: A→s; B→q; C→r; D→p

Total mechanical energy is conserved

$U(x) + KE =$ constant

At $x = 5$

$U(x) = 29$

$KE = 20$

Total ME at any $x = 49$

maximum P.E. $= 49$

maximum K.E. $= 49 - U_{min} = 49 - 20 = 29$

When $U(x)$ is maximum then x will take extreme values.

$20 + (x - 2)^2 = 49$

$(x - 2)^2 = 29$

$x - 2 = +\sqrt{29}$; $x - 2 = -\sqrt{29}$

$x = 7.38$; $x = -3.38$

Sol 34: (A→r; B-p, C-q, D-s)

Work done by gravity

$w = +mgh = 0.72 \times 10 \times 15$

$w_2 = 10800$

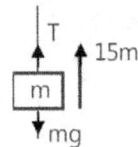

By Newton's second law

$T - mg = ma$; $T = m(g + a)$

$$T = 72\left(R + \frac{g}{10}\right) = \frac{11g}{10} \times 72 = 72 \times 11 = 792$$

Work done $w_1 = 792 \times 15 = 11880$

By work energy theorem

$w_{string} + w_{gravity} = \frac{1}{2}mv^2$

$11880 - 10800 = KE$

$KE = 1080$

$KE = \frac{1}{2}mv^2 = 1080$

$v = 5.47$ m/s

Previous Years' Questions

Sol 1: Given $t = \sqrt{x} + 3$

or $\sqrt{x} = (t - 3)$ ...... (i)

$\therefore x = (t - 3)^2 = t^2 - 6t + 9$ ....(ii)

Differentiating this equation with respect to time, we get

Velocity $v = \dfrac{dx}{dt} = 2t - 6$ .... (iii)

(a) $v = 0$ when $2t - 6 = 0$ or $t = 3s$

Substituting in Eq. (i), we get

$\sqrt{x} = 0$ or $x = 0$

i.e., displacement of particle when velocity is zero is also zero.

(b) From eq. (iii) speed of particle

At $t = 0$ is $v_i = |v| = 6$ m/s

At $t = 6$ s is $v_f = |v| = 6$ m/s

From work energy theorem,

Work done = change in kinetic energy

$$= \frac{1}{2}m[v_f^2 - v_i^2] = \frac{1}{2}m[(6)^2 - (6)^2] = 0$$

Sol 2: $s = vt = 2 \times 5 = 10$ m

Q = work done against friction

$= \mu mgs = 0.2 \times 2 \times 9.8 \times 10 = 39.2$ J $= 9.33$ cal

Sol 3: Normal reaction between blocks A and C will be zero. Therefore, there will be no friction between them.

Both A and B are moving with uniform speed. Therefore net force on them should be zero.

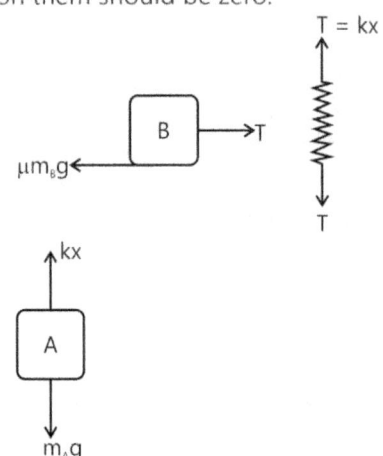

For equilibrium of A

$m_A g = kx$

$\therefore x = \dfrac{m_A g}{k} = \dfrac{(2)(9.8)}{1960} = 0.01$ m

For equilibrium of B $\mu m_B g = T = kx = m_A g$

$m_B = \dfrac{m_A}{\mu} = \dfrac{2}{0.2} = 10$ kg

Energy stored in spring

$U = \dfrac{1}{2}kx^2 = \dfrac{1}{2}(1960)(0.01)^2 = 0.098$ J

Sol 4: Let M strikes with speed v. Then, velocity of m at this instant will be v cos θ or $\dfrac{2}{\sqrt 5}$ v. Further M will fall a distance of 1 m while m will rise up by $(\sqrt 5 - 1)$ m. From energy conservation: decrease in potential energy of M = increase in potential energy of m + increase in kinetic energy of both the blocks.

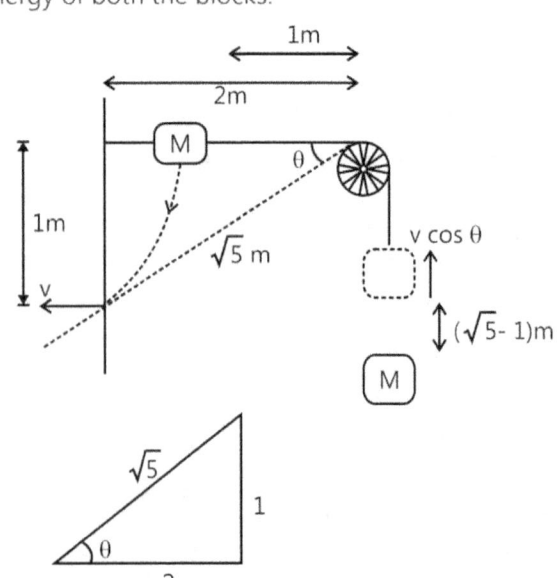

or $(2)(9.8)(1) = (0.5)(9.8)(\sqrt 5 - 1)$

$+ \dfrac{1}{2} \times 2 \times v^2 + \dfrac{1}{2} \times 0.5 \times \left(\dfrac{2v}{\sqrt 5}\right)^2$

Solving this equation, we get v = 3.29 m/s

Sol 5: (a) At the highest point, velocity of bullet is 50cosθ. So, by conservation of linear momentum

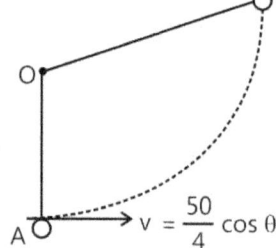

M(50 cos θ) = 4 Mv

$\therefore v = \left(\dfrac{50}{4}\right)\cos\theta$ ... (i)

At point B. T = 0 but v ≠ 0

Hence, $4 Mg \cos 60° = \dfrac{(4M)v^2}{\ell}$

or $v^2 = \dfrac{g}{2}\ell = \dfrac{50}{3}$ ... (ii)

$\left(\text{as } \ell = \dfrac{10}{3}\text{m amd } g = 10 m/s^2\right)$

Also, $v^2 = u^2 - 2gh = u^2 - 2g\left(\dfrac{3}{2}\ell\right) = u^2 - 3(10)\left(\dfrac{10}{3}\right)$

or $v^2 = u^2 - 100$

or solving eqs. (i), (ii) and (iii), we get

cos θ = 0.86 or θ = 30°

(b) $x = \dfrac{\text{Range}}{2} = \dfrac{1}{2}\left(\dfrac{u^2 \sin 2\theta}{g}\right)$

$= \dfrac{50 \times 50 \times \sqrt 3}{2 \times 10 \times 2} = 108.25$ m

$y = H = \dfrac{u^2 \sin^2 \theta}{2g} = \dfrac{50 \times 50 \times 1}{2 \times 10 \times 4} = 31.25$ m

Hence, the desired coordinates are (108.25 m, 31.25 m).

Sol 6: Let the string slack s at point Q as shown in figure. From P to Q path is circular and beyond Q path is parabolic. At point C, velocity of particle becomes horizontal therefore. QD = half the range of the projectile

Now, we have following equations

(1) $T_Q = 0$. Therefore, mg sin θ $= \dfrac{mv^2}{L}$ ... (i)

1.64

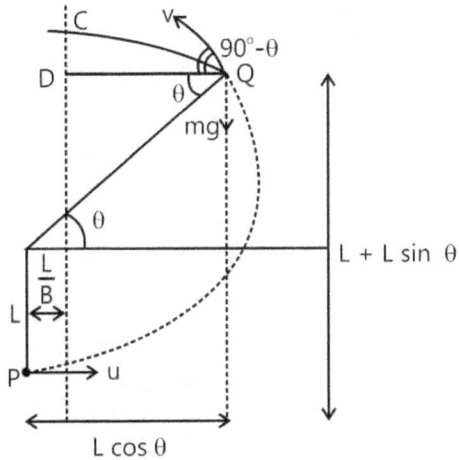

(2) $v^2 = u^2 - 2gh = u^2 - 2gL(1 + \sin\theta)$ ... (ii)

(3) $QD = \dfrac{1}{2}$ (Range)

$\Rightarrow \left(L\cos\theta - \dfrac{L}{g}\right) = \dfrac{v^2 \sin 2(90° - \theta)}{2g} = \dfrac{v^2 \sin 2\theta}{2g}$ ... (iii)

Eq. (iii) can be written as

$\left(\cos\theta - \dfrac{1}{g}\right) = \left(\dfrac{v^2}{gL}\right) \sin\theta \cos\theta$

Substituting value of $\left(\dfrac{v^2}{gL}\right) = \sin\theta$ from Eq. (i), we get

$\left(\cos\theta - \dfrac{1}{8}\right) = \sin^2\theta \cos\theta = (1 - \cos^2\theta) \cos\theta$

or $\cos\theta - \dfrac{1}{8} = \cos\theta - \cos^3\theta$

$\therefore \cos^3\theta = \dfrac{1}{8}$ or $\cos\theta = \dfrac{1}{2}$ or $\theta = 60°$

$\therefore$ From Eq. (i) $v^2 = gL \sin\theta = gL \sin 60°$

or $v^2 = \dfrac{\sqrt{3}}{2} gL$

$\therefore$ Substituting this value of v^2 in eq. (ii)

$u^2 = v^2 + 2gL (1 + \sin\theta)$

$= \dfrac{\sqrt{3}}{2} g L + 2gL \left(1 + \dfrac{\sqrt{3}}{2}\right) = \dfrac{3\sqrt{3}}{2} gL + 2gL$

$= gL\left(2 + \dfrac{3\sqrt{3}}{2}\right)$

$u = \sqrt{gL\left(2 + \dfrac{3\sqrt{3}}{2}\right)}$

Sol 7: $a = \dfrac{\text{Net pulling force}}{\text{Total mass}} = \dfrac{0.72g - 0.36g}{0.72 + 0.36} = \dfrac{g}{3}$

$S = \dfrac{1}{2} at^2 = \dfrac{1}{2}\left(\dfrac{g}{3}\right)(1)^2 = \dfrac{g}{6}$

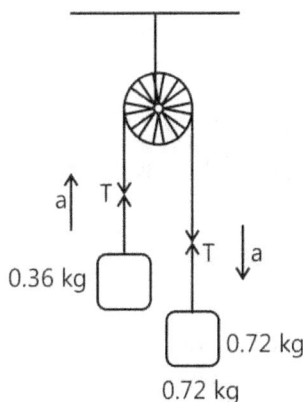

0.36 kg

0.72 kg

0.72 kg

$T - 0.36 g = 0.36 a = 0.36 \dfrac{g}{3}$

$\therefore T = 0.48 g$

Now, $W_T = TS \cos 0°$ (on 0.36 kg mass)

$= (0.48 g)\left(\dfrac{g}{6}\right)(1) = 0.08(g^2) = 0.08(10)^2 = 8$ J

Sol 8: (D) Decrease in mechanical energy

= work done against friction

$\therefore \dfrac{1}{2} mv^2 - \dfrac{1}{2} kx^2 = \mu\, mgx$

or $v = \sqrt{\dfrac{2\mu\, mgx + kx^2}{m}}$

Substituting the values, we get

$v = 0.4$ m/s $= \left(\dfrac{4}{10}\right)$m/s

$\therefore$ Answer is D

Sol 9: (D)

$dw = F.dr = F.(dx\,\hat{i} + dy\hat{j})$

$= K\displaystyle\int \dfrac{xdx}{\left(x^2 + y^2\right)^{3/2}} + \dfrac{ydy}{\left(x^2 + y^2\right)^{3/2}}$

$x^2 + y^2 = a^2$

$w = \dfrac{K}{a^3}\displaystyle\int_a^0 xdx + \int_0^a ydy = \dfrac{K}{a^3}\left(\dfrac{-a^2}{2} + \dfrac{a^2}{2}\right) = 0$

Sol 10: (5) The initial speed of 1st bob (suspended by a string of length l_1) is $\sqrt{5gl_1}$.

The speed of this bob at highest point will be $\sqrt{gl_1}$.

When this bob collides with the other bob there speeds will be interchanged.

$$\sqrt{gl_1} = \sqrt{5gl_2} \Rightarrow \frac{l_1}{l_2} = 5$$

Sol 11: (5) Power $= \dfrac{dW}{dt} \Rightarrow W = 0.5 \times 5 = 2.5 = KE_f - KE_i$

$$2.5 = \frac{M}{2}\left(v_f^2 - v_i^2\right) \Rightarrow v_f = 5$$

Sol 12: (B) $\dfrac{d(KE)}{dt} = mv\dfrac{dv}{dt}$

Sol 13: (D) Condition for not sliding,

$f_{max} > (m_1 + m_2)\, g \sin\theta$

$\mu N > (m_1 + m_2)\, g \sin\theta$

$0.3\, m_2\, g \cos\theta \geq 30 \sin\theta$

$6 \geq 30 \tan\theta$

$1/5 \geq \tan\theta$

$0.2 \geq \tan\theta$

$\therefore$ for P, Q

$f = (m_1 + m_2)\, g \sin\theta$

For R and S

$F = f_{max} = \mu m_2 g \sin\theta$

Sol 14: (A, C)

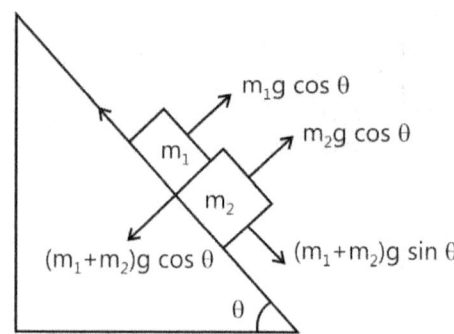

As $I_1 = I_2$

$n_1 w_1 d_1 v_1 = n_2 w_2 d_2 v_2$

Now, $\dfrac{V_2}{V_1} = \dfrac{B_2 v_2 w_2}{B_2 v_1 w_1} = \left(\dfrac{B_2 w_2}{B_1 w_1}\right)\left(\dfrac{n_1 w_1 d_1}{n_2 w_2 d_2}\right) = \dfrac{B_2 n_1}{B_1 n_2}$

2. CENTRE OF MASS AND THE LAW OF CONSERVATION OF MOMENTUM

1. INTRODUCTION

In this chapter we will study the motion of system of particles or bodies. The individual particles or bodies comprising the system in the general case move with different velocities and accelerations and exert forces on each other and are influenced by external or surrounding bodies as well. We will learn the techniques to simplify the analysis of complicated motion of such a system. We will also learn about the dynamics of extended bodies whose shape and/or mass changes during their motion. We define the linear momentum of a system of particles and introduce the concept of center of mass of a system. The dynamics of center of mass and the law of conservation of linear momentum are important tools in the study of system of particles.

2 CENTER OF MASS

When we study the dynamics of the motion of a system of particles as a whole, then we need not bother about the dynamics of individual particles of the system, but only focus on the dynamics of a unique point corresponding to that system. The motion of this unique point is identical to the motion of a single particle whose mass is equal to the sum of the masses of all the individual particles of the system and the resultant of all the forces exerted on all the particles of the system, by the surrounding bodies, or due to action of a field of force, is exerted directly to that particle. This point is called the center of mass (COM) of the system of particles. The COM behaves as if the entire mass of the system is concentrated there. The concept of COM is very useful in analyzing complicated motion of system of objects, in particular, when two or more objects collide or an object explodes into fragments.

2.1 Center of Mass of a System of Particles

For a system of n particles, having masses m_1, m_2, m_3 ..... m_n and position vectors $\vec{r}_1$, $\vec{r}_2$, $\vec{r}_3$,........$\vec{r}_n$ respectively with respect to the origin in a certain reference frame, the position vector of center of mass, $\vec{r}_{cm}$ with respect to the origin is given by

$$\vec{r}_{cm} = \frac{m_1\vec{r}_1 + m_2\vec{r}_2 + ... + m_n\vec{r}_n}{m_1 + m_2 + ... + m_n}$$

If the total mass of the system is M, then $M\vec{r}_{cm} = m_1\vec{r}_1 + m_2\vec{r}_2 + ... + m_n\vec{r}_n$

Co-ordinates of center of mass are

$$x_{cm} = \frac{x_1 m_1 + x_2 m_2 + \ldots\ldots x_m m_n}{m_1 + m_2 + \ldots\ldots m_n}; \qquad y_{cm} = \frac{y_1 m_1 + y_2 m_2 + \ldots\ldots y_m m_n}{m_1 + m_2 + \ldots\ldots m_n}; \qquad z_{cm} = \frac{z_1 m_1 + z_2 m_2 + \ldots\ldots z_m m_n}{m_1 + m_2 + \ldots\ldots m_n}$$

For a system comprising of two particles of masses m_1, m_2, positioned at co-ordinates (x_1, y_1, z_1) and (x_2, y_2, z_1), respectively, we have

$$X_{com} = \frac{m_1 x_1 + m_2 x_2}{m_1 + m_2}; \qquad Y_{com} = \frac{m_1 y_1 + m_2 y_2}{m_1 + m_2}; \qquad Z_{com} = \frac{m_1 z_1 + m_2 z_2}{m_1 + m_2}$$

NOMORECLASS CONCEPTS

For a two-particle system, COM lies closer to the particle having more mass, which is rather obvious. If COM's co-ordinates are made zero, we would clearly observe that distances of individual particles are inversely proportional to their masses.

Illustration 1: Two particles of masses 1 kg and 2 kg are located at x = 0 and x = 3 m respectively. Find the position of their center of mass. **(JEE MAIN)**

Sol: For the system of particle of masses m_1 and m_2, if the distance of particle from the center of mass are r_1 and r_2 respectively then it is seen that $m_1 r_1 = m_2 r_2$.

Since, both the particles lie on x-axis, the COM will also lie on the x-axis. Let the COM be located at x = x, then r_1 = distance of COM from the particle of mass 1 kg = x

Figure 6.1

and r_2 = distance of COM from the particle of mass 2 kg = (3 − x)

Using $\quad \dfrac{r_1}{r_2} = \dfrac{m_2}{m_1} \quad$ or $\quad \dfrac{x}{3-x} = \dfrac{2}{1} \quad$ or x = 2 m

Thus, the COM of the two particles is located at x = 2 m.

Illustration 2: Four particles A, B, C and D having masses m, 2m, 3m and 4m respectively are placed in order at the corners of a square of side a. Locate the center of mass. **(JEE ADVANCED)**

Figure 6.2

Sol: The co-ordinate of center of mass of n particle system are given as

$$X = \frac{\sum\limits_i m_i x_i}{\sum\limits_i m_i}, \quad Y = \frac{\sum\limits_i m_i y_i}{\sum\limits_i m_i}$$

Take the x and y axes as shown in Fig. 6.2. The coordinates of the four particles are as follows:

Particle	Mass	x-coordinate	y-coordinate
A	m	0	0 (taking A as origin)
B	2m	a	0
C	3m	a	a
D	4m	0	a

Hence, the coordinates of the center of mass of the four-particle system are

$$X = \frac{m \cdot 0 + 2ma + 3ma + 4m \cdot 0}{m + 2m + 3m + 4m} = \frac{a}{2}; \quad Y = \frac{m \cdot 0 + 2m \cdot 0 + 3ma + 4ma}{m + 2m + 3m + 4m} = \frac{7a}{10}$$

The center of mass is at $\left(\dfrac{a}{2}, \dfrac{7a}{10}\right)$.

2.2. Center of Mass of a Continuous Body

For continuous mass distributions, the co-ordinates of center of mass are determined by following formulae,

$$X_{cm} = \frac{\int x\,dm}{\int dm}; \qquad Y_{cm} = \frac{\int y\,dm}{\int dm}; \qquad Z_{cm} = \frac{\int z\,dm}{\int dm}$$

where x, y and z are the co-ordinates of an infinitesimal elementary mass dm taken on the continuous mass distribution. The integration should be performed under proper limits, such that the elementary mass covers the entire body.

NOMORECLASS CONCEPTS

Many people have misconception that the center of mass of a continuous body must lie inside the body. Center of mass of a continuous body may lie outside that body also. e.g. Ring.

(a) **Center of Mass of a Uniform Straight Rod**

Center of mass of a rod of length L is at (L/2, 0, 0).

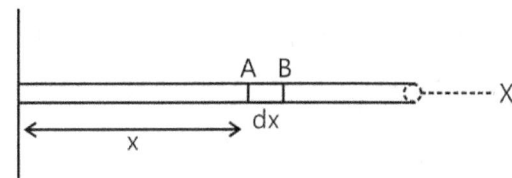

Figure 6.3

(b) **Center of Mass of a Uniform Semicircular Wire**

Center of mass of a Uniform Semicircular Wire of radius R is $(0, 2R/\pi)$.

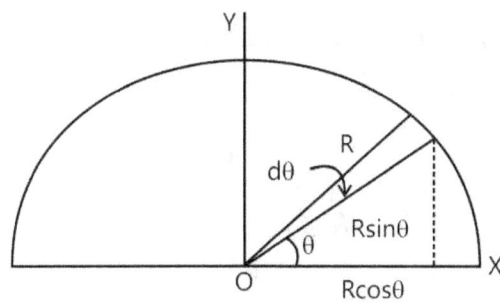

Figure 6.4

(c) **Center of Mass of a Uniform Semicircular Plate**

Center of mass of a uniform semicircular plate of radius R is $(0, 4R/3\pi)$

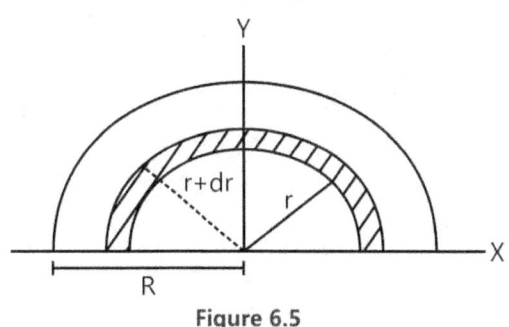

Figure 6.5

(d) Center of mass of a uniform hollow cone

Center of mass of a uniform hollow cone of height H lies on the axis at a distance of H/3 from the center of the bottom.

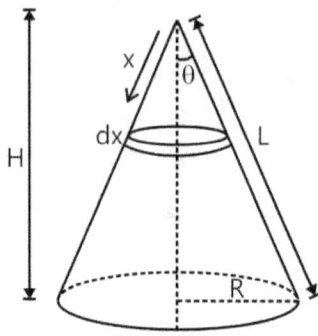

Figure 6.6

NOMORECLASS CONCEPTS

Student must solve the above integrations to get a better view of how we take infinitesimal segment of a body and the corresponding limits to integrate over whole body.

Please solve the integrations for hollow cone and solid cone to note the difference.

Illustration 3: A rod of length L is placed along the x-axis between x = 0 and x = L. The linear density (mass/length) ρ of the rod varies with the distance x from the origin as ρ = a + bx. Here, a and b are constants. Find the position of center of mass of this rod. **(JEE MAIN)**

Sol: To find C.O.M of continuous mass distributions consider a small element of distribution of mass dm. Then the co-ordinate of C.O.M. is given as

$\therefore x_{COM} = \dfrac{\int x\,dm}{M}$ the limits of integration should be chosen such that the small elements covers entire mass distribution.

Choose an infinitesimal element of the rod of length dx situated at co-ordinates (x, 0, 0) (see Fig.6.7) The linear mass density can be assumed to be constant along the infinitesimal length dx.

Thus the mass of the element dm = rdx = (a + bx) dx

As x varies from 0 to L the element covers the entire rod.

Therefore, x-coordinate of COM of the rod will be

Figure 6.7

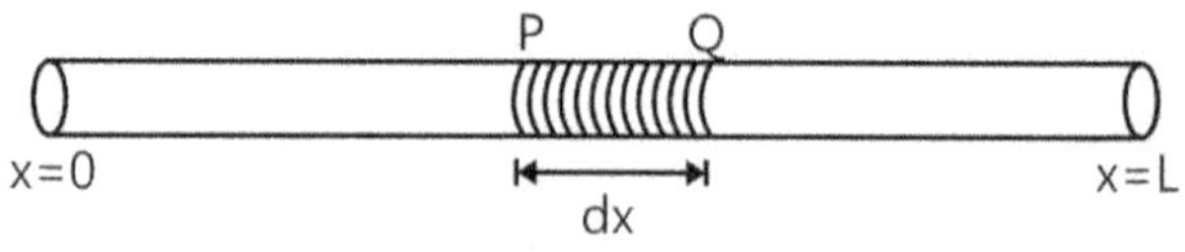

$$x_{COM} = \frac{\int_0^L x\,dm}{\int_0^L dm} = \frac{\int_0^L (x)\,(a+bx)dx}{\int_0^L (a+bx)dx} = \frac{\left[\frac{ax^2}{2}+\frac{bx^3}{3}\right]_0^L}{\left[ax+\frac{bx^2}{2}\right]_0^L} = \frac{3aL^2+2bL^3}{6} \times \frac{2}{2aL+bL^2}$$

$$x_{COM} = \frac{3aL+2bL^2}{6a+3bL}\text{ m}$$

Illustration 4: Determine the center of mass of a uniform solid cone of height h and semi angle α, as shown in Fig. 6.8 **(JEE MAIN)**

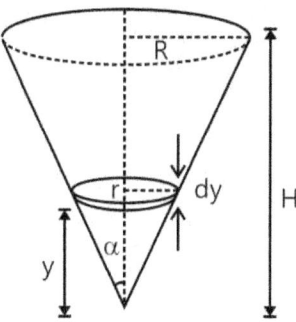

Figure 6.8

Sol: To find C.O.M of continuous mass distributions consider a small element of distribution of mass dm. Then the co-ordinate of C.O.M. is given as

$\therefore Y_{COM} = \dfrac{\int y\, dm}{M}$ the limits of integration should be chosen such that the small elements covers entire mass distribution.

We place the apex of the cone at the origin and its axis along the y-axis. As the cone is a right circular cone then by symmetry it is clear that the center of mass will lie on its axis i.e. on the y-axis. We consider an elementary disk of radius r and infinitesimal thickness dy whose center is on the y-axis at distance y from the origin as shown in Fig. 6.8. The volume of such a disk is

$dV = \pi r^2\, dy = \pi (y \tan \alpha)^2\, dy$

The mass of this elementary disk is dm = rdV. As y varies from 0 to H, the total height of the cone, the elementary disc covers the entire cone. The total mass M of the cone is given by,

$M = \int dm = \pi\rho \tan^2 \alpha \int_0^H y^2 dy = \pi\rho \tan^2 \alpha \dfrac{H^3}{3}$ ....(i)

The position of the center of mass is given by

$y_{com} = \dfrac{1}{M}\int_0^H y\, dm = \dfrac{1}{M}\pi\rho \tan^2 \alpha \int_0^H y^3 dy$

$= \dfrac{1}{M}\pi\rho \tan^2 \alpha \dfrac{H^4}{4}$ ....(ii)

From equations (i) and (ii), we have $y_{com} = \dfrac{3H}{4}$

3. CENTER OF GRAVITY

Definition: Center of gravity is a point, near or within a body, at which its entire weight can be assumed to act when considering the motion of the body under the influence of gravity. This point coincides with the center of mass when the gravitational field is uniform.

Note: The center of mass and center of gravity for a continuous body or a system of particles will be different when there is non- uniform gravitational field.

NOMORECLASS CONCEPTS

You can find the center of gravity and center of mass for a very thin cylinder extending from the surface of earth to the height equal to radius of earth to get the difference. Just sum up all the individual weights of infinitesimal size disks and find the position where gravity will make the same weight of body. This will give center of gravity.

4. CENTER OF MASS OF THE BODY WHEN A PORTION OF THE BODY IS TAKEN OUT

Suppose there is a body of total mass m and a mass m_1 is taken out from this body. The remaining body will have mass $(m - m_1)$ and its center of mass will be at coordinates,

$$x_{cm} = \frac{mx - m_1 x_1}{(m - m_1)}; \quad y_{cm} = \frac{my - m_1 y_1}{(m - m_1)}; \quad z_{cm} = \frac{mz - m_1 z_1}{(m - m_1)}$$

where (x, y, z) are coordinates of center of mass of original (whole) body and (x_1, y_1, z_1) are coordinates of center of mass of the portion taken out.

Illustration 5: A circular plate of uniform thickness has a diameter of 56 cm. A circular portion of diameter 42 cm is removed from one edge of the plate as shown in Fig. 6.9. Find the center of mass of the remaining position.

(JEE MAIN)

Sol: Let O be the center of circular plate and, O_1, the center of circular portion removed from the plate. The COM of the whole plate will lie at O and the COM of the circular cavity will lie at O_1. Let O be the origin. So OO_1 = 28cm – 21cm = 7cm.

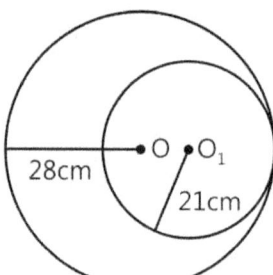

The center of mass of the remaining portion will be given as

$$x_{cm} = \frac{mx - m_1 x_1}{(m - m_1)} = \frac{\sigma(Ax - A_1 x_1)}{\sigma(A - A_1)} = \frac{\pi((28)^2 0 - (21)^2 7)}{\pi((28)^2 - (21)^2)}$$

$$x_{cm} = -9 \text{ cm} = -0.09 \text{ m}.$$

This means that center of mass of the remaining plate is at a distance 9 cm from the center of given circular plate opposite to the removed portion i.e. in this questioon, the new Centre of Mass will shift 9 cm left.

Figure 6.9

5. MOTION OF THE CENTER OF MASS

For a n-particle system of total mass M and individual particles having mass m_1, m_2, ... m_n, from the definition of center of mass we can write,

$$M\vec{r}_{cm} = m_1 \vec{r}_1 + m_2 \vec{r}_2 + ... + m_n \vec{r}_n$$

where $\vec{r}_{cm}$ is the position vector of the center of mass, and $\vec{r}_1$, $\vec{r}_2$, ... $\vec{r}_n$ are the position vectors of the individual particles relative to the same origin in a particular reference frame.

If the mass of each particle of the system remains constant with time, then, for our system of particles with fixed mass, differentiating the above equation with respect to time, we obtain.

$$M\frac{d\vec{r}_{cm}}{dt} = m_1 \frac{d\vec{r}_1}{dt} + m_2 \frac{d\vec{r}_2}{dt} + ... + m_n \frac{d\vec{r}_n}{dt} \qquad ....(i)$$

$$\text{or } M\vec{V}_{cm} = m_1 \vec{v}_1 + m_2 \vec{v}_2 + ... + m_n \vec{v}_n$$

Where $\vec{v}_1$, $\vec{v}_2$, ... $\vec{v}_n$ are the velocities of the individual particles, and $\vec{V}_{cm}$ is the velocity of the center of mass. Again differentiating with respect to time, we obtain

$$M\frac{d\vec{V}_{cm}}{dt} = m_1 \frac{d\vec{v}_1}{dt} + m_2 \frac{d\vec{v}_2}{dt} + ... + m_n \frac{d\vec{v}_n}{dt}$$

$$M\vec{a}_{cm} = m_1 \vec{a}_1 + m_2 \vec{a}_2 + ... + m_n \vec{a}_n \qquad ....(ii)$$

Where $\vec{a}_1$, $\vec{a}_2$, ... $\vec{a}_n$ are the accelerations of the individual particles, and $\vec{a}_{cm}$ is the acceleration of the center of mass. Now, from Newton's second law, the force F_i acting on the ith particle is given by $F_i = m_i a_i$. Then, above equation can be written as

$$M\vec{a}_{cm} = \vec{F}_1 + \vec{F}_2 + ... + \vec{F}_n = \vec{F}_{internal} + \vec{F}_{external}$$ ....(iii)

Internal forces are the forces exerted by the particles of the system on each other. However, from Newton's third law, these internal forces occur in pairs of equal and opposite forces, so their net sum is zero. $\therefore M\vec{a}_{cm} = \vec{F}_{ext}$

This equation states that the center of mass (C.O.M) of a system of particles behaves as if all the mass of the system were concentrated there and the resultant of all the external forces acting on all the particles of the system was applied to it (at C.O.M).

Concept: Whatever may be the rearrangement of the bodies in a system, due to **internal forces** (such as different parts of the system moving away or towards each other or an internal explosion taking place, breaking a body into fragments) provided net F_{ext}=0, we have two possibilities:

(a) If the system as a whole was originally at rest, i.e. the C.O.M was at rest, then the C.O.M. will continue to be at rest.

(b) If before the change, the system as a whole had been moving with a constant velocity (C.O.M was moving with a consant velocity), it will continue to move with a constant velocity.

In presence of a net external force if the C.O.M had been moving with certain acceleration at the instant of an explosion, in a particular trajectory, the C.O.M. will continue to move in the same trajectory, with the same acceleration, as if the system had never exploded at all.

Briefly saying, any internal changes of the body do not effect the motion of C.O.M.

Illustration 6: A man of mass m is standing on a platform of mass M kept on smooth ice (see Fig. 6.10). If the man starts moving on the platform with a speed v relative to the platform, with what velocity relative to the ice does the platform recoil ? **(JEE MAIN)**

Figure 6.10

Sol: When net external force on system is zero, the C.O.M. will either remain at rest or continue the state of motion. i.e. V_{cm} = constant

Let velocity of platform be $\vec{V}$. If velocity of man relative to platform is $\vec{v}$ then the velocity of man in reference frame of ice is $\vec{w} = \vec{v} + \vec{V}$.

Center of mass of the system comprising of "man and the platform" is initially at rest and as no horizontal external force acts on this system (ice is smooth), the center of mass will continue to remain at rest.

$$\vec{V}_{cm} = 0 = \frac{M\vec{V} + m(\vec{v} + \vec{V})}{M + m}$$

or $(M + m)\vec{V} + m\vec{v} = 0$

or $\vec{V} = -\dfrac{m\vec{v}}{M + m}$ ms^{-1}

Negative sign shows that the platform will move in the opposite direction of relative velocity of man.

Illustration 7: Two block of masses m_1 and m_2 connected by a weightless spring of stiffness k rest on a smooth horizontal plane (see Fig 6.11). Block 2 is shifted by a small distance x to the left and released. Find the velocity of the center of mass of the system after block 1 breaks off the wall. **(JEE ADVANCED)**

Sol: Elastic potential energy stored in spring will get converted in kinetic energy of the blocks. If we consider the FBD of mass m_1 at the instant when it breaks off the wall, the normal reaction from the wall is zero, but normal reaction from the wall is equal to is equal to force exerted by spring on mass m_1 so at this instant, force by spring is also zero.

The initial potential energy of compression is $= \frac{1}{2}kx^2$

When the block m_1 breaks off from the wall, the normal reaction from the wall is zero, which in turn means that the tension in the spring is zero. Thus the spring has its natural length at this instant and the kinetic energy of the block m_2 is given by

$$\frac{1}{2}m_2 v_2^2 = \frac{1}{2}kx^2$$

$$v_2^2 = \frac{kx^2}{m_2}$$

$$v_2 = x\sqrt{\frac{k}{m_2}}$$

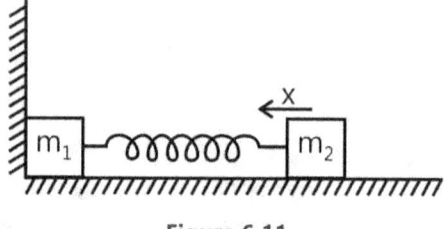

Figure 6.11

Velocity of center of mass is

$$V_{cm} = \frac{m_1 v_1 + m_2 v_2}{m_1 + m_2}$$

At start $v_1 = 0$

$$\therefore V_{cm} = \frac{m_2}{m_1 + m_2}v_2 = \frac{m_2 x}{m_1 + m_2}\sqrt{\frac{k}{m_2}}$$

$\therefore$ Velocity of center of mass of system $V_{cm} = \dfrac{x\sqrt{km_2}}{m_1 + m_2}$ ms^{-1}

6. LINEAR MOMENTUM

The quantity momentum (denoted as $\vec{P}$) is a vector defined as the product of the mass of a particle and its velocity $\vec{v}$, i.e. $\qquad \vec{P} = m\vec{v}$ ....(i)

From Newton's second law of motion, if mass of a particle is constant

$$\vec{F} = m\vec{a} = m\frac{d\vec{v}}{dt} = \frac{d}{dt}(m\vec{v}) = \frac{d\vec{P}}{dt}$$

Thus, for constant m, the rate of change of momentum of a body is equal to the resultant force acting on the body and is in the direction of that force.

For a system of n particles with masses m_1, m_2 ..... etc., and velocities v_1, v_2 ..... etc. respectively, the total momentum $\vec{P}$ in a particular reference frame is,

$$\vec{P} = \vec{P}_1 + \vec{P}_2 + \dots\dots + \vec{P}_n = m_1\vec{v}_1 + m_1\vec{v}_2 + \dots\dots + m_n\vec{v}_n;$$

or $\vec{P} = M\vec{V}_{cm}$

Also, $\qquad \dfrac{d\vec{P}}{dt} = M\dfrac{d\vec{V}_{cm}}{dt} = M\vec{a}_{cm} = \vec{F}_{ext}$

$\therefore \qquad \dfrac{d\vec{P}}{dt} = \vec{F}_{ext}$

The magnitude of linear momentum may be expressed in terms of the kinetic energy as well.

$$p = mv$$

or
$$p^2 = m^2v^2 = 2m\left(\frac{1}{2}mv^2\right) = 2mK$$

Thus,
$$p = \sqrt{2Km} \text{ or } K = \frac{p^2}{2m}$$

6.1. Law of Conservation of Linear Momentum

"The law of conservation of linear momentum states that if no external forces act on a system of particles, then the vector sum of the linear momenta of the particles of the system remains constant and is not affected by their mutual interaction. In other words the total linear momentum of a closed system remains constant in an inertial reference frame."

Proof: For a system of fixed-mass particles having total mass m we have

$$\vec{F}_{ext} = m\vec{a}_{cm} = m\frac{d\vec{v}_{cm}}{dt} = \frac{d(m\vec{v}_{cm})}{dt} = \frac{d\vec{P}}{dt}, \text{ where } \vec{P} \text{ is the total momentum of the system.}$$

Thus,
$$\vec{F}_{ext} = \frac{d\vec{P}}{dt}$$

In case the net external force applied to the system is zero, we have

$$\vec{F}_{ext} = \frac{d\vec{P}}{dt} = 0 \quad \text{or} \quad \vec{P} = \text{constant}$$

Thus for a closed system, the total linear momentum of the system remains constant in an inertial frame of reference.

NOMORECLASS CONCEPTS

Both linear momentum and kinetic energy are dependent on the reference frame since velocity is inclusively dependent on the frame of reference.

Illustration 8: A gun (mass = M) fires a bullet (mass = m) with speed v_r relative to barrel of the gun which is inclined at an angle of 60° with horizontal. The gun is placed over a smooth horizontal surface. Find the recoil speed of gun. **(JEE MAIN)**

Sol: When a bullet is fired, gun recoils in backward direction. Using law of conservation of linear momentum we can find the recoil velocity of gun.

Let the recoil velocity of gun be $\vec{v}$. The relative velocity of the bullet is $\vec{v}_r$ at an angle of 60° with the horizontal. Taking gun + bullet as the system the net external force on the system in horizontal direction is zero. Let x-axis be along the horizontal and bullet be fired towards the positive direction of x-axis. Initially the system was at rest. Therefore, applying the principle of conservation of linear momentum along x-axis, we get

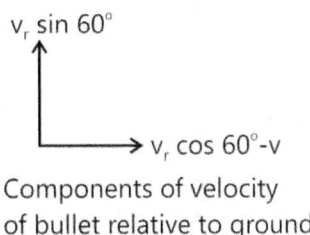

Components of velocity
of bullet relative to ground

Figure 6. 12

$$Mv_x + m(v_{rx} + v_x) = 0$$

$$-Mv + m(v_r \cos 60° - v) = 0$$

$$v = \frac{mv_r \cos 60°}{M+m}$$

or $\quad v = \dfrac{mv_r}{2(M+m)} \ ms^{-1}$

Illustration 9: The block of mass $m_1 = 2kg$ and $m_2 = 5kg$ are moving in the same direction along a frictionless surface with speeds 10 ms^{-1} and 3 ms^{-1}, respectively m_2 being ahead of m_1 as shown in Fig. 6.13. An ideal spring with spring constant K = 1120 N/m is attached to the back side of m_2. Find the maximum compression of the spring when the blocks move together after the collision. **(JEE ADVANCED)**

Figure 6.13

Sol: As frictional force on the blocks is zero the total momentum of blocks can be conserved during collision. At the instant of maximum compression some part of initial total K.E. of blocks is stored as elastic P.E. in the spring.

Let v be the final velocity of the system after collision when the blocks move together.

Applying the law of conservation of momentum, we have

$$m_1u_1 + m_2u_2 = (m_1 + m_2)v$$

Substituting the values,

$$(2 \times 10) + (5 \times 3) = (2 + 5)v$$

$$v = \frac{35}{7} = 5 \ m/s$$

Applying the law of conservation of energy we get

$$\frac{1}{2}m_1u_1^2 + \frac{1}{2}m_2u_2^2 = \frac{1}{2}(m_1 + m_2)v_2^2 + \frac{1}{2}Kx^2$$

$$m_1u_1^2 + m_2u_2^2 = (m_1 + m_2)v_2 + Kx^2$$

$$2 \times (10)^2 + 5 \times (3)^2 = [(2+5) \times (5)^2] + 1120\,x^2$$

$$x^2 = \frac{70}{1120} = \frac{1}{16} \Rightarrow x = 0.25m$$

NOMORECLASS CONCEPTS

In the above questions, note that the compression would be maximum when the relative velocity between the blocks is zero.

7. VARIABLE MASS

From Newton's second law, $\vec{F}_{ext} = m\vec{a}$ is applicable to a system whose total mass m is constant. If total mass of the system is not constant, then this form of Newton's second law is not applicable. If at a certain moment of time the total mass of a system is m and a mass dm is added (or separated) to the system, then we apply $\vec{F}_{ext} = \dfrac{d\vec{p}}{dt}$ to the system comprising "m+dm" to get

$$\vec{F}_{ext}.dt = d\vec{p} = \vec{p}_{final} - \vec{p}_{initial} = (m+dm)(\vec{v}+d\vec{v}) - [m\vec{v} + dm(\vec{v}+\vec{u})]$$

$$\text{or } \vec{F}_{ext}.dt = md\vec{v} - dm\vec{u}; \qquad (dm.d\vec{v} \approx 0)$$

$$\text{or } \vec{F}_{ext} = m\dfrac{d\vec{v}}{dt} - \dfrac{dm}{dt}\vec{u}$$

$$\text{or } m\dfrac{d\vec{v}}{dt} = \vec{F}_{ext} + \dfrac{dm}{dt}\vec{u}$$

where u is velocity of adding or separating mass dm relative to the system having instantaneous mass m and instantaneous velocity v with respect to an inertial reference frame. The term $\dfrac{dm}{dt}$ can be positive or negative depending upon whether mass is added to the system or mass is separating from the system.

Problems related to variable mass can be solved in following three steps.

(a) Make a list of all the external forces acting on the main mass and draw its FBD.

(b) Apply an additional thrust force or reaction force $\vec{F}_t$ on the main mass, due to the action of the added(separated) mass on the main mass, the magnitude of which is $\left|\vec{u}\left(\pm\dfrac{dm}{dt}\right)\right|$ and direction is given by the direction of $\vec{u}$ in case the mass is being added or the direction of $-\vec{u}$ if mass is being separated.

(c) Apply the equation

$$m\dfrac{d\vec{v}}{dt} = \vec{F}_{ext} + \dfrac{dm}{dt}\vec{u} \quad (m = \text{instantaneous mass})$$

Illustration 10: A flat cart of mass m_0 at t=0 starts moving to the left due to a constant horizontal force F. The sand spills on the flat cart from a stationary hopper. The rate of loading is constant and equal to μ kg/s. Find the time dependence of the velocity and the acceleration of the flat cart in the process of loading. The friction is negligibly small. **(JEE ADVANCED)**

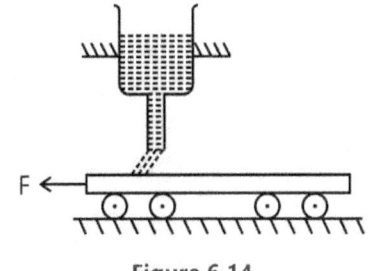

Figure 6.14

Sol: The hopper is at rest in K frame, so in the frame of the cart its initial velocity will be u=-v, where v is velocity of cart in K frame. Here we have used the equation of motion of variable mass $m\dfrac{dv}{dt} = F_{ext} + \dfrac{dm}{dt}u$

The rate of increase of mass of the flat car $\dfrac{dm}{dt} = \mu$ kgs^{-1}

The hopper is stationary and so its relative velocity is u = 0 - v = - v

The equation of motion is given by

$$m\dfrac{dv}{dt} = F + \dfrac{dm}{dt}u = F - \mu v \quad \left[\because \dfrac{dm}{dt} = \mu \text{ and } u = -v\right]$$

At the instant t, $m = m_0 + \mu t$

$$\therefore \dfrac{dv}{F-\mu v} = \dfrac{dt}{m} = \dfrac{dt}{m_0 + \mu t} \quad \Rightarrow \quad \int_0^v \dfrac{dv}{F-\mu v} = \int_0^t \dfrac{dt}{m_0 + \mu t}$$

$$\text{or, } -\dfrac{1}{\mu}\log_e \dfrac{F-\mu v}{F} = \dfrac{1}{\mu}\log_e \dfrac{m_0 + \mu t}{m_0}$$

or $\quad \log_e \dfrac{F}{F - \mu v} = \log_e \dfrac{m_0 + \mu t}{m_0} \Rightarrow \quad v = \dfrac{Ft}{m_0 + \mu t}\ ms^{-1}$

The acceleration a is given by

$$a = \dfrac{dv}{dt} = \dfrac{Fm_0}{\left(m_0 + \mu t\right)^2}\ ms^{-2}$$

Alternative:

$$m\dfrac{dv}{dt} = F + \dfrac{dm}{dt}u = F - \dfrac{dm}{dt}v \qquad \text{or } m\dfrac{dv}{dt} + \dfrac{dm}{dt}v = F$$

$$\text{or } \dfrac{d}{dt}(mv) = F \Rightarrow \int_0^{mv} d(mv) = \int_0^t F\,dt$$

$$\text{or } mv = Ft \Rightarrow v = \dfrac{Ft}{(m_0 + \mu t)}; \quad (\because m = m_0 + \mu t)$$

8. ROCKET PROPULSION

The propulsion of rocket is an example of a system of variable mass. In the combustion chamber of a rocket, the fuel is burnt in the presence of an oxidizing agent due to which a jet of gases emerges from the tail of the rocket. Thus the mass of the rocket is continuously decreasing. This action due to emission of gases in the backward direction produces a reaction force in the forward direction due to which the rocket moves forward.

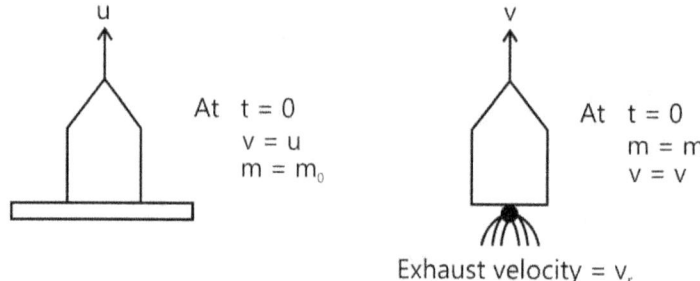

Figure 6.15: Rocket propulsion

Let m_0 be the mass of the rocket and u be its velocity at time t = 0, and m be its mass and v be its velocity at any time t. (see Fig. 6.15)

The mass of the gas ejected per unit time or the rate of change of mass of the rocket is $-\dfrac{dm}{dt}$ and v_r be the exhaust velocity of the gases relative to the rocket. Usually $-\dfrac{dm}{dt}$ and v_r are assumed constant throughout the journey of the rocket.

Now using the equation of motion for a system of variable mass derived in the previous article we get,

$$m\dfrac{d\vec{v}}{dt} = m\vec{g} + \vec{v}_r\dfrac{dm}{dt}$$

$$\text{or } \dfrac{d\vec{v}}{dt} = \vec{g} + \dfrac{\vec{v}_r}{m}\dfrac{dm}{dt}$$

$$\text{or } d\vec{v} = \vec{v}_r\dfrac{dm}{m} + \vec{g}.dt$$

This is a vector equation and we do not assume any sign of $\dfrac{dm}{dt}$. It is taken to be positive. After evaluating the definite integrals, when we substitute the scalar components of the vectors with proper signs we get the correct result.

Integrating on both sides, we get $\int\limits_{\vec{u}}^{\vec{v}} d\vec{v} = \vec{v}_r \int\limits_{m_0}^{m} \dfrac{dm}{m} + \vec{g}.\int\limits_{0}^{t} dt$

or $\vec{v} - \vec{u} = \vec{v}_r \ln\dfrac{m}{m_0} + \vec{g}.t$

or $\vec{v} = \vec{u} + \vec{g}.t + \vec{v}_r \ln\dfrac{m}{m_0}$

Now taking upwards direction as positive and downwards as negative (g and v_r are downwards and u is upwards) we get,

$v = u - g.t + (-v_r)\ln\dfrac{m}{m_0}$

Thus, $v = u - gt + v_r \ln\left(\dfrac{m_0}{m}\right)$

Now if $-\dfrac{dm}{dt} = \mu$ (constant), then $m = m_0 - mt$

Thus, $v = u - gt + v_r \ln\left(\dfrac{m_0}{m_0 - \mu t}\right)$

If the initial velocity of the rocket u = 0, and the weight of the rocket is ignored as compared to the reaction force

of the escaping gases, the above equation reduces to $v = v_r \ln\left(\dfrac{m_0}{m}\right)$

NOMORECLASS CONCEPTS

The concept of variable mass can also be physically visualized by changing the reference frame to the instantaneous velocity of body. In that case mass is either being added by constant speed or being removed by a constant speed. Considering the dm mass and the body as a system, and writing the equations of conservation of momentum one can see the magic!

Illustration 11: (a) A rocket set for vertical firing weighs 50 kg and contains 450 kg of fuel. It can have a maximum exhaust velocity of 2000 m/s. What should be its minimum rate of fuel consumption?

(i) To just lift it off the launching pad?

(ii) To give it an acceleration of 20 m/s²?

(b) What will be the speed of the rocket when the rate of consumption of fuel is 10 kg/s after whole of the fuel is consumed? (Take g = 9.8 m/s²) **(JEE ADVANCED)**

Sol: To lift the rocket upward against gravity, the thrust force in the upward direction due to exiting gases should be greater than or equal to the gravitational force. During motion the mass of rocket decreases till whole of its fuel is consumed. Final velocity of rocket is

$v = u - gt + v_r \ln\left(\dfrac{m_0}{m}\right)$.

(a) (i) To just lift the rocket off the launching pad

Initial weight = thrust force

or $\quad m_0 g = v_r \left(-\dfrac{dm}{dt} \right)$; or $-\dfrac{dm}{dt} = \dfrac{m_0 g}{v_r}$

Substituting the values, we get $-\dfrac{dm}{dt} = \dfrac{(450+50)(9.8)}{2 \times 10^3} = 2.45$ kg/s

(ii) Net acceleration $a = 20$ m/s^2

$\therefore \quad ma = F_t - mg$

or $\quad m(a+g) = F_t = v_r \left(-\dfrac{dm}{dt} \right)$

This gives, $\left(-\dfrac{dm}{dt} \right) = \dfrac{m(g+a)}{v_r}$

Substituting the values, we get $\left(-\dfrac{dm}{dt} \right) = \dfrac{(450+50)(9.8+20)}{2 \times 10^3} = 7.45$ kg/s

(b) The rate of fuel consumption is 10 kg/s. So, the time for the consumption of entire fuel is

$$t = \dfrac{450}{10} = 45 \text{ s}$$

The formula for speed of the rocket at time t is, $v = u - gt + v_r \ln \left(\dfrac{m_0}{m} \right)$

Here, $u = 0$, $v_r = 2 \times 10^3$ m/s, $m_0 = 500$ kg and $m = 50$ kg

Substituting the values, we get $v = 0 - (9.8)(45) + (2 \times 10^3) \ln \left(\dfrac{500}{50} \right)$

or $v = -441 + 4605.17$; \qquad or $v = 4164.17$ m/s; \qquad or $v = 4.164$ km/s

9. COLLISION

An event in which two or more bodies exert forces on each other for a relatively short time is called collision. If net external force acting on the system of bodies is zero, then according to the law of conservation of linear momentum, the total momentum of the system of bodies before and after the collision remains constant.

9.1 Classification of Collisions

Collisions are classified into following types on the basis of the degree of conservation of kinetic energy in a collision:

(a) **Elastic Collision**: If the total kinetic energy of the colliding particles is conserved before and after the collision, the collision is said to be an elastic collision. If two bodies of masses m_1 and m_2 moving with velocities u_1 and u_2 respectively, collide with each other so that their final velocities after collision are v_1 and v_2 respectively, then the collision will be perfectly elastic if,

$$\dfrac{1}{2}m_1 u_1^2 + \dfrac{1}{2}m_2 u_2^2 = \dfrac{1}{2}m_1 v_1^2 + \dfrac{1}{2}m_2 v_2^2$$

(b) **Inelastic Collision**: If the total kinetic energy of the colliding particles is not conserved before and after the collision, the collision is said to be inelastic collision. The kinetic energy is partially converted into other forms of energy like sound, heat, deformation energy etc.

(c) **Perfectly Inelastic Collision**: The collision is said to be perfectly inelastic if the collision results in "sticking together" of the colliding particles after which they move as a single unit with the same velocity.

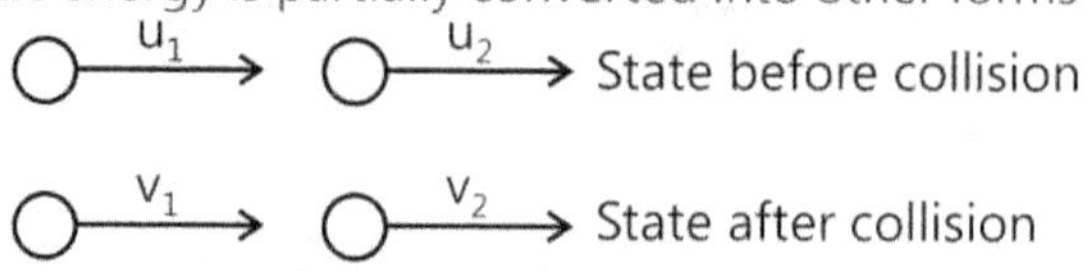

Figure 6.16: Collision in one dimension

Collisions can also be classified on the basis of the line of action of the forces of interaction.

(i) Head- on collisions: A collision is said to be head-on if the direction of the velocities of each of the colliding bodies are along the line of action of the forces of interaction acting on the bodies at the instant of collision.

(ii) Oblique collisions: A collision is said to be oblique if the direction of the velocities of the colliding bodies are not along the line of action of the forces of interaction acting on the bodies at the instant of collision. Just after collision, at least one of the colliding bodies moves in a direction different from the initial direction of motion.

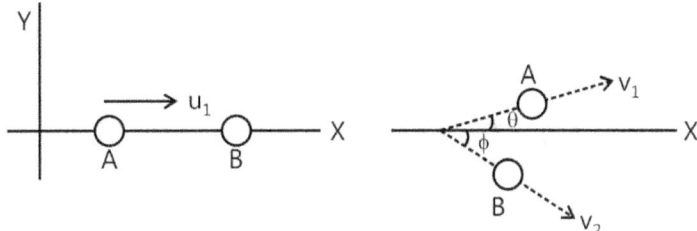

Figure 6.17: Collision in two dimensions

10. COEFFICIENT OF RESTITUTION

Coefficient of restitution is a measure of the elasticity of a collision between two particles. It is defined as the ratio of relative velocity of one of the particles with respect to the other particle after the collision to the relative velocity of the same particle before the collision and the ratio is negative.

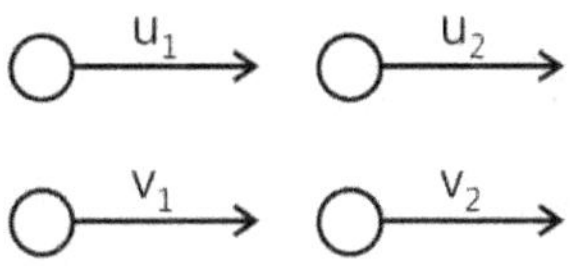

Figure 6.18: Velocities before and after collision

If the velocities of two particles before the collision are u_1 and u_2 respectively and their velocities after the collision are v_1 and v_2 respectively (see Fig. 6.18), then $\dfrac{v_1 - v_2}{u_1 - u_2} = -e$

The coefficient of restitution is also expressed as the ratio of velocity of separation after collision to the velocity of approach before collision.

$$e = \frac{v_2 - v_1}{u_1 - u_2} = \frac{\text{velocity of separation}}{\text{velocity of approach}}$$

NOMORECLASS CONCEPTS

- For perfectly inelastic collision $e = 0$.
- For perfectly elastic collision $e = 1$.
- For partially inelastic collision $0 < e < 1$.

In elastic and inelastic collisioins, momentum is conserved whereas in inelastic collisions, kinetic energy is not conserved.

11. ELASTIC COLLISION

Consider the collision of two small smooth spheres of masses m_1 and m_2 moving with velocities u_1 and u_2 respectively in the same direction along the line joining their centers. Suppose m_1 is following m_2 with $u_1 > u_2$ i.e. m_1 tries to overtake m_2 but as the line of motion is same as the line joining the centers of the spheres, head-on ellastic collision takes place. Let their velocities after the elastic collision are v_1 and v_2 respectively, with $v_2 > v_1$ as shown in the Fig. 6.19

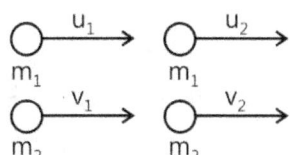

Figure 6.19: Head-on collision between two particles

Conserving momentum before and after collision we get

$$m_1u_1 + m_2u_2 = m_1v_1 + m_2v_2$$

$$m_1(u_1 - v_1) = m_2(v_2 - u_2) \qquad \qquad ...(i)$$

Conserving kinetic energy before and after collision we get

$$\frac{1}{2}m_1u_1^2 + \frac{1}{2}m_2u_2^2 = \frac{1}{2}m_1v_1^2 + \frac{1}{2}m_2v_2^2$$

$$m_1(u_1^2 - v_1^2) = m_2(v_2^2 - u_2^2) \qquad \qquad ...(ii)$$

Dividing (ii) by (i)

$$u_1 + v_1 = v_2 + u_2$$

So $v_1 = -u_1 + u_2 + v_2 \qquad \qquad ...(iii)$

Substitute v_1 in equation (i)

$$m_1(u_1 + u_1 - u_2 - v_2) = m_2v_2 - m_2u_2$$

$$2m_1u_1 + (m_2 - m_1)u_2 = (m_1 + m_2)v_2$$

$$v_2 = \left[\frac{2m_1}{m_1 + m_2}\right]u_1 + \left[\frac{m_2 - m_1}{(m_1 + m_2)}\right]u_2 \qquad \qquad ...(iv)$$

Similarly, $v_1 = \left[\frac{m_1 - m_2}{m_1 + m_2}\right]u_1 + \left[\frac{2m_2}{m_1 + m_2}\right]u_2 \qquad \qquad ...(v)$

Special Cases

(i) When $m_1 = m_2$,

From equation (i)

$$u_1 - v_1 = v_2 - u_2 \qquad \text{or} \qquad v_1 + v_2 = u_1 + u_2 \qquad \qquad ...(vi)$$

Equation (iii) gives

$$v_1 - v_2 = u_2 - u_1 \qquad \qquad ...(vii)$$

Solving (vi) and (vii) we get

$$v_1 = u_2 \text{ and } v_2 = u_1$$

∴ In one dimensional elastic collision of two bodies of equal masses, the bodies exchange there velocities after collision.

(ii) When m_2 is at rest i.e. $u_2 = 0$.

$$v_2 = \frac{2m_1u_1}{m_1 + m_2} + \left(\frac{m_2 - m_1}{m_2 + m_1}\right)u_2 \quad \Rightarrow \quad v_2 = \frac{2m_1u_1}{m_1 + m_2}$$

Now there are three possibilities in this case:

(a) If $m_1 = m_2 = m$; $\quad v_2 = \frac{2mu_1}{2m} = u_1, v_1 = 0$.

The first body stops after collision. Both the momentum and the kinetic energy of the first body are completely transferred to the second body.

(b) If $m_2 \gg m_1$, $v_1 \simeq -u_1$, $v_2 \simeq 0$

Thus when a light body collides with a much heavier stationary body, the velocity of light body is reversed and heavier body almost remains at rest.

(c) If $m_2 << m_1$, $v_2 \simeq u_1$ and $v_2 \simeq 2u_1$

Thus when a heavy body collides with a much lighter stationary body, the velocity of heavier body remains almost unchanged. The lighter body moves forward with approximately twice the velocity of the heavier body.

12. INELASTIC COLLISION

Consider the situation similar to previous article wherein m_1 is following m_2 with $u_1 > u_2$ i.e. m_1 tries to overtake m_2 but as the line of motion is same as the line joining the centers of the spheres, head-on collision takes place. Let their velocities after the collision are v_1 and v_2 respectively, with $v_2 > v_1$. Now suppose that the collision is inelastic, i.e. kinetic energy is not conserved.

Conserving momentum we get,

$m_1u_1 + m_2u_2 = m_1v_1 + m_2v_2$

Restitution equation gives,

$v_1 - v_2 = -e(u_1 - u_2)$

The loss in kinetic energy ΔE in this case, is given by

$$\Delta E = \frac{1}{2}\left(\frac{m_1 m_2}{m_1 + m_2}\right)(e^2 - 1)(u_1 - u_2)^2$$

Putting e = 0 in this equation, it is clear that the loss of kinetic energy is maximum in case of pefectly inelastic collision.

Illustration 12: A block of mass m moving at a velcoity v collides head on with another block of mass 2m at rest. If the coefficient of restitution is 0.5, find the velocities of the blocks after the collision. **(JEE MAIN)**

Sol: Solve using law of conservation of momentum, before and after collision and the equation of restitution.

Suppose after the collision the block of mass m moves at a velocity u_1 and the block of mass 2m moves at a velocity u_2. By conservation of momentum,

$mv = mu_1 + 2mu_2$ ... (i)

The velocity of sepration is $u_2 - u_1$ and the velocity of approach is v.

So, $u_2 - u_1 = \dfrac{v}{2}$ ... (ii)

Solving (i) and (ii) we get, $u_1 = 0 \text{ ms}^{-1}$ and $u_2 = \dfrac{v}{2} \text{ ms}^{-1}$.

Illustration 13: A ball is moving with velocity 2 m/s towards a heavy wall moving towards the ball with speed 1 m/s as shown in Fig. 6.20. Assuming collision to be elastic, find the velocity of ball immediately after the collision. **(JEE MAIN)**

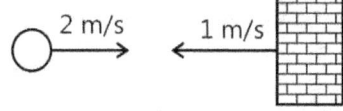

Figure 6.20

Sol: The equation of conservation of momentum will not give us any fruitful result because the mass of the wall is very large and remains at rest before and after the collision. This problem has to be solved by using equation of restitution

Before collision   **After collision**

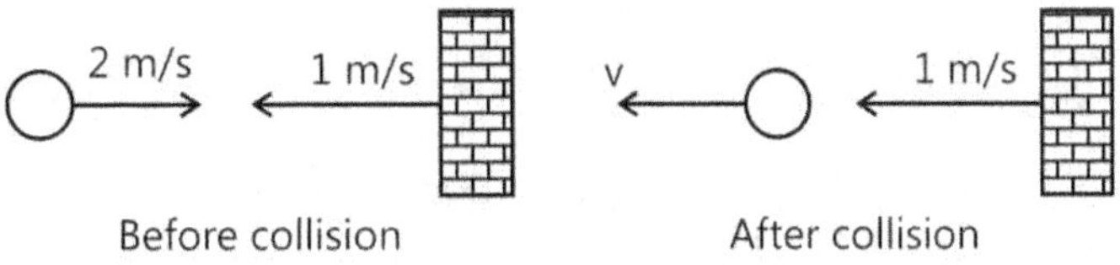

Figure 6.21

The speed of wall will not change after the collision. So, let v be the velocity of the ball after collision in the direction shown in Fig. 6.21. Since, collision is elastic (e = 1).

velocity of Separation=velocity of approach

or $v - 1 = 2 - (-1)$

or $v = 4 \, m/s$

Illustration 14: A ball of mass m is projected vertically up from smooth horizontal floor with a speed V_0. Find the total momentum delivered by the ball to the surface, assuming e as the coefficient of restitution of impact.

(JEE MAIN)

Sol: By Newton's third law the impulse delivered by the ball to the surface at each collision will be equal in magnitude to the impulse delivered to the ball by the surface i.e. change in momentum of ball at each collision. The total impulse will be the sum of DP due to all the collisions.

The momentum delivered by the ball at first, second, third impact etc. can be given as the corresponding change in its momentum (ΔP) at each impact.

$$(\Delta \vec{P})_1 = \left| (mV_1)\hat{j} - m(-V_0)\hat{j} \right| \quad \Rightarrow \quad \Delta P_1 = m(V_1 + V_0)$$

Similarly $\Delta P_2 = m(V_1 + V_2)$, $\Delta P_3 = m(V_2 + V_3)$, ... and so on.

$\Rightarrow$ The total momentum transferred $\Delta P = \Delta P_1 + \Delta P_2 + \Delta P_3 + ...$

Putting the values of ΔP_1, ΔP_2 etc., we obtain,

$$\Delta P = m\left[V_0 + 2(V_1 + V_2 + V_3 + ...) \right]$$

Putting $V_1 = eV_0$, $V_2 = e^2 V_0$, $V_3 = e^3 V_0$

We obtain,

$$\Delta P = mV_0 \left[1 + 2(e + e^2 + e^3 + ...) \right] \quad \Rightarrow \quad \Delta P = mV_0 \left(1 + 2\frac{e}{1-e} \right) = mV_0 \left(\frac{1+e}{1-e} \right)$$

Illustration 15: A stationary body explodes into four identical fragments such that three of them fly off mutually perpendicular to each other, each with same K.E. Find the energy of explosion.   **(JEE ADVANCED)**

Sol: As the body is initially at rest, the vector sum of momentum of all fragments will be zero. The energy of explosion will appear as K.E. of fragments.

Let the three fragments move along X, Y and Z axes. Therefore their velocities can be given as

$$\vec{V}_1 = V\hat{i}, \quad \vec{V}_2 = V\hat{j} \text{ and } \vec{V}_3 = V\hat{k},$$

where V = speed of each of the three fragments. Let the velocity of the fourth fragment be $\vec{V}_4$ Since, in explosion no net external force is involved, the net momentum of the system remains conserved just before and after explosion. Initially the body is a rest,

$$\Rightarrow \qquad m\vec{V}_1 + m\vec{V}_2 + m\vec{V}_3 + m\vec{V}_4 = 0$$

Putting the values of $\vec{V}_1$, $\vec{V}_2$ and $\vec{V}_3$, we obtain, $\vec{V}_4 = -V\left(\hat{i} + \hat{j} + \hat{k}\right)$

Therefore, $V_4 = \sqrt{3} \, V$

The energy of explosion

$$\therefore E = KE_f - KE_i = \left(\frac{1}{2}mV_1^2 + \frac{1}{2}mV_2^2 + \frac{1}{2}mV_3^2 + \frac{1}{2}mV_4^2 \right) - (0)$$

Putting $V_1 = V_2 = V_3 = V$ and setting $\frac{1}{2}mV^2 = E_0$, we obtain, $E = 6E_0$.

13. OBLIQUE COLLISION

Let us now consider the case when the velocities of the two colliding spheres are not directed along the line of action of the forces of interaction or the line of impact (line joining the centers). As already discussed this kind of impact is said to be oblique.

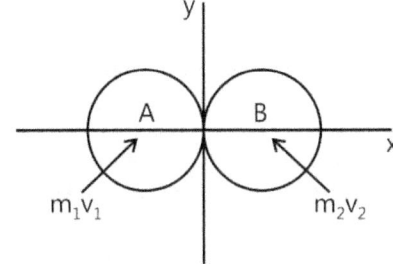

Let us consider the collision of two spherical bodies. Since velocities v'_1 and v'_2 of the bodies after impact are unknown in direction and magnitude, their determination will require the use of four independent equations.

We choose the x-axis along the line of impact, i.e. along the common normal to the surfaces in contact, and the y-axis along their common tangent as shown in Fig. 6.22. Assuming the spheres to be perfectly smooth and frictionless, the impulses exerted on the spheres during the collision are along the line of impact i.e., along the x-axis. So,

Figure 6.22: Oblique collision of two particles

(i) the component of the momentum of each sphere along the y-axis, considered separately is conserved; hence the y component of the velocity of each sphere remains unchanged. Thus we can write

$$(v_1)_y = (v'_1)_y \qquad ....(i)$$
$$(v_2)_y = (v'_2)_y \qquad ....(ii)$$

(ii) the component of total momentum of the two spheres along the x-axis is conserved. Thus we can write

$$m_1(v_1)_x + m_2(v_2)_x = m_1(v'_1)_x + m_2(v'_2)_x \qquad ....(iii)$$

(iii) The component along the x-axis of the relative velocity of the two spheres after impact i.e. the velocity of separation along x-axis is obtained by multiplying the x-component of their velocity of approach before impact by the coefficient of restitution. Thus we can write

$$(v'_2)_x - (v'_1)_x = e\left[(v_1)_x - (v_2)_x\right] \qquad ....(iv)$$

Now the four equations obtained above can be solved to find the velocities of the spheres after collision.

NOMORECLASS CONCEPTS

It is not advised to break the components of velocity in any other direction even though they are still valid. The only problem will be in using the coefficient of restitution.

Definition of coefficient of restitution can be applied in the normal direction in the case of oblique collision.

Illustration 16: After perfectly inelastic collision between two identical particles moving with same speed in different directions, the speed of the particles becomes half the initial speed. Find the angle between the two before collision. **(JEE MAIN)**

Sol: In case of an oblique collision, the momentum of individual particles are added vectorially in the equation of conservation of linear momentum.

Let θ be the desired angle. Linear momentum of the system will remain conserved.

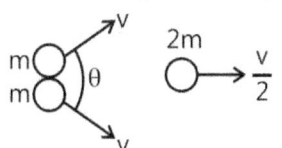

Figure 6.23

Hence $P^2 = P_1^2 + P_2^2 + 2P_1P_2 \cos\theta$

or $\left\{2m\left(\dfrac{v}{2}\right)\right\}^2 = (mv)^2 + (mv)^2 + 2(mv)(mv)\cos\theta$

or $1 = 1 + 1 + 2\cos\theta$ or $\cos\theta = -\dfrac{1}{2}$

∴ $\theta = 120°$

Illustration 17: A ball of mass m hits the floor with a speed v making an angle of incidence θ with the normal. The coefficient of restitution is e. Find the speed of the reflected ball and the angle of reflection of the ball.

(JEE MAIN)

Sol: In case of an oblique collision with fixed surface the component of velocity of colliding particle parallel to surface doesn't change. The impulse will act along the normal to the surface so use the equation of restitution along the normal.

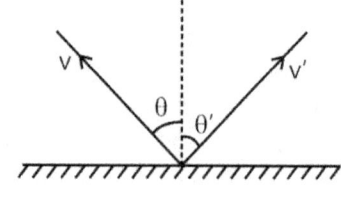

See Fig. 6.24. Let the angle of reflection is θ' and the speed after the collision is v'. The impulse on the ball is along the normal to the floor during the collision. There is no impulse parallel to the floor. Thus, the component of the velocity of the ball parallel to the surface remains unchanged before and after the collision. This gives

Figure 6.24

$v'\sin\theta' = v\sin\theta$ ...(i)

As the floor is stationary before and after the collision, the equation of conservation of momentum in the direction normal to the floor will not give any result. We have to use the formula for coefficient of restitution along the direction normal to the floor.

The velocity of separation along the normal $= v'\cos\theta'$

The velocity of approach along the normal $= v\cos\theta$

Hence, $v'\cos\theta' = ev\cos\theta$ ...(ii)

From (i) and (ii),

$v' = v\sqrt{\sin^2\theta + e^2\cos^2\theta}$ and $\tan\theta' = \dfrac{\tan\theta}{e}$

For elastic collision, e = 1 so that $\theta' = \theta$ and $v' = v$

NOMORECLASS CONCEPTS

Inelastic collision doesn't always mean that bodies will stick, which is very clear from the concept of oblique collision. Only the velocities along n-axis become same and may be different in t-direction.

14. CENTER OF MASS FRAME

We can rigidly fix a frame of reference to the center of mass of a system. This frame is called the C-frame of reference and in general is a non-inertial reference frame. Relative to this frame, the center of mass is at rest $(\vec{V}_{com,C} = 0)$ and according to equation $\vec{P} = M\vec{V}_{com}$ the total momentum of a system of particle in the C-frame of reference is always zero.

$\vec{P} = \Sigma\vec{P}_i = 0$ in the C-frame of reference.

Note : When the net external force acting on the system is zero, the C-frame becomes an inertial frame.

14.1 A System of Two Particles

Suppose the masses of the particles are equal to m_1 and m_2 and their velocities in the given reference frame K be $\vec{v}_1$ and $\vec{v}_2$ respectively. Let us find the expressions defining their momentum and total kinetic energy in the C-frame. The velocity of C-frame relative to K-frame is $\vec{v}_c$.

The momentum of the first particle in the C-frame is $\vec{P}_{1/c} = m_1 \vec{v}_{1/c} = m_1(\vec{v}_1 - \vec{v}_c)$ where $\vec{v}_c$ is the velocity of the center of mass of the system in the K frame. Substituting the expression for $\vec{v}_c$

$$\vec{v}_c = \frac{m_1\vec{v}_1 + m_2\vec{v}_2}{m_1 + m_2}$$

we get $\qquad \vec{P}_{1/c} = \mu(\vec{v}_1 - \vec{v}_2)$

where μ is the reduced mass of the system, given by $\mu = \dfrac{m_1 m_2}{m_1 + m_2}$

Similarly, the momentum of the second particle in the C-frame is $\vec{P}_{2/c} = \mu(\vec{v}_2 - \vec{v}_1)$

Thus, the momenta of the two particles in the C-frame are equal in magnitude and opposite in direction; the modulus of the momentum of each particle is

$$P_{1/c} = \mu v_{rel}$$

where $v_{rel} = |\vec{v}_1 - \vec{v}_2|$ is the modulus of velocity of one particle relative to another.

Finally, let us consider total kinetic energy. The total kinetic energy of the two particles in the C-frame is

$$K_{sys/c} = K_{1/c} + K_{2/c} = \frac{1}{2}m_1 v_{1/c}^2 + \frac{1}{2}m_2 v_{2/c}^2 = \frac{P_{1/c}^{\,2}}{2m_1} + \frac{P_{1/c}^{\,2}}{2m_2}$$

Now, $\qquad \mu = \dfrac{m_1 m_2}{m_1 + m_2} \quad$ or $\quad \dfrac{1}{m_2} + \dfrac{1}{m_2} = \dfrac{1}{\mu}$

Then $\qquad K_{sys/c} = \dfrac{\vec{P}_{1/c}^{\,2}}{2\mu} = \dfrac{\mu v_{rel}^2}{2}$

The total kinetic energy of the partices of the system in the K-frame is related to the total kinetic energy in C-frame. The velocity of the ith particle of the system in K-frame can be expressed as:

$$\vec{v}_i = \vec{v}_{i/c} + \vec{v}_c$$

So we can write $K_{sys} = \dfrac{1}{2}\sum m_i v_i^2 = \dfrac{1}{2}\sum m_i(\vec{v}_{i/c} + \vec{v}_c)^2 = \dfrac{1}{2}\sum m_i v_{i/c}^2 + \vec{v}_c \sum m_i \vec{v}_{i/c} + \dfrac{v_c^2}{2}\sum m_i$

In the C-frame, the summation $\sum m_i \vec{v}_{i/c} = M\vec{V}_{com,C} = 0$.

So we get $K_{sys} = \dfrac{1}{2}\sum m_i v_{i/c}^2 + \dfrac{v_c^2}{2}\sum m_i = K_{sys/c} + \dfrac{v_c^2}{2}\sum m_i$

For a two-particle system, we get

$$K_{sys} = \frac{\mu v_{rel}^2}{2} + \frac{Mv_c^2}{2} \quad \text{(where } M = m_1 + m_2)$$

Illustration 18: Two blocks of mass m_1 and m_2 connected by an ideal spring of spring constant k are kept on a smooth horizontal surface. Find maximum extension of the spring when the block m_2 is given an initial velocity of v_0 towards right as shown in Fig. 6.25.   **(JEE ADVANCED)**

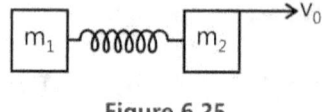

Figure 6.25

Sol: In absence of frictional forces on block, the total mechanical energy of the system comprising the blocks and spring will be conserved. At the time of maximum expansion of spring, the mechanical energy in C frame will be totally stored as elastic P.E. of the spring

This problem can be best solved in the C-frame or the reference frame rigidly fixed to the center of mass of the system of two blocks.

Initially at t=0 when the block m_2 is given velocity v_0, the total kinetic energy of the blocks in C-frame is related to the total kinetic energy in the given frame K by the relation,

$$K_{sys} = \frac{\mu v_{rel}^2}{2} + \frac{(m_1 + m_2)v_c^2}{2} = K_{sys/c}(0) + \frac{(m_1 + m_2)v_c^2}{2} \qquad \left(\mu = \frac{m_1 m_2}{m_1 + m_2}; \ v_{rel} = v_0; \ v_c = \frac{m_2 v_0}{m_1 + m_2} \right)$$

where the first term on the right hand side of this relation is the total kinetic energy in C-frame at t=0, $K_{sys/c}(0)$, and the second term is the kinetic energy associated with the motion of the system of blocks as a whole in the K-frame. As there are no dissipative external forces acting on the system, the total mechanical energy will remain constant, both in the C-frame and the K-frame. In the C-frame the blocks will oscillate under the action of spring force and the kinetic energy in the C-frame will get converted into the elastic potential energy of the spring and vice-versa, the total mechanical energy remaining constant at each instant, equal to the total kinetic energy in C-frame at t=0, $K_{sys/c}(0)$.

Initially at t=0 when the block m_2 is given velocity v_0, the mechanical energy in C frame will be totally kinetic ($K_{sys/c}(0)$), and at the instant of maximum extension of the spring, the mechanical energy in C-frame will be totally converted into elastic potential energy of the spring. So we have,

$$K_{sys/c}(0) = \frac{1}{2}kx_{max}^2 \Rightarrow \frac{1}{2}\frac{m_1 m_2}{m_1 + m_2}v_0^2 = \frac{1}{2}kx_{max}^2$$

Thus, maximum extension is $x_{max} = v_0 \sqrt{\dfrac{m_1 m_2}{k(m_1 + m_2)}}$

15. IMPULSE AND MOMENTUM

When two bodies collide during a very short time period, large impulsive forces are exerted between the bodies along the line of impact. Common examples are a hammer striking a nail or a bat striking a ball. The line of impact is a line through the common normal to the surfaces of the colliding bodies at the point of contact.

When two bodies collide, the momentum of each body is changed due to the force on it exerted by the other. On an ordinary scale, the time duration of the collision is very small and yet the change in momentum is sizeable. This means that the magnitude of the force of interaction must be very large on an ordinary scale. Such large forces acting for a very short duration are called impulsive forces. The force may not be uniform during the interaction.

We know that the force is related to momentum as $\vec{F} = \dfrac{d\vec{P}}{dt} \quad \Rightarrow \quad \vec{F}dt = d\vec{P}$

We can find the change in momentum of the body during a collision (from $\vec{P}_i$ to $\vec{P}_f$) by integrating over the time of collision and assuming that the force during collision has a constant direction, $\vec{P}_f - \vec{P}_i = \int_{P_i}^{P_f} d\vec{P} = \int_{t_i}^{t_f} \vec{F}dt$;

Here the subscripts i (= initial) and f (= final) refer to the times before and after the collision. The integral of a force over the time interval during which the force acts is called impulse.

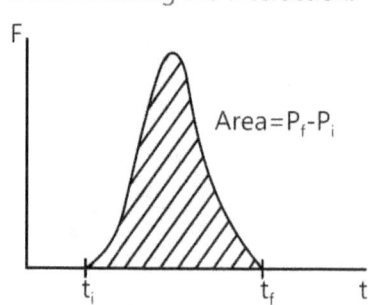

Figure 6.26: Impulse imparted to the particle

Thus the quantity $\int_{t_i}^{t_f} \vec{F} \cdot dt$ is the impulse of the force $\vec{F}$ during the time interval t_i and t_f and is equal to the change in the momentum of the body in which it acts.

The magnitude of impulse $\int_{t_i}^{t_f} \vec{F} dt$ is the area under the force-time curve as shown in Fig. 6.26

Illustration 19: A block of mass m and a pan of equal mass are connected by a string going over a smooth light pulley as shown in Fig. 6.27. Initially the system is at rest when a particle of mass m falls on the pan and sticks to it. If the particle strikes the pan with a speed v find the speed with which the system moves just after the collision.

(JEE MAIN)

Sol: By Newton's third law, the impulse imparted to the particle in upward direction will be equal in magnitude to the total impulse imparted to the system of block and the pan.

Let N be the contact force between the particle and the pan during the collision.

Consider the impulse imparted to the particle. The force N will be in upward direction and the impulse imparted to it will be $\int N \, dt$ in the upward direction. This should be equal to the change in momentum imparted to it in the upward direction.

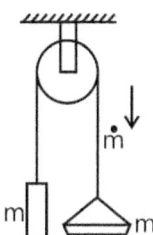

Thus, $\int N \, dt = P_f - P_i = -mV - (-mv) = mv - mV$ ....(i)

Similarly considering the impulse imparted to the pan. The forces acting on it are tension T upwards and contact force N downwards. The impulse imparted to it in the downward direction will be, $\int (N-T) dt = mV - 0 = mV$ ....(ii)

Figure 6.27

Impulse imparted to the block by the tension T will be upwards,

$\int T \, dt = mV - 0 = mV$ ....(iii)

Adding (ii) and (iii) we get, $\int N dt = 2mV$ ....(iv)

Comparing (i) and (iv) we get, $mv - mV = 2mV$ or $V = \dfrac{v}{3}$ ms^{-1}

PROBLEM-SOLVING TACTICS

Applying the principle of Conservation of Linear Momentum

(a) Decide which objects are included in the system.

(b) Relative to the system, identify the internal and external forces.

(c) Verify that the system is isolated.

(d) Set the final momentum of the system equal to its initial momentum. Remember that momentum is a vector.

(e) Always check whether kinetic energy is conserved or not. If it is conserved, it gives you an extra equation. Otherwise use work-energy theorem, carefully.

(f) Try to involve yourself physically in the question, imagine various events. This would help in some problems where some parameters get excluded by conditions. This will also help in checking your answer.

Impulse

(g) Ignore any finite-value forces, while dealing with impulses.

(h) Write impulse equations carefully, because integration which we are unable to calculate will always cancel out.

Collisions

(i) Remembering special cases of collisions would be nice.

FORMULAE SHEET

Position of center of mass of a system: $\vec{r}_{com} = \dfrac{\sum\limits_i m_i \vec{r}_i}{M}$

$$\vec{r}_{COM} = x_{COM}\hat{i} + y_{COM}\hat{j} + z_{COM}\hat{k}$$

$$x_{COM} = \frac{m_1 x_1 + m_2 x_2 + \ldots + m_n x_n}{m_1 + m_2 + \ldots + m_n} = \frac{\sum\limits_i m_i x_i}{\sum\limits_i m_i}$$

For continuous bodies $x_{COM} = \dfrac{\int x\,dm}{\int dm} = \dfrac{\int x\,dm}{M}$

For a two-particle system, we have

$$r_1 = \left(\frac{m_2}{m_2 + m_1}\right)d \quad \text{and} \quad r_2 = \left(\frac{m_1}{m_1 + m_2}\right)d$$

where d is the separation between the particles.

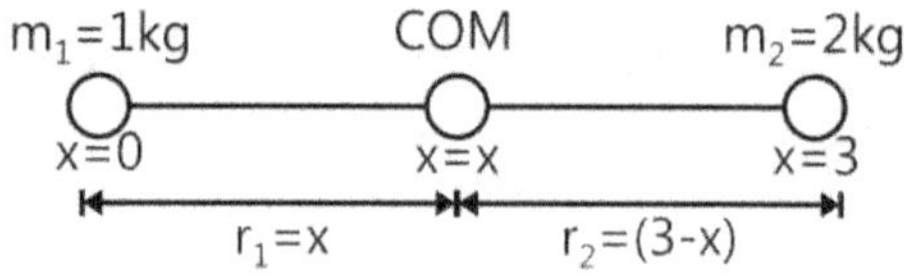

Center of Mass of a Uniform Rod $\left(\dfrac{L}{2}, 0, 0\right)$

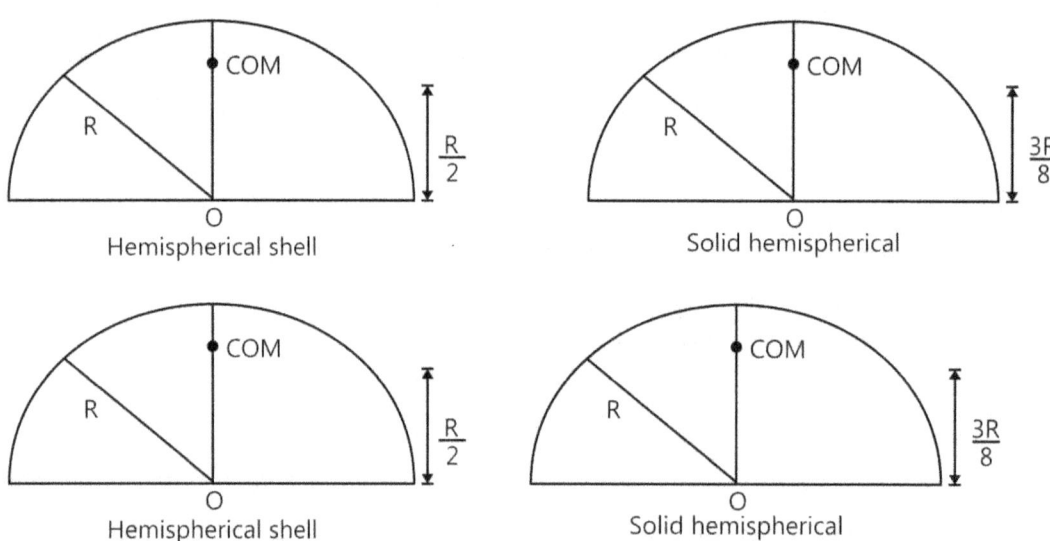

Figure 6.28

Figure 6.29

If some mass or area is removed from a rigid body, then the position of center of mass of the remaining portion is obtained from the following formula:

$$\vec{r}_{COM} = \frac{m_1 \vec{r}_1 - m_2 \vec{r}_2}{m_1 - m_2}$$

Where m_1 is the mass of the body after filling all cavities with same density and m_2 is the mass filled in the cavity. Cavity mass is assumed negative.

Velocity of COM $\vec{v}_{COM} = \dfrac{m_1 \vec{v}_1 + m_2 \vec{v}_2 + \ldots + m_n \vec{v}_n}{m_1 + m_2 + \ldots + m_n} = \dfrac{\sum\limits_i m_i \vec{v}_i}{\sum\limits_i m_i}$

Total momentum of a n-particle system $\vec{P}_{COM} = \vec{P}_1 + \vec{P}_2 + ... + \vec{P}_n = M\vec{v}_{COM}$

Acceleration of COM $\vec{a}_{COM} = \dfrac{m_1\vec{a}_1 + m_2\vec{a}_2 + ... + m_n\vec{a}_n}{m_1 + m_2 + ... + m_n} \dfrac{\sum\limits_i m_i \vec{a}_i}{\sum\limits_i m_i}$

Net force acting on the system $\vec{F}_{COM} = \vec{F}_1 + \vec{F}_2 + ... + \vec{F}_n$

Net external force on center of mass is $M\vec{a}_{cm} = \vec{F}_{ext}$

If net force on the system $\vec{F} = \vec{F}_1 + \vec{F}_2 + \vec{F}_3 + ... + \vec{F}_n = 0$ then, $\vec{P}_1 + \vec{P}_2 + \vec{P}_3 + ... + \vec{P}_n = $ constant

Equation of motion of a body with variable mass is:

$$m\left(\dfrac{d\vec{v}}{dt}\right) = \vec{F} + \left(\dfrac{dm}{dt}\right)\vec{u}$$

Where $\vec{u}$ is the velocity of the mass being added(separated) relative to the given body of instantaneous mass m and $\vec{F}$ is the external force due to surrounding bodies or due to field of force.

In case of reducing mass of a system $\dfrac{dm}{dt} = \mu$ kgs^{-1}

For a rocket we have, $m\left(\dfrac{d\vec{v}}{dt}\right) = m\vec{g} + \left(\dfrac{dm}{dt}\right)\vec{v}_r$

Where $\vec{v}_r$ is the velocity of the ejecting gases relative to the rocket.

In scalar form we can write

$$m\left(\dfrac{dv}{dt}\right) = -mg + v_r\left(-\dfrac{dm}{dt}\right)$$

Here $-\dfrac{dm}{dt} = $ rate at which mass is ejecting and $v_r\left(-\dfrac{dm}{dt}\right) = $ Thrust force.

Final velocity of rocket $v = u - gt + v_r \ln\left(\dfrac{m_0}{m}\right)$

Impulse of a force: $\vec{J} = \int \vec{F} dt = \Delta\vec{p} = \vec{p}_f - \vec{p}_i$

Collision

(a) In the absence of any external force on the system the linear momentum of the system will remain conserved before, during and after collision, i.e.,

$$m_1v_1 + m_2v_2 = (m_1 + m_2)v = m_1v'_1 + m_2v'_2 \qquad ...(i)$$

(b) In the absence of any dissipative forces, the mechanical energy of the system will also remain conserved, i.e.

$$\dfrac{1}{2}m_1v_1^2 + \dfrac{1}{2}m_2v_2^2 = \dfrac{1}{2}(m_1 + m_2)v^2 + \dfrac{1}{2}kx_m^2 = \dfrac{1}{2}m_1v_1'^2 + \dfrac{1}{2}m_2v_2'^2 \qquad ...(ii)$$

Head on Elastic Collision

$$v'_1 = \left(\dfrac{m_1 - m_2}{m_1 + m_2}\right)v_1 + \left(\dfrac{2m_2}{m_1 + m_2}\right)v_2$$

$$v'_2 = \left(\dfrac{m_2 - m_1}{m_1 + m_2}\right)v_2 + \left(\dfrac{2m_1}{m_1 + m_2}\right)v_1$$

$$\frac{\text{separation speed after collision}}{\text{approach speed before collision}} = e$$

$$v'_1 = \left(\frac{m_1 - em_2}{m_1 + m_2}\right)v_1 + \left(\frac{m_2 + em_2}{m_1 + m_2}\right)v_2$$

$$v'_2 = \left(\frac{m_2 - em_1}{m_1 + m_2}\right)v_2 + \left(\frac{m_1 + em_1}{m_1 + m_2}\right)v_1$$

The C-frame: Total kinetic energy of system in K-frame is related to total kinetic energy in C-frame as:

$$K_{sys} = K_{sys/c} + \frac{Mv_c^2}{2}; M = \sum m_i$$

For a two-particle system: $\vec{P}_{1/c} = -\vec{P}_{2/c} = \frac{m_1 m_2}{m_1 + m_2}(\vec{v}_1 - \vec{v}_2)$

Or $P_{1/c} = P_{2/c} = \mu v_{rel} = \frac{m_1 m_2}{m_1 + m_2}|\vec{v}_1 - \vec{v}_2|$ and $K_{sys/c} = \frac{\mu v_{rel}^2}{2} = \frac{P_{1/c}^2}{2}$

Solved Examples

JEE Main/Boards

Example 1: The linear mass density of rod of a length l=2 m varies from A as (2+x) kg/m. What is the position of center of mass from end A.

Sol: To find C.O.M of continuous mass distributions consider a small element of distribution of mass dm. Then the co-ordinate of C.O.M. is given as

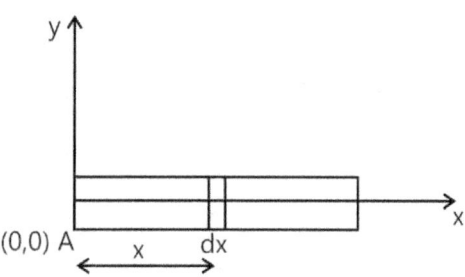

(0,0) A   x   dx

$\therefore x_{COM} = \dfrac{\int x \, dm}{M}$ the limits of integration should be chosen such that the small elements covers entire mass distribution.

Take an element of the rod of infinitesimal length dx at distance x from point A. The mass of the element will be

$dm = \lambda \, dx = (2 + x)dx$ As x varies from o to l the element covers the entire rod.

Center of mass of rod $X_{cm} = \dfrac{\int x \, dm}{\int dm}$

$$X_{cm} = \frac{\int_0^l x(2+x)dx}{\int_0^l (2+x)dx} = \frac{\left|(x^2 + \frac{x^3}{3})\right|_0^l}{\left|2x + \frac{x^2}{2}\right|_0^l}$$

$$X_{cm} = \frac{l^2 + \dfrac{l^3}{3}}{2l + \dfrac{l^2}{2}} = \frac{6l + 2l^2}{12 + 3l}$$

For l = 2 m, $X_{cm} = \dfrac{6 \times 2 + 2 \times 4}{12 + 3 \times 2} = \dfrac{20}{18} = \dfrac{10}{9}$ m

So center of mass is at a distance $\dfrac{10}{9}$ m from A.

Example 2: One fourth of the mass of square lamina is cut off (see figure). Where does the center of mass of the remaining part of the square shift.

Sol: To find the C.O.M. of a body having a cavity we first fill the cavity with the same density as body and find the C.O.M. (x,y,z) of the whole body. Then we consider the cavity as second body having negative mass and find the C.O.M (x_1, y_1, z_1) of the cavity. The C.O.M. of the body with cavity is

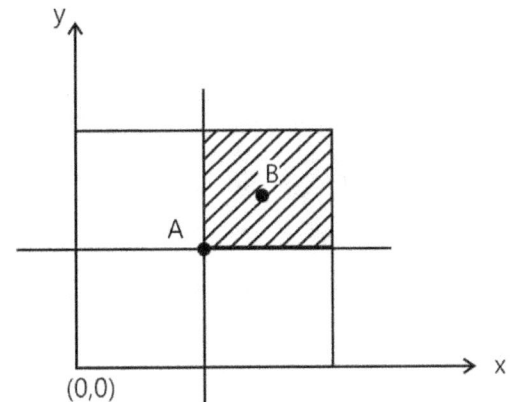

(0,0)

$$x_{cm} = \frac{mx - m_1 x_1}{(m - m_1)}; \quad y_{cm} = \frac{my - m_1 y_1}{(m - m_1)}; \quad z_{cm} = \frac{mz - m_1 z_1}{(m - m_1)}$$

Part of the lamina cut-off is taken as negative mass. Coordinates of center of mass of whole lamina are $\left(\frac{a}{2}, \frac{a}{2}\right)$ and the coordinates of center of mass of cut-off part are $\left(\frac{3a}{4}, \frac{3a}{4}\right)$. So the center of mass of remaining part is given as,

$$X_{cm} = \frac{m\frac{a}{2} - \frac{m}{4} \cdot \frac{3a}{4}}{m - \frac{m}{4}} = \frac{\frac{a}{2} - \frac{3a}{16}}{\frac{3}{4}} = \frac{5a}{12} m$$

$$Y_{cm} = \frac{m\frac{a}{2} - \frac{m}{4} \cdot \frac{3a}{4}}{m - \frac{m}{4}} = \frac{5a}{12} m$$

Example 3: The magnitude and direction of the velocities of two identical frictionless balls before they strike each other are as shown in figure. Assuming e=0.90, determine the magnitude and direction of the velocity of each ball after the impact. $v_A = 30$ ms^{-1}, $v_B = 40$ ms^{-1}

Sol: In case of an oblique collision, the momentum of individual particles are added vectorially in the equation of conservation of linear momentum. The equation of restitution is used along line of impact

The component of velocity of each ball along the common tangent at the point of impact will remain the same before and after the collisions. Let x and y axes be along the common normal and common tangent respectively.

So $v'_{Ay} = v_{Ay} = v_A \sin 30° = 15$ ms^{-1} ...(i)

$v'_{By} = v_{By} = v_B \sin 60° = 20\sqrt{3} = 34.6$ ms^{-1} ...(ii)

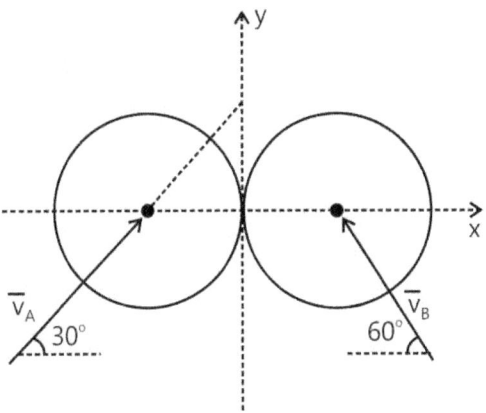

Along the x axis we conserve the momentum to get

$$m_A v_{Ax} + m_B v_{Bx} = m_A v'_{Ax} + m_B v'_{Bx}$$

$$m v_A \cos 30 + m(-v_B \cos 60°) = mv'_{Ax} + mv'_{Bx}$$

$$v'_{Ax} + v'_{Bx} = 15\sqrt{3} - 20 \quad \text{... (iii)}$$

Velocity of separation = e (velocity of approach)

$$\Rightarrow v'_{Bx} - v'_{Ax} = e(v_{Ax} - v_{Bx})$$

$$v'_{Bx} - v'_{Ax} = 0.9 \, (15\sqrt{3} + 20) \quad \text{... (iv)}$$

Solving (iii) and (iv) are get

$$v'_{Ax} = -17.7 \text{ ms}^{-1}$$

$$v'_{Bx} = 23.68 \text{ ms}^{-1}$$

$$\Rightarrow v'_A = \sqrt{v'^2_{Ax} + v'^2_{Ay}} = 23.2 \text{ ms}^{-1}$$

and $v'_B = \sqrt{v'^2_{Bx} + v'^2_{By}} = 41.92 \text{ ms}^{-1}$

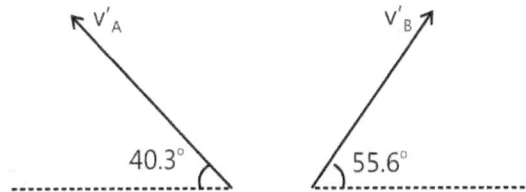

Example 4: The mass of a rocket is 2.8 x 10⁶ kg at launch time of this 2x10⁶kg is fuel. The exhaust speed is 2500m/s and the fuel is ejected at the rate of 1.4x10⁴kg/sec.

(a) Find thrust on the rocket.

(b) What is initial acceleration at launch time?

Ignore air resistance.

Sol: To lift the rocket upward against gravity, the thrust force in the upward direction due to exiting gases should be greater than or equal to the gravitational force. The equation of motion of the rocket can be written in terms of force as $m\dfrac{dv}{dt} = F_g + v_r\left(\dfrac{dm}{dt}\right)$

Thrust force

$F_{th} = v_r\left(\dfrac{dm}{dt}\right) = 2500\ ms^{-1} \times 1.4 \times 10^4\ kgs^{-1}$

$F_{th} = 3.5 \times 10^7\ N$

Equation of motion of rocket is

$m\dfrac{dv}{dt} = -W + v_r\left(\dfrac{dm}{dt}\right) = -mg + v_r\left(\dfrac{dm}{dt}\right)$

$\Rightarrow \left.\dfrac{dv}{dt}\right|_{t=0} = -g + \dfrac{F_{th}}{m_0} = -9.8 + \dfrac{3.5 \times 10^7\ N}{2.8 \times 10^6\ kg}$

$\Rightarrow a_0 = \left(\dfrac{350}{28} - 9.8\right) ms^{-2}$

$\Rightarrow a_0 = 12.5 - 9.8 = 2.7\ ms^{-2}$

Example 5: A projectile is fired at a speed of 100 m/s at an angle of 37° above horizontal (see figure) At the highest point the projectile breaks into two parts of mass ratio 1:3. Find the distance from the launching point to the point where the heavier piece lands. The smaller mass has zero velocity with respect to the earth immediately after explosion.

Sol: The range of center of mass of the system during projectile motion is given by $X_{CM} = R = \dfrac{u^2 \sin 2\theta}{g}$. This range is not effected by any internal changes in the system.

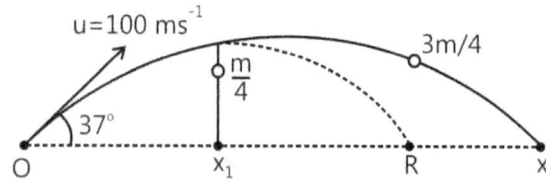

The C.O.M of the projectile will hit the horizontal plane at the same point where it would have hit without any explosion i.e. the range of COM will not change. Both the smaller and larger mass will reach the ground together because the vertical components of their velocity are equal to zero after the explosion. (Explosion took place at the highest point of the trajectory and the smaller mass comes to rest just after the explosion). At highest

point $x_1 = \dfrac{R}{2}$.

$X_{cm} = R = \dfrac{2u^2 \sin\theta\cos\theta}{g} = \dfrac{2 \times 10^4 \times 0.6 \times 0.8}{10}$

$X_{cm} = 960 = \dfrac{\dfrac{m}{4}\cdot\dfrac{960}{2} + \dfrac{3m}{4}\cdot x_2}{m}$

$\Rightarrow 960 = \dfrac{960}{8} + \dfrac{3}{4}x_2$

$\Rightarrow x_2 = \dfrac{4}{3} \times 960 \times \dfrac{7}{8} = 160 \times 7 = 1120\ m$

Example 6: A bullet of mass m strikes a block of mass M connected to a light spring of stiffness k, with a speed v_0 and gets embedded into mass M. Find the loss of K.E. of the system just after impact

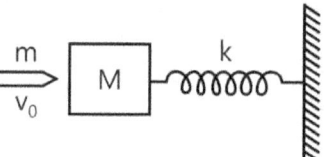

Sol: During collision as there is no net external force acting on the bullet-block system, hence the momentum of the system can be conserved. As the bullet hits block in-elastically, some of its initial K.E. is lost during the collision.

As the bullet of mass m hits the block of mass M, and gets embedded into it, we can write the equation of conservation of linear momentum at the instant of collision, assuming the force due to spring to be negligible, as at this instant the block M has just started moving and the compression in the spring is negligible (see figure)

$mV_0 = (m + M)V \Rightarrow V = \dfrac{mV_0}{m+M}$

where V is the velocity of block just after collision.

Loss in kinetic energy of the system of bullet and block is,

$\Delta K = \dfrac{1}{2}mv_0^2 - \dfrac{1}{2}(M + m)V^2$

$$= \frac{1}{2}\left[mv_0^2 - (M+m)\frac{m^2v_0^2}{(M+m)^2}\right]$$

$$= \frac{mv_0^2}{2}\left[1 - \frac{m}{M+m}\right]$$

$$\Delta K = \frac{mMv_0^2}{2(M+m)} \, J$$

Example 7: Two blocks B and C of mass m each connected by a spring of natural length I and spring constant k rest on an absolutely smooth horizontal surface as shown in figure A third block A of same mass collides elastically to block B with velocity v. Calculate the velocities of blocks, when the spring is compressed as much as possible and also the maximum compression.

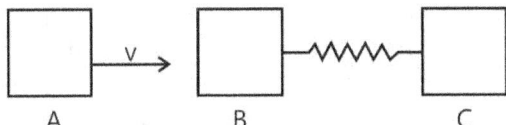

A   B   C

Sol: In absence of frictional forces on block, the total mechanical energy of the system comprising the blocks B and C and spring will be conserved. At the time of maximum compression of spring, the mechanical energy of this system in C-frame will be totally stored as elastic potential energy of the spring.

Block A collides with block B elastically. So conserving momentum between A and B we get, (spring force is negligible at the instant of collision)

$$mv = mv_A = mv_A + mv_B$$

or $v = v_A + v_B$ ...(i)

$v = v_B - v_A$ ...(ii) (restitution equation)

Solving (i) and (ii), we get

$v_A = 0$ and $v_B = v$ ...(iii)

For system comprising blocks B and C, the velocity of center of mass after collision is,

$$V_{cm} = \frac{mv_B}{m+m} = \frac{v}{2} \, ms^{-1} \quad \text{...(iv)}$$

As there are no dissipative forces in the horizontal direction the velocity of COM will remain constant. Let us consider the motion of B and C in the C-frame. At the instant of maximum compression the blocks B and C will come to rest in the C-frame. So there velocity in K- frame will become equal to the velocity of COM.

$$\therefore \ v_B = v_C = \frac{v}{2} \, ms^{-1} \quad \text{...(v)}$$

Kinetic energy of system "B + C" just after collision in K frame is,

$$K_{sys} = \frac{1}{2}m \, v_B^2 + 0 = \frac{1}{2}mv^2$$

Now $K_{sys} = K_{sys/c} + \dfrac{2mv_{cm}^2}{2}$

$$\Rightarrow \frac{1}{2}mv^2 = K_{sys/c} + m\frac{v^2}{4}$$

$$\Rightarrow K_{sys/c} = \frac{1}{4}mv^2 \quad \left[K_{sys/c} = \frac{\mu v_{rel}^2}{2} = \frac{m}{2}\cdot\frac{v^2}{2}\right]$$

Initially the potential energy of spring is zero and when the compression is maximum the energy in C-frame will be entirely converted into potential energy of the spring, thus we can write

$$\frac{mv^2}{4} = \frac{1}{2}k\,x_{max}^2$$

$$\Rightarrow x_{max} = v\sqrt{\frac{m}{2k}}$$

JEE Advanced/Boards

Example 1: A body of mass M with a small disc of mass m placed on it rests on a smooth horizontal plane as shown in figure. The disc is set in motion in the horizontal direction with velocity v. To what height relative to the initial level will the disc rise after breaking off the body M? The friction is assumed to be absent.

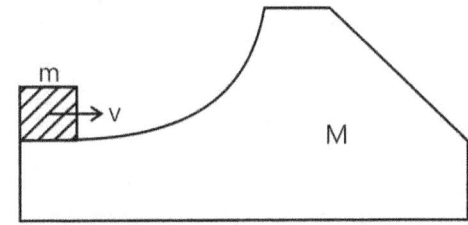

Sol: As there are no external forces acting on the system comprising m and M in the horizontal direction, the momentum is conservative in the horizontal direction. At the instant the disk m breaks – off from block M, it has a component of velocity u in vertical direction and the disk m and block M have a common velocity V in the horizontal direction.

In horizontal direction we can write,

$$mv = (m+M)V \Rightarrow V = \frac{mv}{m+M} \, ms^{-1} \quad \text{... (i)}$$

Once the disc breaks–off the block, then its horizontal velocity will not change and at the highest point of its trajectory, the vertical component of its velocity becomes zero.

Using law of conservation of energy, we get

$$\frac{1}{2}mv^2 = \frac{1}{2}mV^2 + mgH + \frac{1}{2}MV^2 \qquad \text{... (ii)}$$

where H is the height raised by the disc.

$$\Rightarrow mgH = \frac{1}{2}m(v^2) - \frac{1}{2}(m+M)V^2$$

$$\Rightarrow mgH = \frac{1}{2}mv^2 - \frac{1}{2}\frac{m^2v^2}{(m+M)} \quad \text{(using (i))}$$

$$\Rightarrow H = \frac{v^2}{2g}\left[1 - \frac{m}{m+M}\right]$$

$$H = \frac{v^2}{2g}\cdot\left(\frac{M}{m+M}\right)m$$

Example 2: A 20 gm bullet pierces through a plate of mass $M_1 = 1$ kg and then comes to rest inside a second plate of mass $M_2 = 2.98$ kg as shown in the figure. It is found that the two plates, initially at rest, now move with equal velocities. Find the percentage loss in the initial velocity of the bullet when it is between M_1 and M_2. Neglect any loss of material of the plates, due to action of bullet.

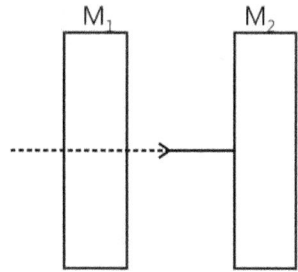

Sol: As the net external force on the system during collision is zero, the momentum of system can be conservative.

Conserve momentum for collision of bullet with fist plate, $mv = mu + M_1V$ ....(i)

Conserve momentum for collision of bullet with second plate, $mu = (m + M_2)V$ ....(ii)

Here the two plates move with equal velocity V after the collision.

Eliminate V from equations (i) and (ii) to get

$$mv = mu + M_1\frac{mu}{(m+M_2)}$$

$$u\left[1 + \frac{M_1}{m+M_2}\right] = v \; ; \; \Rightarrow u = \frac{v(m+M_2)}{m+M_1+M_2}$$

$$\Rightarrow \frac{v-u}{v} = 1 - \frac{u}{v} = 1 - \frac{m+M_2}{m+M_1+M_2}$$

$$\Rightarrow \frac{\Delta v}{v}\times 100\% = \frac{M_1}{m+M_1+M_2}\times 100\%$$

Putting the values of m, M_1 and M_2 we get

$$\% \text{ loss} = \frac{1}{0.020+1+2.98}\times 100\%$$

$$\% \text{ loss} = 25\%$$

Example 3: Two bodies A and B of masses m and 2m respectively are placed on a smooth floor. They are connected by a spring. A third body C of mass m moves with a velocity v_0 along the line joining A and B and collides elastically with A, as shown in figure. At a certain time t_0, it is found that the instantaneous velocities of A and B are the same. Further, at this instant the compression of the spring is found to be x_0.

Find: (a) The common velocity of A and B at the time t_0. (b) The spring constant.

Sol: The collision between the blocks A and C is elastic. In C-frame at time of maximum compression of spring, the total mechanical energy will be stored as elastic potential energy of spring.

Masses of bodies C and A are same and C

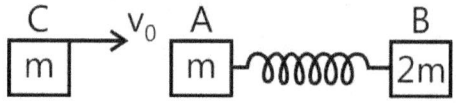

collides elastically with body A initially at rest. So after collision C will come to rest and A will take up the velocity of C (spring force during collision is negligible.)

The velocity of center of mass (COM) of the system comprising blocks A and B just after collision is,

$$v_{cm} = \frac{m.v_0}{3m} = \frac{v_0}{3} \text{ ms}^{-1}$$

As there are no external forces acting in horizontal direction, the velocity of COM will be constant.

In the C-frame when the compression in the spring is maximum the blocks will come to rest momentarily. Thus there velocity in K – frame will be equal to the velocity of COM.

$$\Rightarrow v_A = v_B = v_{cm} = \frac{v_0}{3} \text{ ms}^{-1}$$

Just after collision the total kinetic energy of blocks A and B in C – frame is,

$$K_{sys/c} = \frac{1}{2} \mu \, v_{rel}^2 = \frac{(m)(2m)}{2 \times 3m} . v_0^2 \Rightarrow K_{sys/c} = \frac{mv_0^2}{3} \, J$$

This energy will get converted into the elastic potential energy of the spring at the instant of maximum compression,

$$\frac{mv_0^2}{3} = \frac{1}{2} k \, x_{max}^2 = \frac{1}{2} k \, x_0^2 \; ; \; \Rightarrow k = \frac{2mv_0^2}{3x_0^2} \, J$$

Example 4: A block of mass M with a semi-circular track of radius R rests on a horizontal frictionless surface. A uniform cylinder of radius r and mass m is released from rest at the point A as shown in the figure. The cylinder slips on the semicircular frictionless track.

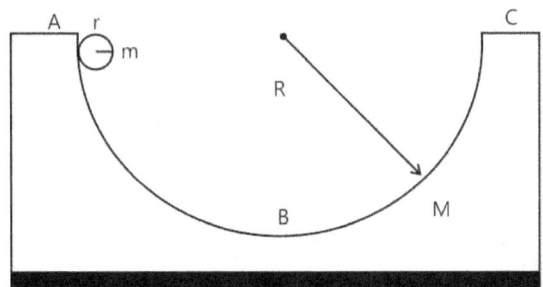

(a) How far has the block moved when the cylinder reaches the bottom point B of the track?

(b) How fast is the block moving when the cylinder reaches the bottom of the track?

Sol: As there are no frictional forces acting on the system comprises cylinder and block, the gravitational potential energy of cylinder is converted into the kinetic energy of cylinder and block.

(a) There are no external forces acting on the system comprising cylinder and the block in the horizontal direction. So we can conserve momentum in the horizontal direction, so when cylinder reaches point B on the block, let its velocity in K-frame be v_1 towards right and velocity of block in K-frame be v_2 towards left. So we get

$$0 = mv_1 - Mv_2 \text{ or } mv_1 = Mv_2 \qquad ....(i)$$

Also the COM of the system was initially at rest and will continue to remain at rest in absence of horizontal external forces. When m moves towards right a distance of (R – r) relative to block M.

We can write,

$$X_{cm} = \frac{mx_1 + Mx_2}{m+M} \Rightarrow \Delta X_{cm} = \frac{m \, \Delta x_1 + M \Delta x_2}{m+M} = 0$$

Now $\Delta x_1 = (R-r) + \Delta x_2$

$$\Rightarrow m(R - r + \Delta x_2) + M\Delta x_2 = 0$$

$$\Rightarrow \Delta x_2 = -\frac{m}{M+m}(R - r) \qquad ...(ii)$$

Now (R – r) is towards right, so Δx_2 will be towards left.

(b) Now as there are no dissipative forces acting on the system, total energy of system is conserved. i.e.

$$mg(R - r) = \frac{1}{2} mv_1^2 + \frac{1}{2} Mv_2^2 \qquad ....(iii)$$

From (i) and (iii) eliminate v_1 to get

$$mg(R - r) = \frac{1}{2} m \frac{M^2 v_2^2}{m^2} + \frac{1}{2} Mv_2^2$$

$$\Rightarrow mg(R - r) = \frac{1}{2} Mv_2^2 \left[\frac{M}{m} + 1 \right]$$

$$\Rightarrow mg(R - r) = \frac{M(M + m)v_2^2}{2m}$$

$$\Rightarrow v_2 = m \sqrt{\frac{2g(R - r)}{M(M + m)}} \; ms^{-1}$$

Example 5: Two balls of masses m and 2m are suspended by two threads of same length l from the same point on the ceiling. The ball m is pulled aside through an angle α and released after imparting to it a tangential velocity v_0 towards the other stationery ball. To what heights will the balls rise after collision, if the collision is perfectly elastic?

Sol: In case of perfectly elastic collision, the kinetic energy of the system is conserved. At the maximum vertical displacement of the ball the total kinetic energy is converted in to gravitational potential energy.

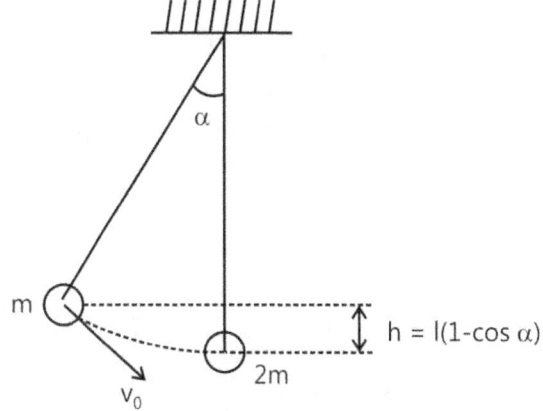

Ball of mass m will collide the ball of mass 2m, which is initially at rest.

The velocity of impact of m be v, then by conserving energy of m we get

$$\frac{1}{2}mv_0^2 + mgl(1-\cos\alpha) = \frac{1}{2}mv^2$$

$$v_0^2 + 2gl(1-\cos\alpha) = v^2 \qquad \ldots(i)$$

Conserve momentum of balls before and after collision to get,

$$mv = mv_1 + 2mv_2 \text{ or } v = v_1 + 2v_2 \qquad \ldots(ii)$$

Equation for coefficient of restitution gives

$$v = v_2 - v_1 \qquad \ldots(iii)$$

Add (ii) and (iii) to get $2v = 3v_2$

$$\text{or } v_2 = \frac{2v}{3}ms^{-1} \text{ and } v_1 = -\frac{v}{3}ms^{-1} \qquad \ldots(iv)$$

Conserve energy for 'm' as it reaches maximum height,

$$\frac{1}{2}m\left(\frac{v}{3}\right)^2 = mg\,h_1$$

$$\text{or } h_1 = \frac{v^2}{18g} = \frac{v_0^2 + 2gl(1-\cos\alpha)}{18g} \text{ [using (i)]}$$

Conserve energy for '2m' as it reaches maximum height,

$$\frac{1}{2}2m\left(\frac{2v}{3}\right)^2 = 2mgh_2$$

$$\Rightarrow h_2 = \frac{1}{2g}\cdot\frac{4v^2}{9}$$

$$\Rightarrow h_2 = \frac{2}{9g}[v_0^2 + 2gl(1-\cos\alpha)] \text{ [using (i)]}$$

Example 6: A gun is mounted on a gun carriage movable on a smooth horizontal plane and the gun is elevated at an angle 45° to the horizontal. A shot is fired and leaves the gun inclined at an angle θ to the horizontal. If the mass of gun and carriage is n times that of the shot, find the value of θ.

Sol: As the frictional force on the cart is zero, the momentum of cart comprising cart and bullet is conserved in horizontal direction.

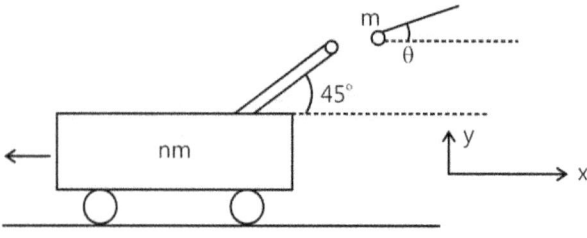

Let the mass of the shot be m and the mass of the gun carriage be nm.

Suppose the velocity of the shot relative to the gun be u and its velocity relative to the ground be V. The gun recoils with a speed v. As the system comprising gun and the shot rests on a smooth horizontal plane, the net horizontal external force will be zero, so conserving momentum in the horizontal direction, taken as the x – axis, we get

$$(nm)v_x + mV_x = 0 \Rightarrow nmv_x + m(u_x + v_x) = 0$$

$$\Rightarrow (nm)(-v) + m(u\cos 45 - v) = 0$$

$$\Rightarrow -(n+1)mv + \frac{mu}{\sqrt{2}} = 0 \qquad \ldots(i)$$

Again, $V_x = -n\,v_x = -n\,(-v)\,;\Rightarrow V_x = nv \qquad \ldots(ii)$

Now the component of the velocity of the gun along the vertical i.e. along the y – axis is zero, so the velocity of the shot along the y – axis will be given by

$$V_y = u\,\sin 45 + 0\,;\, V_y = \frac{u}{\sqrt{2}} \qquad \ldots(iii)$$

$$\Rightarrow \tan\theta = \frac{V_y}{V_x} = \frac{u/\sqrt{2}}{n\,v} \text{ (using (ii) & (iii))}$$

$$\Rightarrow \tan\theta = \frac{u}{\sqrt{2}\,n\,v} \qquad \ldots(iv)$$

From (i) we get $v = \dfrac{u}{\sqrt{2}(n+1)}$

$$\Rightarrow \frac{u}{v} = \sqrt{2}\,(n+1) \qquad \ldots(v)$$

From (iv) and (v) we get

$$\tan\theta = \frac{n+1}{n} \Rightarrow \theta = \tan^{-1}\left(\frac{n+1}{n}\right)$$

Exercise 1

Q.1 Show that center of mass of an isolated system moves with a uniform velocity along a straight line path.

Q.2 Locate the center of mass of uniform triangular lamina and a uniform cone.

Q.3 Explain what is meant by center of gravity.

Q.4 Obtain an expression for the position vector of center of mass of a two particle system.

Q.5 Obtain an expression for the position vector of the center of mass of a system of n particle.

Q.6 Prove that center of mass of an isolated system moves with a uniform velocity along a straight line path.

Q.7 Find the center of mass of three particles at the vertices of an equilateral triangle. The masses of the particles are 0.10 kg, 0.15 kg and 0.20 kg respectively. Each side of the quilateral triangle is 0.5 m long.

Q.8 Find the center of mass of a triangular lamina.

Q.9 Two bodies of masses 0.5 kg and 1 kg are lying in XY plane at (−1, 2) and (3, 4) respetively. What are the co-ordinates of the center of mass ?

Q.10 Three point masses of 1 kg, 2 kg and 3 kg lie at (1, 2), (0, −1) and (2, −3) respectively. Calculate the co-ordinates of the center of mass of the system.

Q.11 Two particles of mass 2 kg and 1 kg are moving along the same straight line with speeds 2 ms^{-1} and 5 ms^{-1} respectively. What is the speed of the center of mass of the system if both the particles are moving (a) in same direction (b) in opposite direction ?

Q.12 Consider a two-particle system with the particles having masses m_1 and m_2. If the first particle is pushed towards the center of mass through a distance d, by what distance should the second particle be moved so as to keep the center of a mass at the same position?

Q.13 Two particles of masses 1 kg and 3 kg are located at $(2\hat{i} + 5\hat{j} + 13\hat{k})$ and $(-6\hat{i} + 4\hat{j} - 2\hat{k})$ meter respectively. Find the position of their center of mass.

Q.14 Four particles of masses $m_1 = 1kg$, $m_2 = 2kg$, $m_3 = 3kg$ and $m_4 = 4kg$ are located at the corners of a rectangle as shown in figure. locate the position of center of mass.

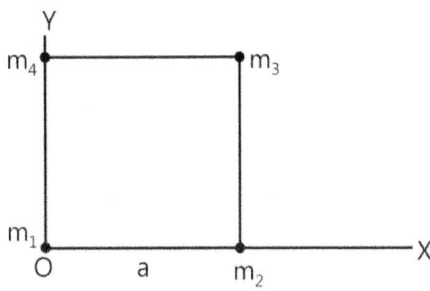

Q.15 Find the center of mass of uniform L shaped (a thin flat plate) with dimensions as shown in figure. The mass of the lamina is 3 kg.

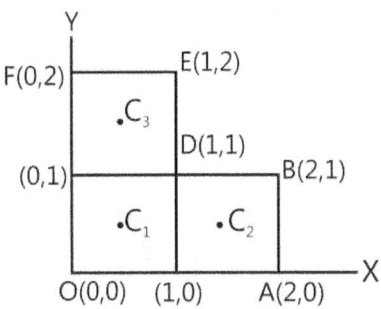

Exercise 2

Single Correct Choice Type

Q.1 A bullet of mass m moving with a velocity v strikes a vertically suspended wooden block of mass M and embedded in it. If the block rises to a height h, the initial velocity of the bullet will be

(A) $\sqrt{2hg}$

(B) $\left(\dfrac{M+m}{m}\right)\sqrt{2hg}$

(C) $\left(\dfrac{m}{M+m}\right)\sqrt{2hg}$

(D) $\left(\dfrac{M+m}{m}\right)\sqrt{hg}$

Q.2 A body of mass 1 kg, which was initially at rest, explodes and breaks into three fragments of masses in the ratio of $1:1:3$.

Both the pieces of equal masses fly off perpendicular to each other with a speed of 30 m/s each. The velocity of the heavier fragment is

(A) $\dfrac{10}{\sqrt{2}}$ ms^{-1}

(B) $10\sqrt{2}$ ms^{-1}

(C) 20 ms^{-1}

(D) $20\sqrt{2}$ ms^{-1}

Q.3 If the linear momentum of a body is increased by 50%, its kinetic energy will increase by

(A) 50%   (B) 100%   (C) 125%   (D) 150%

Q.4 Two perfectly elastic particles A and B of equal masses travelling along the line joining them with velcity 25 ms^{-1} and 20 ms^{-1} respectively collide. Their velocities after the elastic collision will be (in ms^{-1}) respectively.

(A) 0 and 45

(B) 5 and 45

(C) 20 and 25

(D) 25 and 20

Q.5 A body of mass 2.9 kg is suspended from a string of length 2.5 m and is at rest. A bullet of mass 0.1 kg, moving horizontally with a speed of 150 ms^{-1} strikes and sticks to it. What is the maximum angle made by the string with the vertical after the impact?

($g = 10$ ms^{-2})

(A) 30°   (B) 45°   (C) 60°   (D) 90°

Q.6 An isolated particle of mass m is moving in a horizontal plane (x-y), along the x-axis, at a certain height above the ground. It suddenly explodes into two fragments of mass $\dfrac{m}{4}$ and $\dfrac{3m}{4}$. An instant later, the smaller fragment is at y = +15 cm. The larger fragment at this instant is at

(A) y = -5 cm

(B) y = +20 cm

(C) y = +5 cm

(D) y = -20cm

Q.7 A ball collides elastically with another ball of the same mass. The collision is oblique and initially one of the ball was at rest. After the collision, the two balls move with same speeds. What will be the angle between the velocity of the balls after the collision?

(A) 30°   (B) 45°   (C) 60°   (D) 90°

Q.8 A body of mass 2kg moving with a velocity of 3 ms^{-1} collides head-on with a body of mass 1 kg moving with a velocity of 4 ms^{-1}. After collision the two bodies stick together and move with a common velocity which in the units m/s is equal to

(A) $\dfrac{1}{4}$ (B) $\dfrac{1}{3}$ (C) $\dfrac{2}{3}$ (D) $\dfrac{3}{4}$

Q.9 Two particles of masses M and 2M are at a distance D apart. Under the mutual gravitational force they start moving towards each other. The acceleration of their center of mass when they are D/2 apart is:

(A) $2\,GM/D^2$

(B) $4\,GM/D^2$

(C) $8\,GM/D^2$

(D) Zero

Previous Year's Questions

Q.1 Two particles A and B initially at rest, move towards each other by mutual force of attraction. At the instant when the speed of A is v and the speed of B is 2v, the speed of the center of mass of the system is   *(1982)*

(A) 3v   (B) v   (C) 1.5v   (D) zero

Q.2 A shell is fired from a cannon with a velocity v (ms^{-1}) at an angle θ with the horizontal direction. At the highest point in its path it explodes into two pieces of equal mass. One of the pieces retraces its path to the cannon and the speed (ms^{-1}) of the other piece immediately after the explosion is   *(1986)*

(A) $3v\cos\theta$

(B) $2v\cos\theta$

(C) $\dfrac{3}{2}v\cos\theta$

(D) $\sqrt{\dfrac{3}{2}}v\cos\theta$

Q. 3 Two particles of masses m_1 and m_2 in projectile motion have velocities $\vec{v}_1$ and $\vec{v}_2$ respectively at time t = 0. They collide at time t_0. Their velocities become $\vec{v}_1'$ and $\vec{v}_2'$ at time $2t_0$ while still moving in air. The value of $|(m_1\vec{v}_1' + m_2\vec{v}_2') - (m_1\vec{v}_1 + m_2\vec{v}_2)|$ is   *(2001)*

(A) Zero

(B) $(m_1 + m_2)g\,t_0$

(C) $2(m_1 + m_2)g\,t_0$

(D) $\dfrac{1}{2}(m_1 + m_2)g\,t_0$

Q.4 Two blocks of masses 10 kg and 4 kg are connected by a spring of negligible mass and placed on a frictionless horizontal surface. An impulse gives a velocity of 14 ms^{-1} to the heavier block in the direction of the lighter block. The velocity of the center of mass is **(2002)**

(A) 30 ms⁻¹   (B) 20 ms⁻¹

(C) 10 ms⁻¹   (D) 5 ms⁻¹

Q.5 Look at the drawing given in the figure, which has been drawn with ink of uniform line-thickness. The mass of ink used to draw each of the two inner circles, and each of the two line segments is m. The mass of the ink used to draw the outer circle is 6m. The coordinates of the center of the different parts are: outer circle (0, 0), left inner circle (-a, a), right inner circle (a, a), vertical line (0, 0) and horizontal line (0, -a). The y-coordinate of the center of mass of the ink in this drawing is **(2009)**

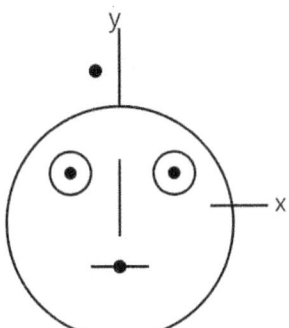

(A) $\dfrac{a}{10}$ (B) $\dfrac{a}{8}$ (C) $\dfrac{a}{12}$ (D) $\dfrac{a}{3}$

Q.6 Two small particles of equal masses start moving in opposite directions from a point A in a horizontal circular orbit. Their tangential velocities are v and 2v respectively, as shown in the figure. Between collisions, the particles move with constant speed. After making how many elastic collisions, other than that at A, these two particles will again reach the point A? **(2009)**

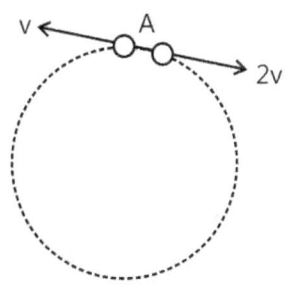

(A) 4   (B) 3   (C) 2   (D) 1

Q.7 A ball of mass 0.2 kg rests on a vertical post of height 5 m. a bullet of mass 0.01 kg, travelling with a velocity v ms⁻¹ in a horizontal direction, hits the center of the ball. After the collision, the ball and bullet travel independently. The ball hits the ground at a distance

of 20 m and the bullet at the distance of 100 m from the foot of the post. The initial velocity v of the bullet is **(2011)**

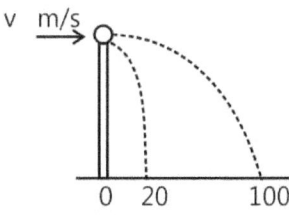

(A) 250 ms⁻¹   (B) $250\sqrt{2}$ ms⁻¹

(C) 400 ms⁻¹   (D) 500 ms⁻¹

Paragraph: Q.8 - Q.10

A small block of mass M moves on a frictionless surface of an inclined plane, as shown in figure. The angle of the incline suddenly changes from 60° to 30° at point B. the block is initially at rest at A. Assume that collisions between the block and the incline are total inelastic (g = 10 ms⁻²)

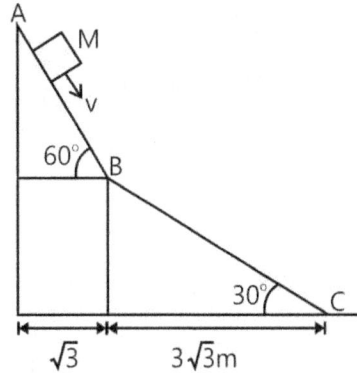

Q.8 The speed of the block at point B immediately after it strikes the second incline is **(2008)**

(A) $\sqrt{60}$ ms⁻¹   (B) $\sqrt{45}$ ms⁻¹

(C) $\sqrt{30}$ ms⁻¹   (D) $\sqrt{15}$ ms⁻¹

Q.9 The speed of the block at point C, immediately before it leaves the second incline is **(2008)**

(A) $\sqrt{120}$ ms⁻¹   (B) $\sqrt{105}$ ms⁻¹

(C) $\sqrt{90}$ ms⁻¹   (D) $\sqrt{75}$ ms⁻¹

Q.10 If collision between the block and the incline is completely elastic, then the vertical (upward) component of the velocity of the block at point B, immediately after it strikes the second incline is **(2008)**

(A) $\sqrt{30}$ ms⁻¹   (B) $\sqrt{15}$ ms⁻¹

(C) Zero   (D) $-\sqrt{15}$ ms⁻¹

Q.11 This question has Statement-I and Statement-II. Of the four choices given after the Statements, choose the one that best describes the two Statements.

Statement-I: A point particle of mass m moving with speed v collides with stationary point particle of mass M. If the maximum energy loss possible is given as $f\left(\dfrac{1}{2}mv^2\right)$ then $f = \left(\dfrac{m}{M+m}\right)$.

Statement-II: Maximum energy loss occurs when the particles get stuck together as a result of the collision.
(2013)

(A) Statement-I is true, statement-II is true, statement-II is not a correct explanation of statement-I.

(B) Statement-I is true, statement-II is false.

(C) Statement-I is false, statement-II is true

(D) Statement-I is true, statement-II is true, statement-II is a correct explanation of statement-I.

Q.12 Distance of the centre of mass of a solid uniform cone from its vertex is z_0. If the radius of its base is R and its height is h then z_0 is equal to: *(2015)*

(A) $\dfrac{3h}{4}$ (B) $\dfrac{5h}{8}$ (C) $\dfrac{3h^2}{8R}$ (D) $\dfrac{h^2}{4R}$

Q.13 A particle of mass m moving in the x direction with speed 2v is hit by another particle of mass 2m moving in the y direction with speed v. If the collision is perfectly inelastic, the percentage loss in the energy during the collision is close to *(2015)*

(A) 50%   (B) 56%   (C) 62%   (D) 44%

JEE Advanced/Boards

Exercise 1

Q.1 A block of mass 10 kg is suspended from a 3 m long weightless string. A bullet of mass 0.2 kg is fired into the block of horizontally with a speed of 20 ms⁻¹ and it gets embedded in the block. Calculate

(a) The speed acquired by the block

(b) The maximum displacement of the block

(c) The energy converted to heat in the collision.

Q.2 A projectile of mass 50 kg shot vertically upwards with an initial velocity of 100 ms⁻¹. After 5 s it explodes into two fragments, one of which having mass 20 kg travels vertically up with a velocity of 150 metres/sec. if g = 9.8 ms⁻².

(a) What is the velcoity of the other fragment at that instant ?

(b) Calculate the sum of the momenta of the two fragments 3 s after the explosion. What would have been the momentum of the projectile at this instant if there had been no explosion?

Q.3 Particle P and Q of mass 20 g and 40 g respectively are simultaneously projected from points A and B on the ground. The initial velocities of P and Q make angle

45° and 135° respectively with line AB. Each particle has an initial speed of 49 ms⁻¹. The separation AB is 245 m. Both particles travel in the same vertical plane and undergo a collision. After the collision, P retraces its path. Taking g = 9.8 ms⁻², determine

(a) The position of Q when it hits the ground.

(b) How much time, after the collision, does Q take to reach the ground.

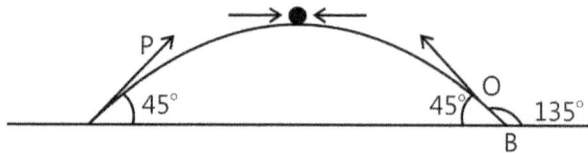

Q.4 A shell of mass 500 kg travelling horizontally at a speed of 100 ms⁻¹ explodes into just three parts. The first part of mass 200 kg travels vertically upwards at a speed of 150 ms⁻¹ and the second part of mass 150 kg travels horizontal with a speed of 60 ms⁻¹, but in a direction opposite to that of the original shell. What is the velocity fo the third part? What is the path of the center of mass of the fragments after the explosion?

Q.5 A small sphere of mass 10 g is attached to a point of smooth vertical wall by a light string of length 1 m. The sphere is pulled out in vertical plane perpendicular to the wall so that the string makes an angle of 60°

with the wall and is then released. It is found that after the first rebound, the string makes a maximum angle of 30° with the wall. Calculate the coefficient of restitution and the loss of kinetic energy due to impact. If all the energy is converted into heat, find the heat produced by the impact.

Q.6 A small ball A slides down the quadrant of a circle as shown in the figure and hits the ball B of equal mass which is initially at rest. Find the velocities of both the balls after collision. Neglect the effect of friction and assume the collision to be elastic.

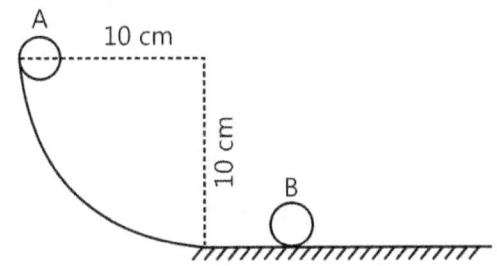

Q.7 Two balls A and B of mass 0.10 kg and 0.25 kg respectively are connected by a stretched spring of negligible mass and spring constant 2 Nm⁻¹. Unstretched length of the spring is 0.6 m and placed on a smooth table. When the balls are released simultaneously the initial acceleration of ball B is 50 cm s⁻² west-ward.

(a) What is the magnitude and direction of the initial acceleration of the ball A?

(b) What is the initial compression of the spring.

(c) What is the maximum distance between balls A and B.

Q.8 Find the center of mass of a uniform disc of radius a from which a circular section of radius b has been removed. The center of the hole is at a distance c from, the center of the disc.

Q.9 A man of mass m climbs a rope of length L suspended below a balloon of mass M. The ballon is stationary with respect to ground,

(a) If the man begins to climb up the rope at a speed v_{rel} (relative to rope) in what direction and with what speed (relative to ground) will the balloon move?

(b) How much has the balloon by climbing the rope.

(c) What is the state of motion after the man stops climbing?

Q.10 Prove that in case of oblique elastic collision of two particles of equal mass out of which one is at rest, the recoiling particles always move off at right angles to each other.

Q.11 A uniform thin rod of mass M and length L is standing vertically along the y-axis on a smooth hroizontal surface, with its lower end at the origin (0,0). A slight disturbance at t = 0 causes the lower end to slip on the smooth surface along the positive x-axis, and the rod starts falling.

(a) What is the path followed by the center of mass of the rod during its fall?

(b) Find the equation of trajectory of a point on the rod located at a distance r from the lower end.

Q.12 Two blocks of masses m_1 and m_2 are connected by a light inextensible string passing over a smooth fixed pulley of negligible mass. Find the acceleration of the center of mass of the system when blocks move under gravity.

Q.13 A block of mass m is resting on the top of a smooth prism of mass M which is resting on a smooth table. Calculate the distance moved by the prism when the block reaches the bottom.

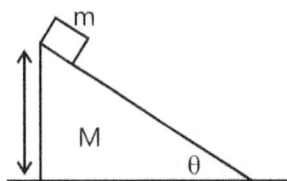

Q.14 A shell is fired from a cannon with a velocity v m/s at an angle θ with the horizontal direction. At the highest point of its path is explodes into two pieces of equal masses. What is the speed of other piece immediately after explosion, if one of the piece retraces its path to the cannon?

Q.15 A particle of mass 4m which is at rest explodes into three fragments. Two of the fragments each of mass m are found to move with a speed v each is mutually perpendicular directions. Calculate the energy released in the process of explosion.

Q.16 A moving particle of mass m makes a head on elastic collision with a particle of mass 2m which is initially at rest. Show that the colliding particle losses (8/9)th of its energy after collision.

Q.17 A ball is dropped on the ground from a height h. If the coefficient of restitution is e, then find the total distance travelled by the ball before coming to rest and the total time elapsed.

Q.18 A block of mass m_1 = 150 kg is at rest on a very long frictionless table, one end of which is terminated in a wall. Another block of mass m_2 is placed between the first block and the wall, and set in motion towards m_1 with constant speed u_2.

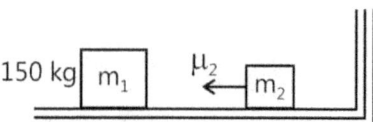

Assuming that all collisions are completely elastic, find the value of m_2 for which both blocks move with the same velocity after m_2 has collided once with m_1 and once with the wall. The wall has effectively infinite mass.

Q.19 A simple pendulum is suspended from a peg on a vertical wall. The pendulum is pulled away from the wall to a horizontal position and released. The ball hits the wall, the coefficient of restitution being $\left(\dfrac{2}{\sqrt{5}}\right)$.

What is the minimum number of collisions after which the amplitude of oscillation becomes less than 60°?

Q.20 A block A of mass 2m is placed on another block b of mass 4 m which in turn in placed on a fixed table. The two blocks have the same length 4d and they are placed as shown in the figure.

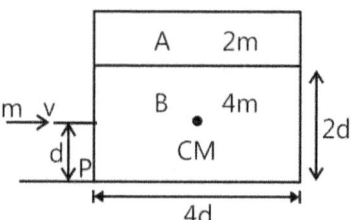

The coefficient of friction (both static and kinetic) between the block B and the table is μ. There is no friction between the two blocks. A small object of mass m moving horizontally along a line passing through the center of mass of the block B and perpendicular to its face with a speed v collides elastically with the block B at a height d above the table.

(a) What is the minimum value of v (call it v_0), required to make the block A topple?

(b) If v = $2v_0$ find the distance (from the point P) at which the mass m falls on the table after collision.

Q.21 A 60 kg man and a 50 kg woman are standing on opposite ends of a platform of mass 20 kg. The platform is placed on a smooth horizontal ground. The man and the woman begin to approach each other. Find the displacement of the platform when the two meet in terms of the displacement x_0 of the man relative to the platform. The length of the platform is 6m.

Q.22 A rope thrown over a pulley has a ladder with a man A on one of its ends and a counter balancing mass M on it other end. The man whose mass is m, climb upwards by $\overrightarrow{\Delta r}$ relative to the ladder and the stops. Ignoring the masses of the pulley and the rope, as well as the friction in the pulley axis, find the displacement of the center of mass of this system.

Q.23 A drinking straw of length $\dfrac{3a}{2}$ and mass 2m is placed on a square table of side 'a' parallel to one of its sides such that one third of its length extends beyond the table. An insect of mass $\dfrac{m}{2}$ lands on the inner end of the straw (i.e., the end which lies on the table) and walks along the straw until it reaches the outer end. It does not topple even when another insect lands on top of the first one. Find the largest mass of the second insect that can have without toppling the straw. Neglect friction.

Q.24 A boy throws a ball with initial speed $2\sqrt{ag}$ at an angle θ to the horizontal. It strikes a smooth vertical wall and returns to his hand. Show that if the boy is standing at a distance 'a' from the wall, the coefficient of restitution between the ball and the wall equals $\dfrac{1}{(4\sin 2\theta - 1)}$. Also show that θ cannot be less than 15°.

Q.25 A ball is projected from a point A on a smooth inclined plane which makes an angle α to the horizontal. The velocity of projection makes an angle θ with the plane upwards. If on the second bounce the ball is moving perpendicular to the plane, find e in terms of α and θ. Here e is the coefficient of restitution between the ball and the plane.

Q.26 Two identical smooth balls are projected toward each ther from points A and B on the horizontal ground with same speed of projection. The angle of projection.

The angle of projection in each case is 30°. The distance between A and B is 100 m. The balls collide in air and return to their respective points of projection. If coefficient of restitution is e = 0.7, find

(a) The speed of projection of either ball.

(b) Coordinates of point with respect to A where the balls collide.(Take g = 10 ms⁻²)

Q.27 Three identical particles A, B and C lie on a smooth horizontal table. (see figure) Light inextensible strings which are just taut connect AB and BC and $\angle ABC$ is 135°. An impulse J is applied to the particle C in the direction BC.

Find the initial speed of each particle. The mass of each particle in m.

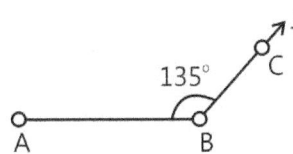

Q.28 A 2 kg sphere A is connected to a fixed point O by an inextensible cord of length 1.2 m (see figure). The sphere is resting on a frictionless horizontal surface at a distance of 0.5 m from O when it is given a velocity v_0 in a direction perpendicular to the line OA. It moves freely until it reaches position A' when the cord becomes taut.

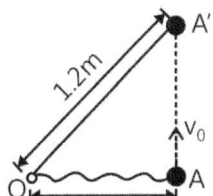

Determine

(a) The maximum allowable velocity v_0 if the impulse of the force exerted on the cord is not to exceed 3 Ns.

(b) The loss of energy as the cord becomes taut, if the sphere is given the maximum allowable velocity v_0.

Q.29 An open car of mass 1000 kg is running at 25 m/s holds three men each of mass 75 kg. Each man runs with a speed of 5 ms⁻¹ relative to the car and jumps off from the back end. Find the speed of the car if the three men jump off.

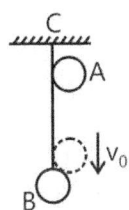

(a) In succession

(b) All together.

Neglect friction between the car and the ground.

Q.30 Ball B is hanging from an inextensible cord BC. An identical ball A is released from rest when it is just touching the cord and acquires a velocity v_0 before striking ball B. Assuming perfectly elastic impact (e = 1) and no friction, determine the velocity of each ball immediately after impact.

Q.31 A particle whose initial mass is m_0 is projected vertically upwards at time t = 0 with speed gT, where T is a constant. The particle gradually acquires mass on its way up and at time t the mass of the particle has increased to $m_0 e^{dT}$. If the added mass is at rest relative to the particle when it is acquired, find the time when the particle is at highest point and its mass at that instant.

Q.32 Two blocks of mass 2kg and M are at rest on an inclined plane and are separated by a distance of 6.0 m as shown. The coefficient of friction between each block and the inclined plane is 0.25. the 2 kg block is given a velocity of 10.0 ms⁻¹ up the inclined plane.

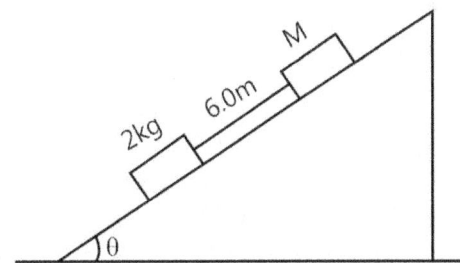

It collides with M, comes back and has a velocity of 1.0 m/s when it reaches its initial position. The other blocks M after the collision moves 0.5 m up and comes to rest. Calculate the coefficient of restitution between the blocks and the mass of the block M.

(Take $\sin\theta \approx \tan\theta = 0.05$ and g = 10 ms⁻²)

Exercise 2

Single Correct Choice Type

Q.1 A bullet of mass m moving with a velocity v strikes a vertically suspended wooden block of mass M and embedded in it. If the block rises to a height h, the initial velocity of the bullet will be

(A) $\sqrt{2hg}$

(B) $\left(\dfrac{M+m}{m}\right)\sqrt{2hg}$

(C) $\left(\dfrac{m}{M+m}\right)\sqrt{2hg}$

(D) $\left(\dfrac{M+m}{m}\right)\sqrt{hg}$

Q.2 Two identical billiard balls A and B of equal mass and radius are in contact on a horizontal table. A similar third ball C strikes these balls symmetrically in the middle and remains at rest after the impact, the coefficient of restitution of the balls is

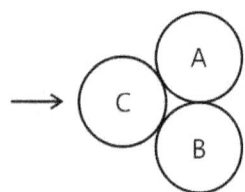

(A) $\dfrac{1}{6}$ (B) $\dfrac{1}{3}$ (C) $\dfrac{2}{3}$ (D) $\dfrac{\sqrt{3}}{2}$

Q.3 A sphere of mass m moving with a constant velocity u hits another stationary sphere of the same mass. If e is the coefficient of the restitution, then the ratio of the velocities of the two spheres after collision will be

(A) $\left(\dfrac{1-e}{1+e}\right)$ (B) $\left(\dfrac{1+e}{1-e}\right)$ (C) $\left(\dfrac{e+1}{e-1}\right)$ (D) $\left(\dfrac{e-1}{e+1}\right)$

Q.4 A cannon ball is fired with a velocity of 200 ms⁻¹ at an angle of 60° with the horizontal. At the highest point it explodes into three equal fragments. One goes vertically upwards with a velocity of 100 ms⁻¹, the second one falls vertically downwards with a velocity of 100 ms⁻¹. The third one moves with a velcoity of

(A) 100 ms⁻¹ horizontally

(B) 300 ms⁻¹ horizontally

(C) 200 ms⁻¹ at 60° with the horizontal

(D) 300 ms⁻¹ at 60° with the horizontal

Q.5 A bullet of mass 0.01 kg, travelling at a speed of 500 m/s, strikes a block of mass 2kg, which is suspended by a string of length 5 m, and emerges out. The block rises by a vertical distance of 0.1 m. The speed of the bullet after it emerges from the block is

(A) 55 ms⁻¹   (B) 110 ms⁻¹

(C) 220 ms⁻¹   (D) 440 ms⁻¹

Q.6 A 1 kg ball, moving at 12 ms⁻¹ collides head-on with a 2 kg ball moving in the opposite direction at 24 m/s. If the coefficient of restitution is $\dfrac{2}{3}$, then the energy lost in the collision is

(A) 60 J   (B) 120 J

(C) 240 J   (D) 480 J

Q.7 A body of mass m_1 and speed v_1 makes a head-on, elastic collision with a body of mass m_2, initially at rest. The velocity of m_1 after the collision is

(A) $\dfrac{m_1+m_2}{m_1 m_2}v_1$ (B) $\dfrac{m_1-m_2}{m_1+m_2}v_1$

(C) $\dfrac{2m_1 v_1}{m_1+m_2}$ (D) $\dfrac{2m_2 v_1}{m_1+m_2}$

Q.8 In the above example, the velocity of mass m_2 after the collision is

(A) $\dfrac{m_1+m_2}{m_1 m_2}v_1$ (B) $\dfrac{m_1-m_2}{m_1+m_2}v_1$

(C) $\dfrac{2m_1 v_1}{m_1+m_2}$ (D) $\dfrac{2m_2 v_1}{m_1+m_2}$

Q.9 A ball of mass m approaches a moving wall of infinite mass with speed v along the normal to the wall. The speed of the wall is u towards the ball. The speed of the ball after an elastic collision with the wall is

(A) u + v away from the wall

(B) 2u + v away from the wall

(C) u – v away from the wall

(D) v – 2u away from the wall.

Q.10 A neutron is moving with velocity u. It collides head on and elastically with an atom of mass number A. If the initial K.E. of the neutron is E, how much K.E. is retained by neutron after collision?

(A) $[A/(A+1)]^2 E$ B) $[A/(A+1)^2]E$

(C) $[(1-A)/(A+1)^2]E$ (D) $[(A-1)/(A+1)^2]E$

Q.11 A ball is dropped from a height h on the ground. If the coefficient of restitution is e, the height to which the ball goes up after it rebounds for the nth time is

(A) $he^2 n$ (B) he^2 (C) $\dfrac{e^2 n}{h}$ (D) $\dfrac{h}{e^{2n}}$

Q.12 Two equal spheres A and B lie on a smooth horizontal circular groove at opposite ends of diameter. A is projected along the groove and at the end of time t impinges on B. If e is coefficient of restitution, the second impact will occur after a time

(A) $\dfrac{2t}{e}$ (B) $\dfrac{t}{e}$ (C) $\dfrac{\pi t}{e}$ (D) $\dfrac{2\pi t}{e}$

Q.13 The center of mass of triangle shown in the figure. has co-ordinates.

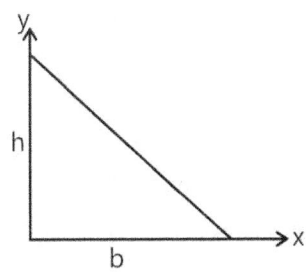

(A) $x = \dfrac{h}{2}; y = \dfrac{b}{2}$ (B) $x = \dfrac{b}{2}; y = \dfrac{h}{2}$

(C) $x = \dfrac{b}{3}; y = \dfrac{h}{3}$ (D) $x = \dfrac{h}{3}; y = \dfrac{b}{3}$

Q.14 A cart of mass M is tied to one end of a massless rope of length 10 m. The other end of the rope is in the hands of a man of mass M, the entire system is on a smooth horizontal surface. The man is at x = 0 and the cart at x = 10 m. if the man pulls the cart by a rope, the man and the cart will meet at the point :

(A) x = 0   (B) x = 5 m

(C) x = 10 m   (D) They will nevemeet

Multiple Correct Choice Type

Q.15 Which one of the following statements does not hold god when two balls of masses m_1 and m_2 undergo elastic collision?

(A) when $m_1 < m_2$ and m_2 at rest, there will be maximum transfer of momentum.

(B) when $m_1 > m_2$ and m_2 at rest, after collision the ball of mass m_2 moves with four times the velocity of m_1

(C) when $m_1 = m_2$ and m_2 at rest, there will be maximum transfer of K.E.

(D) when collision is oblique and m_2 at rest with $m_1 = m_2$, after collision the ball moves in opposite directions.

Assertion Reasoning Type

Each of the questions given below consists of two statements, an assertion (A) and reason (R). Select the number corresponding to the appropriate alternative as follows.

(A) If both A and R are true and R is the correct explanation of A.

(B) If both A and R are true but R is not the correct explanation of A.

(C) If A is true but R is false

(D) If A is flase but R is true

Q.16 Assertion: When two bodies of different masses are just released from different position above the ground, then acceleration of their center of mass is zero.

Reason: When bodies move, their center may change position but is not accelerated.

Q.17 Assertion: The center of mass of a proton and an electron, released from their respective positions remains at rest.

Reason: The proton and electron attract and move towards each other. No external force is applied, therefore, their center of mass remains at rest.

Q.18 Assertion: The center of mass of a body may lie where there is no mass.

Reason: Center of mass of a body is a point, where the whole mass of the body is supposed to be concentrated.

Q.19 Assertion: When a body dropped from a height explodes in mid air, the pieces fly in such a way that their center of mass keeps moving vertically downwards.

Reason: Explosion occurs under internal forces only. External force = 0.

Q.20 Assertion: The center of mass of a circular disc lies always at the center of the disc.

Reason: Circular disc is a symmetrical body.

Q.21 Assertion: At the center of earth, a body has center of mass, but no center of gravity.

Reason: This is because g = 0 at the center of earth.

Q.22 Assertion: The center of mass of a body may lie where there is no mass.

Reason: The center of mass has nothing to do with the mass.

Comprehension Type

In physics, we come across many examples of collisions. The molecules of a gas collide with one another and with the container. The collisions of a neutron with an atom is well known. In a nuclear reactor, fast neutrons produced in the fission of uranium atom have to be slowed down. They are, therefore, made to collide with hydrogen atom. The term collision does not necessarily mean that a particle or a body must actually strike another. In fact, two particles may not even touch each other and yet they are said to collide if one particle influences the motion of the other. When two bodies collide, each body exerts an equal and opposite force on the other. The fundamental conservation law of physics are used to determine the velocities of the bodies after the collision. Collision may be elastic or inelastic. Thus a collision may be defined as an event in which two or more bodies exert relatively strong forces on each other for a relatively short time. The forces that the bodies exert on each other are internal to the system.

Almost all the knowledge about the sub-atomic particles such as electrons, protons, neutrons, muons, quarks, etc. is obtained from the experiments involving collisions.

There are certain collisions called nuclear reactions in which new particles are formed. For example, when a slow neutron collides with a U^{235} nucleus, new nuclei barium-141 and Kr^{92} are formed. This collisioin is called nuclear fission. In nuclear fusion, two nuclei deuterium and tritium collide (or fuse) to form a helium nucleus with the emission of a neutron.

Q.23 Which one of the following collisions is not elastic?

(A) A hard steel ball dropped on a hard concrete floor and rebounding to its original height.

(B) Two balls moving in the same direction collide and stick to each other

(C) Collision between molecules of an ideal gas.

(D) Collisions of fast neutrons with hydrogen atoms in a fission reactor.

Q.24 Which one of the following statemnts is true about inelastic collision?

(A)The total kinetic energy of the particles after collision is equal to that before collision.

(B) The total kinetic energy of the particle after collision is less than that before collision.

(C) The total momentum of the particles after collision is less than that before collision.

(D) Kinetic energy and momentum are both conserved in the collision.

Q.25 In elastic collision

(A) Only energy is conserved.

(B) Only momentum is conserved.

(C) Neither energy nor momentum is conserved.

(D) Both energy and momentum are conserved.

Previous Years' Questions

Q.1 A body of mass m moving with a velocity v in the x-direction collides with another body of mass M moving in the y-direction with a velocity V. They coalesce into one body during collsion. Find

(a) The direction and magnitude of the momentum of the composite body.

(b) The fraction of the initial kinetic energy transformed into heat during the collision. **(1978)**

Q.2 A 20 g bullet pierces through a plate of mass $M_1 = 1 kg$ and then comes to rest inside a second plate of mass $M_2 = 2.98 kg$ as shown in the figure. It is found that the two plates initially at rest, now move with equal velocities. Find the percentage loss in the initial velocity of the bullet when it is between M_1 and M_2. Neglect any loss of material of the plates due to the action of bullet. Both plates are lying on smooth table. **(1979)**

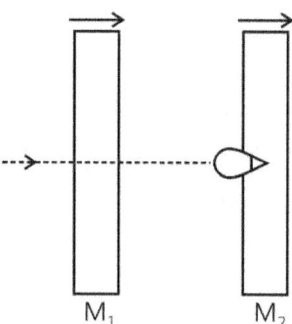

Q.3 A circular plate of uniform thickness has a diameter of 56 cm. A circular portion of diameter 42 cm is removed from one edge of the plate as shown in figure. Find the position of the center of mass of the remaining portion. **(1980)**

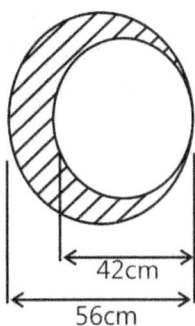

Q.4 Three particles A, B and C of equal mass move with equal speed v along the medians of an equilateral triangle as shown in figure. They collide at the centroid G of the triangle. After the collision, A comes to rest, B retraces its path with speed v. What is the velocity of C?
(1982)

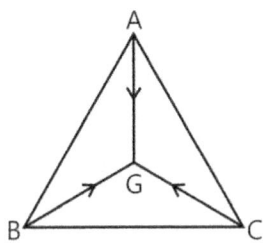

Q.5 Two bodies A and B of masses m and 2m respectively are placed on a smooth floor. They are connected by a spring. A third body C of mass m moves with velocity v_0 along the line joining A and B and collides elastically with A as shown in figure. At a certain instant of time t_0 after collision, it is found that the instantaneous velocities of A and B are the same. Further at this instant the compression of the spring is found to be x_0. Determine (a) the common velocity of A and B at time t_0 and (b) the spring constant.
(1984)

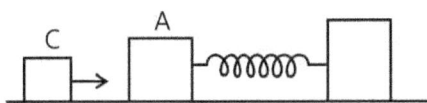

Q.6 A simple pendulum is suspended from a peg on a vertical wall. The pendulum is pulled away from the wall to a horizontal position (see figure) and released. The ball hits the wall, the coefficient of restitution being $\frac{2}{\sqrt{5}}$.

What is the minimum number of collisions after which the amplitude of oscillations becomes less than 60 degrees?
(1987)

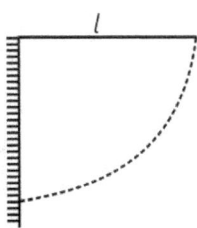

Q.7 A uniform thin rod of mass M and length L is standing vertically along the y-axis on a smooth horizontal surface, with its lower end at the origin (0, 0). A slight disturbance at t = 0 causes the lower end to slip on the smooth surface along the positive x-axis, and the rod starts falling.

(a) What is the path followed by the center of mass of the rod during its fall?

(b) Find the equation of the trajectory of a point on the rod located at a distance r from the lower end. What is the shape of the path of this point?
(1993)

Q.8 A wedge of mass m and triangular cross-section (AB = BC = CA = 2R) is moving with a constant velocity $(-v\hat{i})$ towards a sphere of radius R fixed on a smooth horizontal table as shown in the figure. The wedge makes an elastic collision with the fixed sphere and returns along the same path without any rotation. Neglect all friction and suppose that the wedge remains in contact with the sphere for a very short time Δt during which the sphere exerts a constant force $\vec{F}$ on the wedge.
(1998)

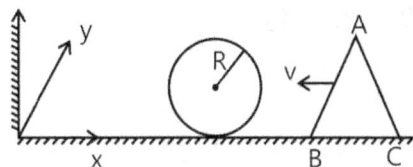

(a) Find the force $\vec{F}$ and also the normal force $\vec{F}$ exerted by the table on the wedge during the time Δt.

(b) Let h denote the perpendicular distance between the center of mass of the wedge and the line of action of force. Find the magnitude of the torque due to the normal force $\vec{N}$ about the center of the wedge during the interval Δt.

Q.9 Three objects A, B and C are kept in a straight line on a frictionless horizontal surface (see figure). These have masses m, 2m and m, respectively. The object A moves towards B with a speed 9ms^{-1} and makes an elastic collision with it. Thereafter, B makes completely inelastic collision with C. All motions occur on the same straight line. Find the final speed (in ms^{-1}) of the object C.
(2009)

m	2m	m
A	B	C

Q.10 A particle of mass m is projected from the ground with an initial speed u_0 at an angle α with the horizontal.

At the highest point of its trajectory, it makes a completely inelastic collision with another identical particle, which was thrown vertically upward from the ground with the same initial speed u_0. The angle that the composite system makes with the horizontal immediately after the collision is   *(2013, 14, 15, 16)*

(A) $\dfrac{\pi}{4}$ (B) $\dfrac{\pi}{4} + \alpha$ (C) $\dfrac{\pi}{4} - \alpha$ (D) $\dfrac{\pi}{2}$

Q.11 A bob of mass m, suspended by a string of length l_1, is given a minimum velocity required to complete a full circle in the vertical plane. At the highest point, it collides elastically with another bob of mass m suspended by a string of length l_2, which is initially at rest. Both the strings are mass-less and inextensible. If the second bob, after collision acquires the minimum speed required to complete a full circle in the vertical plane, the ratio l_1 / l_2 is   *(2013)*

Important Questions

JEE Main/Boards

Exercise 1

Q. 7   Q.9   Q.16

Exercise 2

Q.1   Q.9

Previous Years' Questions

Q.2   Q.3   Q.5
Q.8   Q.9   Q.10

JEE Advanced/Boards

Exercise 1

Q.3   Q.6   Q.7
Q.10   Q.20   Q.21
Q.28   Q.32

Exercise 2

Q.1   Q.2   Q.3
Q.4   Q.9

Previous Years' Questions

Q.2   Q.5   Q.8

JEE Main/Boards

Exercise 1

Q.9 $\dfrac{5}{3}, \dfrac{10}{3}$

Q.10 $\dfrac{7}{6}, -\dfrac{3}{2}$

Q.11 (a) 3 ms⁻¹ (b) $\dfrac{1}{3}$ ms⁻¹ in the direction of motion of 1 kg

Q.12 $\dfrac{m_1}{m_2}d$

Q.13 $-\hat{i} + \dfrac{17}{4}\hat{j} + \dfrac{7}{4}\hat{k}$

Q.14 $0.5a\hat{i} + 0.7b\hat{j}$

Q.15 $\dfrac{5}{6}m; \dfrac{5}{6}m$

Exercise 2

Single Correct Choice Type

Q.1 B   **Q.2** B   **Q.3** C   **Q.4** C   **Q.5** C   **Q.6** A

Q.7 D   **Q.8** C   **Q.9** D

Previous Years Questions

Q.1 D   **Q.2** A   **Q.3** C   **Q.4** C   **Q.5** A   **Q.6** C

Q.7 D   **Q.8** B   **Q.9** B   **Q.10** C   **Q.11** C   **Q.12** A

Q.13 B

JEE Advanced/Boards

Exercise 1

Q.1 (a) 0.39 m/s (b) 0.220 m (c) 39.32 J

Q.2 15 m/s, 1080 kg ms⁻¹

Q.3 (a) 122.5 m (b) $5\sqrt{2}$ second.

Q.4 441.25 m/s, −27°

Q.5 0.518, 0.0359 J, 0.0085 cals

Q.6 $v_A = 0$, $v_B = 1.4$ m/s

Q.7 (a) 1.25 cm/s² (eastwards) (b) 6.25 cm (c) 66.25 cm

Q.8 At a distance $\dfrac{cb^2}{a^2 - b^2}$ from O on the other side of the hole.

Q.9 (a) $-m\bar{v}_{rel}/(M+m)$ (b) $L\dfrac{m}{M+m}$ (c) system is stationary

Q.11 (a) Straight line (b) $\dfrac{x^2}{[L/2-r]^2} + \dfrac{y^2}{r^2} = 1$

Q.12 $\left(\dfrac{m_1 - m_2}{m_1 + m_2}\right)^2 g$

Q.13 $\dfrac{mh\cot\theta}{M+m}$

Q.14 $3v\cos\theta$

Q.15 $\dfrac{3}{2}mv^2$

Q.17 $\dfrac{h(1+e^2)}{1-e^2}$, $\sqrt{\dfrac{2h}{g}}\left[\dfrac{1+e}{1-e}\right]$

Q.18 50 kg

Q.19 4

Q.20 (a) $\dfrac{5}{2}\sqrt{6\mu gd}$ (b) $= -6d\sqrt{3\mu}$

Q.21 $\dfrac{30-11x_0}{13}$

Q.22 $\dfrac{m}{2M}\overrightarrow{\Delta r}$

Q. 23 $m'-m = \dfrac{m}{2}$

Q.25 $\dfrac{\cot\theta\cot\alpha}{2}-1$

Q.27 $\dfrac{\sqrt{2}}{7m}$, $\dfrac{\sqrt{10}}{7m}$, $\dfrac{3J}{7m}$

Q.28 (a) 1.65 m/s (b) 2.25 J

Q.29 (a) 25.97 m/s (b) 25.92 m/s

Q.30 $|v'_B| = 0.721v_0$, $|v'_A| = 0.693v_0$

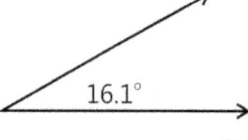

16.1°

Q.31 T ln (2), $2m_0$

Q.32 0.84, 15.011 kg

Exercise 2

Single Correct Choice Type

Q.1 B	**Q.2** C	**Q.3** A	**Q.4** B	**Q.5** C	**Q.6** C
Q.7 B	**Q.8** C	**Q.9** B	**Q.10** C	**Q.11** A	**Q.12** A
Q.13 C	**Q.14** B				

Multiple Correct Choice Type

Q.15 C, D

Assertion Reasoning Type

Q.16 D	**Q.17** A	**Q.18** B	**Q.19** A	**Q.20** D	**Q.21** A
Q.22 B				

Comprehension Type

Q.23 B	**Q.24** B	**Q.25** D

Previous Years' Questions

Q.1 (a) $\theta = \tan^{-1}\dfrac{MV}{mv}$, $P = \sqrt{m^2v^2 + M^2V^2}$ (b) $\dfrac{\Delta K}{K_i} = \dfrac{Mm(v^2+V^2)}{(M+m)(mv^2+MV^2)}$

Q.2 25%

Q.3 9 cm

Q.4 Opposite to velocity of B

Q.5 (a) $v_0/3$ (b) $\dfrac{2mv_0^2}{3x_0^2}$

Q.6 4

Q.7 (a) a straight line (b) $\dfrac{x^2}{\left(\dfrac{L}{2}-r\right)^2}+\dfrac{y^2}{r^2}=1$

Q.8 (a) $\overline{F} = \dfrac{2mv}{\Delta t}\hat{i} - \dfrac{2mv}{\sqrt{3}\Delta t}\hat{k}$, $\overline{N} = \left(\dfrac{2mv}{\sqrt{3}\Delta t}+mg\right)\hat{k}$ (b) $\Rightarrow |\overline{\tau}_N| = \dfrac{4\,mvh}{\sqrt{3}\Delta t}$

Q.9 4 ms^{-1}

Q.10 A

Q.11 5

JEE Main/Boards

Exercise 1

Sol 1: Isolated system, so external force = 0

F_{ext} = 0, therefore acceleration of centre of mass = 0.

So the centre of mass moves with a constant velocity along a straight line path (Ist law of motion).

Sol 2: (i) Lamina: mass per unit area

$$= \frac{M}{\frac{\sqrt{3}a^2}{4}} = \rho \text{ (say)}$$

then $X_{COM} = \int_0^{\frac{\sqrt{3}}{2}a} \frac{(\rho.dA).x}{M}$

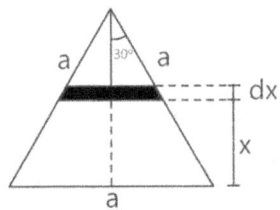

where dA = area of the strip of thickness dx (shaded reg.)

$$= 2\left(\frac{\sqrt{3}}{2}a - x\right).\tan 30° \, dx$$

$$dA = 2\left(\frac{a}{2} - \frac{x}{\sqrt{3}}\right).dx$$

so, $X_{COM} = \int_0^{\frac{\sqrt{3}a}{2}} \frac{\rho \times 2 \times \left(\frac{a}{2} - \frac{x}{\sqrt{3}}\right).x.dx}{M}$

$$= \frac{\rho}{M} \times 2 \times \int_0^{\frac{\sqrt{3}a}{2}} \left(\frac{ax}{2} - \frac{x^2}{\sqrt{3}}\right)dx$$

$$= \frac{4}{\sqrt{3}a^2} \times 2 \times \left[\frac{ax^2}{4} - \frac{x^2}{\sqrt{3}}\right]_0^{\frac{\sqrt{3}a}{2}}$$

$$= \frac{4}{\sqrt{3}a^2} \times 2 \times \left[\frac{a}{4} \times \frac{3}{4}a^2 - \frac{1}{3\sqrt{3}} \times \frac{3\sqrt{3}a^3}{8}\right]$$

$$= \frac{4}{\sqrt{3}a^2} \times 2 \times \left[\frac{3a^3}{16} - \frac{a^3}{8}\right] = \frac{4}{\sqrt{3}a^2} \times \frac{3a^3}{16}$$

$$= \frac{a}{2\sqrt{3}} \text{ (from bottom)}$$

(ii)

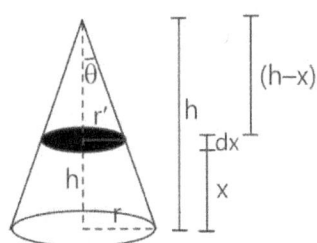

$$\tan \theta = \frac{r}{h}$$

Let $\rho = \frac{3M}{\pi r^2 h} = \frac{mass}{volume}$

then dm = ρdV, where dV = volume of shaded region

dV = $\pi. r'^2. dx$

$$= \pi \times [(h - x).\tan\theta]^2 \, dx = \frac{\pi.(h-x)^2.r^2}{h^2}.dx$$

so, $X_{COM} = \frac{\int_0^h x \, dm}{M} = \int_0^h \frac{x.\rho.dV}{M} = \rho.\int_0^h \frac{x.dV}{M}$

$$= \frac{3M}{\pi r^2 h.M}\int_0^h x.dV = \frac{3}{\pi r^2 h}\int_0^h \frac{\pi.x.(h-x)^2.r^2}{h^2}dx$$

$$= \frac{3}{h^3}\int_0^h x.(h-x)^2 dx = \frac{3}{h^3}.\frac{h^4}{12}; \quad X_{COM} = \frac{h}{4}$$

Sol 3: Centre of gravity is the point at which all the force of gravity is assumed to be applied i.e., there is a force of gravity on each point of the body and hence the complications are reduced by finding a point where all the force is assumed to be applied, this point is centre of gravity.

Sol 4: Now we have

$$M_{tot}\vec{a}_{COM} = m_1\vec{a}_1 + m_2\vec{a}_2$$

where m = mass

a = acceleration

so $M_{Tot.} \dfrac{\partial \vec{V}_{COM}}{dt} = m_1 \dfrac{\partial \vec{V}_1}{dt} + m_2 \dfrac{\partial \vec{V}_2}{dt}$

x dt, on integrating w. r. t dt, we get

$M_{Tot.} \vec{V}_{COM} = m_1 \vec{V}_1 + m_2 \vec{V}_2$

$\Rightarrow M_{Tot.} \dfrac{\partial \vec{x}_{COM}}{dt} = m_1 \dfrac{\partial \vec{x}_1}{dt} + m_2 \dfrac{\partial \vec{x}_2}{dt}$

$\Rightarrow$ On multiplying by dt, and integrating

$M_{Tot} \vec{x}_{COM} = m_1 \vec{x}_1 + m_2 \vec{x}_2$

So $\vec{x}_{COM} = \dfrac{m_1 \vec{x}_1 + m_2 \vec{x}_2}{M_{tot.}}$

Sol 5: $M_{Tot.} \cdot \vec{a}_{COM} = m_1 \vec{a}_1 + m_2 \vec{a}_2 + \dots m_n \vec{a}_n$

(Now, just like above question, question-4, we can find that

$\vec{x}_{COM} = \dfrac{m_1 \vec{x}_1 + m_2 \vec{x}_2 + \dots m_n \vec{x}_n}{M_{tot.}}$

Sol 6: We have

$M_{tot.} \cdot \vec{a}_{COM} = m_1 \vec{a}_1 + m_2 \vec{a}_2 + \dots m_n \vec{a}_n$

Now, $\vec{a}_{COM} = 0$, so we have

$0 = m_1 \dfrac{d\vec{V}_1}{dt} + m_2 \dfrac{d\vec{V}_2}{dt} + \dots m_n \dfrac{d\vec{V}_n}{dn}$

x dt, and integrating, we get

$C = m_1 \vec{V}_1 + m_2 \vec{V}_2 + \dots m_n \vec{V}_n = M_{Tot.} V_{COM}$

So $V_{COM} = \dfrac{C}{M_{Tot.}} = $ constant

Hence proved.

Sol 7:

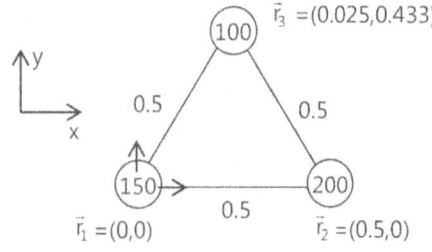

$\vec{r}_3 = (0.025, 0.433)$

$\vec{r}_1 = (0, 0)$

$\vec{r}_2 = (0.5, 0)$

So $\vec{x}_{COM} = \dfrac{m_1 \vec{x}_1 + m_2 \vec{x}_2 + m_3 \vec{x}_3}{M_{Tot.}}$

$= \dfrac{150 \times (10) + 200 \times (0.5) + 100 \times (0.25)}{450}$

$= \dfrac{25 + 180}{450} = \dfrac{125}{450}$

$\vec{x}_{COM} = 0.277\ \hat{i}$

$\vec{y}_{COM} = \dfrac{m_1 \vec{y}_1 + m_2 \vec{y}_2 + m_3 \vec{y}_3}{M_{Tot.}}$

$= \dfrac{150 \times 0 + 200 \times (0) + 100 \times 0.433}{450} = \dfrac{43.3}{450}$

$\vec{y}_{COM} = 0.096\ \hat{j}$

So $\vec{r}_{COM} = (0.277\hat{i} + 0.096\hat{j})$

Sol 8:

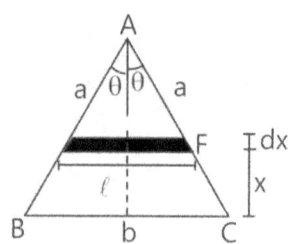

$\sin\theta = \dfrac{b}{2a}$, $\cos\theta = \dfrac{\sqrt{4a^2 - b^2}}{2a}$,

$\tan\theta = \dfrac{b}{\sqrt{4a^2 - b}}$

Now, $\rho = \dfrac{M}{\dfrac{1}{2}a^2 \sin 2\theta} = \dfrac{M}{a^2 \sin\theta \cos\theta}$

$\ell = \overbrace{2(a\cos\theta - x)}^{EF} . \tan\theta$
$\underbrace{\quad}_{\substack{AE \\ (AD - DE)}}$

So, $dA = \ell . dx$

So, $dm = \rho . dA = \rho . \ell . dx$

So $\vec{x}_{COM} = \dfrac{\displaystyle\int_0^{a\cos\theta} x . dx}{M}$

$= \dfrac{\rho}{M} . \displaystyle\int_0^{a\cos\theta} x \times 2x(a\cos\theta - x) . \tan\theta . dx$

$= \dfrac{1 \times 2}{a^2 \sin\theta . \cos\theta} . \displaystyle\int_0^{a\cos\theta} x(a\sin\theta - x\tan\theta) dx$

$= \dfrac{2}{a^2 \sin\theta . \cos\theta} . \left[\dfrac{x^2}{2} a\sin\theta - \dfrac{x^3 \tan\theta}{3} \right]_0^{a\cos\theta}$

$= \dfrac{2}{a^2 \sin\theta . \cos\theta} . \left[\dfrac{a^3 \sin\theta . \cos 2\theta}{2} - \dfrac{a^3 \sin\theta . \cos^2\theta}{3} \right]$

$$= \frac{a^3 \sin\theta \cdot \cos^2\theta}{3a^2 \sin\theta \cdot \cos\theta} = \frac{a\cos\theta}{3}$$

$$= \frac{\sqrt{4a^2 - b^2}}{6}$$

Sol 9: $\vec{r}_{COM} = \dfrac{m_1\vec{r}_1 + m_2\vec{r}_2}{M_{Tot.}}$

$$= \frac{0.5(-1, 2) + 1(3, 4)}{0.5 + 1}$$

$$= \frac{(-0.5, 1) + (3, 4)}{1.5} = \frac{(2.5, 5)}{1.5}$$

$$\vec{r}_{COM} = \left(\frac{5}{3}, \frac{10}{3}\right)$$

Sol 10: Same as question (9)

$$\vec{x}_{COM} = \frac{m_1\vec{x}_1 + m_2\vec{x}_2 + m_3\vec{x}_3}{m_1 + m_2 + m_3} \text{ and }$$

$$\vec{y}_{COM} = \frac{m_1\vec{y}_1 + m_2\vec{y}_2 + m_3\vec{y}_3}{m_1 + m_2 + m_3}$$

Sol 11: (a) ②$\longrightarrow$ ------ ①$\longrightarrow$
       2m/s   5m/s

$$\vec{v}_{COM} = \frac{\vec{m}_1\vec{v}_1 + \vec{m}_2\vec{v}_2}{m_1 + m_2}$$

$$= \frac{2 \times 2 + 5 \times 1}{2 + 1} = \frac{9}{3} = 3 \text{ m/s}$$

(b) $y \downarrow\!\!\!_{\longrightarrow x}$ ②$\longrightarrow$ $\longleftarrow$① ------
        2m/s −5m/s

$$\vec{v}_{COM} = \frac{2 \times 2 - 5 \times 1}{2 + 1} = -1/3 \text{ m/s}$$

(−ve direction ⇒ velocity is negative direction)

Sol 12:

$\vdash\!\!\!\!\!\bullet\!\!-\!\!-\!\!-\!\!-\!\!-\!\!-\!\!-\!\!\bullet$
m_1 x_{COM} m_2
$(-a,0)$ $(0,0)$ $(b,0)$

     m_1 $\overbrace{\quad}^{d_2}$
$\vdash\!\!-\!\!\odot\!\!-\!\!-\!\!\bullet\!\!-\!\!-\!\!\odot\!\!-\!\!⊕$
$(-a+d)$ x_{COM} $(b-d_2)$ m_2

and $m_2 b - m_1 a = 0$

$\Rightarrow m_2 b = m_1 a$ ...(i)

Take origin at x_{COM} for simplicity.

Assuming the x_{COM} at origin, we have

$m_1(-a + d) + m_2(b - d_2) = 0$

$\Rightarrow m_2(b - d_2) = m_1(a - d)$

$\Rightarrow m_2 b - m_2 d_2 = m_1 a - m_1 d$ ...(2)

from (1), $m_2 b = m_1 a$, putting this in (2)

$m_1 a - m_2 d_2 = m_1 a - m_1 d$

$$\Rightarrow \boxed{\frac{m_1 d}{m_2} = d_2}$$

Sol 13: $\vec{r}_{COM} = \dfrac{m_1\vec{r}_1 + m_2\vec{r}_2}{m_1 + m_2}$

$$\vec{x}_{COM} = \frac{2 \times 1 + 3(-6)}{1 + 3} = \frac{-18 + 2}{4} = -1$$

$$\vec{y}_{COM} = \frac{5 \times 1 + 4 \times 3}{1 + 3} = \frac{12 + 5}{4} = \frac{17}{4}$$

$$\vec{z}_{COM} = \frac{13 \times 1 + 3 \times (-2)}{1 + 3} = \frac{13 - 6}{1 + 3} = \frac{7}{4}$$

So, $\vec{r}_{COM} = \left(-1, \dfrac{17}{4}, \dfrac{7}{4}\right)$

Sol 14: $x_{COM} = \dfrac{m_1 x_1 + m_2 x_2 + m_3 x_3 + m_4 x_4}{m_1 + m_2 + m_3 + m_4}$

$$= \frac{1 \times 0 + 2 \times a + 3 \times a + 4 \times 0}{1 + 2 + 3 + 4} = \frac{5a}{10} = \frac{a}{2}\hat{i}$$

$$y_{COM} = \frac{m_1 y_1 + m_2 y_2 + m_3 y_3 + m_4 y_4}{m_1 + m_2 + m_2 + m_4}$$

$$= \frac{1 \times 0 + 2 \times 0 + 3 \times b + 4 \times b}{10} = \frac{7b}{10}\hat{j}$$

So $\vec{r}_{COM} = \dfrac{a}{2}\hat{i} + \dfrac{7b}{10}\hat{j}$

Sol 15: Divide the lamina in 3 equal parts with centre of masses as C_1, C_2, C_3 respectively.

So, the centre of mass of the whole plate can be found using the centre of mass of these three plates.

Now, from symmetry, we can say that the centre of mass of square plate lies at its centre.

So, $\vec{r}_{C_1} = (0.5, 0.5)$

$\vec{r}_{C_2} = (1.5, 0.5)$ and $\vec{r}_{C_3} = (0.5, 1.5)$

mass of each plate = 1 kg

So $x_{COM} = \dfrac{0.5 \times 1 + 1.5 \times 1 + 0.5 \times 1}{(1 + 1 + 1)} = \dfrac{2.5}{3}$

$y_{COM} = \dfrac{0.5 \times 1 + 0.5 \times 1 + 1.5 \times 1}{3} = \dfrac{2.5}{3}$

So $\vec{r}_{CM} = \left(\dfrac{2.5}{3}, \dfrac{2.5}{3}\right) = \left(\dfrac{5}{6}, \dfrac{5}{6}\right)$

Exercise 2

Single Correct Choice Type

Sol 1: (B)

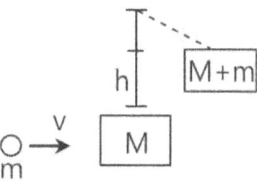

velocity of system after collision = $\sqrt{2gh}$

so using momentum constant

$(M + m)\sqrt{2gh} = mv$

$\Rightarrow v = \dfrac{(M + m)}{m}\sqrt{2gh}$

Sol 2: (B) $m_1 = 200g$, $m_2 = 200$ g, $m_3 = 600$ g

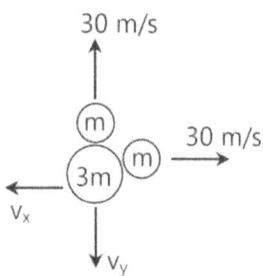

from momentum conservation

$x \Rightarrow 30 \times m = 3m \times v_x = 10$ m/s

$y \Rightarrow 30 \times m = 3m \times v_y = 10$ m/s

$\Rightarrow v = \sqrt{10^2 + 10^2} = 10\sqrt{2}$ m/s

Sol 3: (C) Momentum = mv (mass = constant)

so new $v_n = \dfrac{3v}{2}$

New, K. E. $= \dfrac{1}{2} \times m \times v_n^2 = \dfrac{1}{2} \times mv^2 \times \left(\dfrac{9}{4}\right)$

So increase $= \dfrac{9}{4} \times \left(\dfrac{1}{2}mv^2\right) - \dfrac{1}{2}mv^2$

$= \dfrac{5}{4} \times \left(\dfrac{1}{2}mv^2\right) = \dfrac{5}{4}$ of initial K. E.

Sol 4: (C)

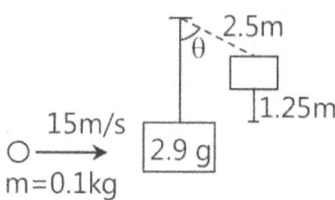

$e = \dfrac{25 - 20}{v_B - v_A} = \dfrac{5}{v_B - v_A} = 1$

$\Rightarrow v_B - v_A = 5$

Momentum conservation $\Rightarrow$

m. (25) + m. (20) = m. (v_A) + m(v_B)

$\Rightarrow v_A + v_B = 45$

$\Rightarrow v_B = 25$ m/s

$v_A = 20$ m/s

Sol 5: (C)

Using momentum conservation

$1.50 \times 0.1 = (2.9 + 0.1) v$

$\Rightarrow 15 = 3 \times v \Rightarrow v = 5$ m/s

so $\dfrac{1}{2}mv^2 = mgh \Rightarrow h = \dfrac{v^2}{2g}$

$\Rightarrow h = \dfrac{25}{2 \times 10} = \dfrac{5}{4} = 1.25$ m

So L. $(1 - \cos\theta) = 1.25$ m

$\Rightarrow \cos\theta = \dfrac{1}{2} \Rightarrow \theta = 60°$

Sol 6: (A) The center of mass will be on x-axis, so $y_{COM} = 0$.

$\Rightarrow 15 \times \dfrac{m}{4} + y \times \dfrac{3m}{4} = 0$

$\Rightarrow y = -5$ cm

Sol 7: (D) Elastic collision $\Rightarrow$ Energy is conserved.

$$\frac{1}{2} \times mv_1^2 = \frac{1}{2} \times m \times v_2^2 + \frac{1}{2} m \times v_2^2$$

$$\Rightarrow v_1^2 = 2v_2^2$$

$$\Rightarrow v_1 = \sqrt{2}\, v'_2$$

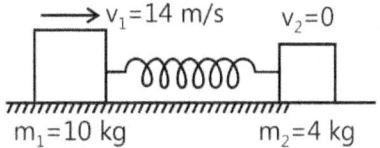

Using momentum conservation

$$\Rightarrow 2mv_2 \cos\theta = mv_1$$

$$\Rightarrow \cos\theta = \frac{v_1}{v'_2} \times \frac{1}{2}$$

$$\boxed{\cos\theta = \frac{1}{\sqrt{2}}} \Rightarrow \theta = 45° \Rightarrow 90°$$

Sol 8: (C) Momentum conservation

$$\Rightarrow 2 \times 3 - 1 \times 4 = (2 + 1)v$$

$$\Rightarrow v = \frac{2}{3} \text{ m/s}$$

Sol 9: (D) No external force $\Rightarrow a_{COM} = 0$

Previous Years' Questions

Sol 1: (D) Net force on centre of mass is zero. Therefore, centre of mass always remains at rest.

Sol 2: (A) Let v' be the velocity of second fragment. From conservation of linear momentum,

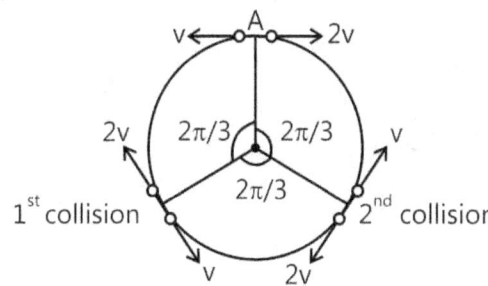

$$2m(v\cos\theta) = mv' - m(v\cos\theta)$$

$$\therefore v' = 3v\cos\theta$$

Sol 3: (C) $|(m_1\vec{v}_1 + m_2\vec{v}_2) - (m_1\vec{v}_1 + m_2\vec{v}_2)|$

= |Change in momentum of the two particle|

= |External force on the system| × time interval

$= (m_1 + m_2)g\,(2t_0) = 2\,(m_1 + m_2)gt_0$

Sol 4: (C) $v_{CM} = \dfrac{m_1v_1 + m_2v_2}{m_1 + m_2}$

$$= \frac{10 \times 14 + 4 \times 0}{10 + 4} = \frac{140}{14} = 10 \text{ m/s}$$

$v_1 = 14$ m/s $v_2 = 0$

$m_1 = 10$ kg $m_2 = 4$ kg

Sol 5: (A) y_{CM}

$$= \frac{m_1y_1 + m_2y_2 + m_3y_3 + m_4y_4 + m_5y_5}{m_1 + m_2 + m_3 + m_4 + m_5}$$

$$= \frac{(6m)(0) + (m)(a) + m(a) + m(0) + m(-a)}{6m + m + m + m + m}$$

$$= \frac{a}{10}$$

Sol 6: (C) At first collision one particle having speed $2v$ will rotate $240°$ $\left(\text{or } \dfrac{4\pi}{3}\right)$ while other particle having speed v will rotate $120°$ $\left(\text{or } \dfrac{2\pi}{3}\right)$. At first collision they will exchange their velocities. Now as shown in figure, after two collisions they will again reach at point A.

Sol 7: (D) $R = u\sqrt{\dfrac{2h}{g}}$

$$\Rightarrow 20 = v_1\sqrt{\frac{2 \times 5}{10}} \text{ and } 100 = v_2\sqrt{\frac{2 \times 5}{10}}$$

$$\Rightarrow v_1 = 20 \text{ m/s}, \ v_2 = 100 \text{ m/s}$$

Applying momentum conservation just before and just after the collision.

$$(0.01)(v) = (0.2)(20) + (0.01)(100)$$

$$v = 500 \text{ m/s}$$

Sol 8: (B) Between A and B, height fallen by block

$$h_1 = \sqrt{3}\tan 60° = 3m.$$

∴ Speed of block just before striking the second incline,

$$v_1 = \sqrt{2gh_1} = \sqrt{2 \times 10 \times 3} = \sqrt{60}\ ms^{-1}$$

In perfectly inelastic collision, component of v_1 perpendicular to BC will become zero, while component of v_1 parallel to BC will remain unchanged.

∴ Speed of block B immediately after it strikes the second inline is,

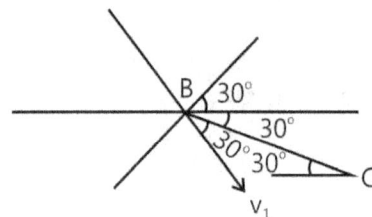

v_2 = component of v_1 along BC

$$= v_1\cos 30° = (\sqrt{60})\left(\frac{\sqrt{3}}{2}\right)$$

$$= \sqrt{45}\ ms^{-1}$$

Sol 9: (B) Height fallen by the block from B to C

$$h_2 = 3\sqrt{3}\ \tan 30° = 3\ m$$

Let v_3 be the speed of block, at point C, just before it leaves the second incline, then:

$$v_3 = \sqrt{v_2^2 + 2gh_2}$$

$$= \sqrt{45 + 2 \times 10 \times 3} = \sqrt{105}\ ms^{-1}$$

Sol 10: (C) In elastic collision, component of v_1 parallel to BC will remain unchanged, while component perpendicular to BC will remain unchanged in magnitude but its direction will be reversed.

B ↘ 30° C   ⇒   B 60° 30° 60° $v_\perp$ $v_\parallel$ 30° C

Just before   Just after

$$v_\parallel = v_1\cos 30° = (\sqrt{60})\left(\frac{\sqrt{3}}{2}\right)$$

$$= \sqrt{45}\ ms^{-1}$$

$$v_\perp = v_1\sin 30° = (\sqrt{60})\left(\frac{1}{2}\right)$$

$$= \sqrt{15}\ ms^{-1}$$

Now vertical component of velocity of block

$$v = v_\perp\cos 30° - v_\parallel\cos 60°$$

$$= (\sqrt{15})\left(\frac{\sqrt{3}}{2}\right) - (\sqrt{45})\left(\frac{1}{2}\right) = 0$$

Sol 11: (C) Loss of energy is maximum when collision is inelastic as in an inelastic collision there will be maximum deformation.

KE in COM frame is $\dfrac{1}{2}\left(\dfrac{Mm}{M+n}\right)v_{rel}^2$

$$KE_i = \frac{1}{2}\left(\frac{Mm}{M+m}\right)v^2 \quad KE_f = 0 \ (\because v_{rel} = 0)$$

Hence loss in energy is $\dfrac{1}{2}\left(\dfrac{Mm}{M+m}\right)v^2$

$$\Rightarrow f = \frac{M}{M+m}$$

Sol 12: (A) $z_0 = h - \dfrac{h}{4} = \dfrac{3h}{4}$

Sol 13: (B) $E_{initial} = \dfrac{1}{2}m(2v)^2 + \dfrac{1}{2}2m(v)^2 = 3mv^2$

$$E_{final} = \frac{1}{2}3m\left(\frac{4}{9}v^2 + \frac{4}{9}v^2\right) = \frac{4}{3}mv^2$$

∴ Fractional loss $= \dfrac{3 - \dfrac{4}{3}}{3} = \dfrac{5}{9} = 56\%$

JEE Advanced/Boards

Exercise 1

Sol 1: (a)

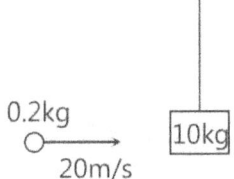

From the conservation of momentum we have,

$$m_1 v_1 + m_2 v_2 = m_{Tot} \times v$$

$$0.2 \times 20 + 10 \times 0 = (10 + 0.2) \times v$$

$$\Rightarrow 4 = 10.2 \times v$$

$$v = \frac{4}{10.2} \text{ m/s} = 0.392 \text{ m/s}$$

(b) From conservation of energy (Force by string is perpendicular to displacement, hence no work done by string)

$$\frac{1}{2} \times m \times v^2 = mgh_{vert.}$$

$$\Rightarrow h = \frac{v^2}{2g} = \frac{(0.392)^2}{2 \times 9.81} = 0.0078 \text{ m}$$

$$= 7.8 \text{ mm} \approx 0.008 \text{ m}$$

In horizontal direction:

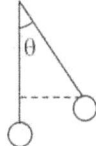

$$R(1 - \cos\theta) = 0.008 \text{ m}$$

$$\Rightarrow \cos\theta = \frac{1 - 0.08}{R} = 1 - 0.0027 = 0.9973$$

So $R \sin\theta = 3 \times \sqrt{(1 - \cos^2\theta)}$

$$= 3 \times \sqrt{1 - (0.9937)^2} = 0.220 \text{ m}$$

So total displacement = 0.220 m

(c) Initial energy:

$$\frac{1}{2} \times m_B \times v_B^2 = \frac{1}{2} \times (0.2) \times (20)^2 = 40 \text{ J}$$

Final energy:

$$\frac{1}{2} \times m_{Tot} \times v_{Tot}^2 = \frac{1}{2} \times 10.2 \times (0.392)^2 = 0.8 \text{ J}$$

So energy lost = 40 J – 0.8 J = 39.2 J

Sol 2:

⓪ ↑100 m/s
㉚ ↑v'

↑100 m/s   ↓g=9.8m/s²

㊿

(a) Now, from equation of motion

$$v = u + at = 100 - 9.8 \times (5) = 51 \text{ m/s}$$

so v = 51 m/s

Now at this velocity, particle exploded, Δt is very small and hence momentum can be conserved.

So $50 \times 51 = 20 \times 150 + 30 \times v'$

$$\Rightarrow 2550 = 3000 + 30 \times v'$$

$$\Rightarrow \boxed{v' = \frac{-450}{30} = -15 \text{m/s}}$$

(b) When no explosion:

$$v = u + at$$

$$\Rightarrow v = 100 - 9.8 \times 8 = 21.6 \text{ m/s}$$

so momentum = m × v

$$= 50 \times 21.6 = 1080 \text{ kg. m/s}$$

When explosion:

For 20 kg v = u + at

$$= 150 - 9.8 \times (s) = 120.6 \text{ m/s}$$

For 30 kg v = u + at

$$= -15 - 9.8 \times 3 = -44.4 \text{ m/s}$$

So total momentum

$$= 20 \times (120.6) - (44.4) \times (30)$$

$$= 2412 - 1332 = 1080 \text{ kg m/s}$$

Sol 3: (a) The particle are meet at the mid-point of the trajectory (i. e. vertical velocity = 0)

u = 49 m/s

So $t = \dfrac{u\sin\theta}{g}$

Thus, v_x = horizontal velocity = $u\cos\theta$

Now to retrace the path velocity of P must be ucosθ in the (–)ve direction, so now using momentum balance

$m_1v_1 + m_2v_2 = m_1v'_1 + m_2v'_2$

$\Rightarrow 20 \times (u\cos\theta) + 40 \times (-u\cos\theta)$

$= 20 \times (u\cos\theta) + 40 \times v'_2$

$\Rightarrow 40 \times v'_2 = 0 \Rightarrow v'_2 = 0$

so the horizontal velocity of particle Q after collision would be 0.

so position of Q would be just below the point of collision

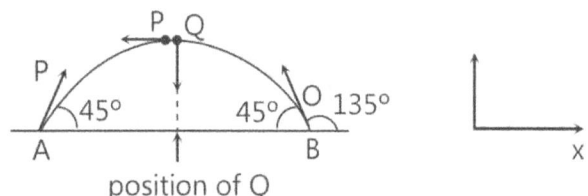

position of Q

So position of Q $= \dfrac{u\cos 45° \times u\sin 45°}{g}$

$= \dfrac{u^2 \sin 90°}{2g} = \dfrac{u^2}{2g} = \dfrac{(49)^2 \times 10}{2 \times 9.8} = 122.5$ m

From position A in the (+)ve x-direction

(b) Time take would be same as the vertical component has not changed, so

$t = \dfrac{u\sin\theta}{g} = \dfrac{49 \times \sin 45°}{9.8} = 3.54$ sec

Sol 4:

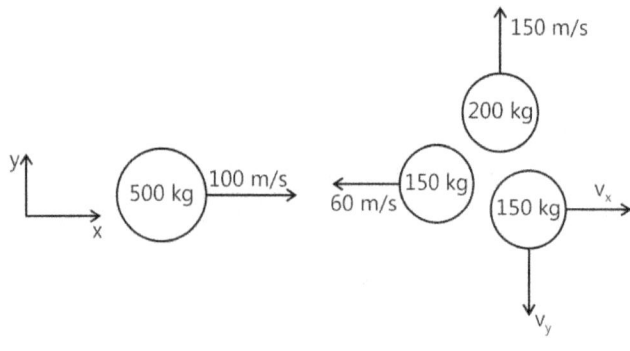

Now, since there is no external force

Using momentum conservation in x-direction

$500 \times 100 = 150 \times v_x + 150 \times (-60)$

$\Rightarrow 150 \times v_x = 5 \times 10^4 + 9000$

$150 \times v_x = 59 \times 10^3$

$\Rightarrow v_x = 393.33$ m/s

Similarly in y-direction

$0 = 150 \times 200 + 150 \times (-v_y)$

$\Rightarrow v_y = 200$ m/s

So $v_{III} = 393.33\ \hat{i} - 200\ \hat{j}$

$\| v_{III} \| = 441.26$ m/s

and $\theta = \tan^{-1}\dfrac{-200}{393.33} = -27°$

Sol 5:

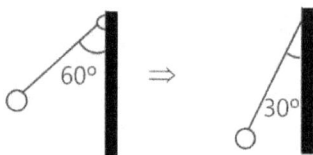

Velocity of ball just before impact

$= \sqrt{2gL(1-\cos\theta)} = \sqrt{2 \times 9.8 \times 1/2}$

$= \sqrt{9.8} = 3.13$ m/s

v of ball after impact $\Rightarrow mg(\Delta h) = \dfrac{1}{2}mv^2$

(energy conservation)

$\Rightarrow \sqrt{2gL.(1-\cos 30°)} = v_f$

$\Rightarrow \sqrt{2 \times 9.8 \times \left(1 - \dfrac{\sqrt{3}}{2}\right)} = v_f$

$\Rightarrow v_f = \sqrt{2.626}$

$\Rightarrow v_f = 1.62$ m/s

so coefficient of rest. $= \dfrac{1.62}{3.13} = 0.517$

Loss of kinetic energy = heat produced

$= \dfrac{1}{2} \times m \times (v_i^2 - v_f^2)$

$= \dfrac{1}{2} \times (0.01) \times [(3.13)^2 - (1.62)^2]$

$= \dfrac{0.0717}{2} = 0.036$ J

$= 0.0085$ cal.

Sol 6: No friction $\Rightarrow$ no torque, so its pure translational motion

Now, conservation of energy

$\Rightarrow \dfrac{1}{2} \times mv^2 = mgh$

$\Rightarrow v = \sqrt{2gh}$

$= \sqrt{2 \times 9.8 \times 0.1}$

$v = 1.4$ m/s

Now, as collision is elastic

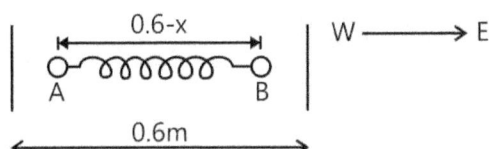

so $v_B - v_A = 1.4$ m/s ...(i)

and conservation of momentum gives:

$m_A v_A + m_B v_B = m_A \times 1.4$ m/s

$\Rightarrow v_A + v_B = 1.4$ m/s ...(ii)

on solving (i) and (ii)

$v_B = 1.4$ m/s

$v_A = 0$ m/s

Sol 7:

(a) Initial acceleration of B = 0.5 m/s^2 W

So, $kx = (0.5)(0.25)$

As the magnitude of force would be the same for A,

Initial acceleration of A $= \dfrac{kx}{0.1} = \dfrac{(0.5)(0.25)}{0.1}$

$\qquad = 1.25$ m/s^2 E

$\qquad = 1.25$ cm/s^2 E

(b) $x = \dfrac{(0.5)(0.25)}{k} = \dfrac{(0.5)(0.25)}{2}$

$\qquad = 0.0625$ m

$\qquad = 6.25$ cm

(c) Max distance would be when spring is fully elongated. And, symmetry of conservation of energy implies that expansion would be equal to companion.

So, Maximum distance between

A and B = 60 cm + 6.25 cm

$= 66.25$ cm

Sol 8:

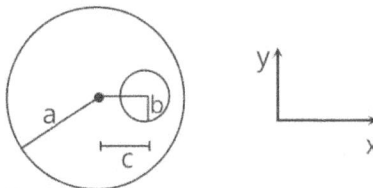

Now $\vec{r}_{COM} = \dfrac{m_1 \vec{r}_1 + m_2 \vec{r}_2}{m_1 + m_2}$

Take the disks as these two bodies and treat m_2 as negative

Given body =

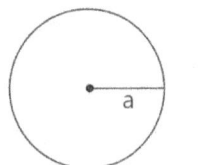

So $x_{COM} = \dfrac{\rho.\pi a^2.(0) - \rho.\pi b^2 \times c}{\rho.(\pi a^2 - \pi b^2)}$

where $\rho = \dfrac{m}{\text{area}}$

$x_{COM} = \dfrac{-b^2 c}{(a^2 - b^2)}$

Sol 9: (a) No external force, hence $v_{COM} = 0$

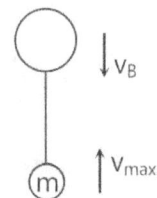

thus,

$m. v_{max} - M. v_B = 0$

$\Rightarrow m v_{max} = M. v_B$ ...(i)

and $v_{rel} = v_{max} + v_B$ ...(ii)

$v_{rel} = \left(\dfrac{M}{m} + 1\right) v_B$

$\Rightarrow v_B = \dfrac{v_{rel}.m}{(m+M)}$

$\Rightarrow v_B = \dfrac{m.v_{rel}}{(m+M)}$ in (–)ve direction

(b) $D_{rel} = L$

Let $x_{COM} = 0$

then $m. x_1 - M. x_2 = 0$

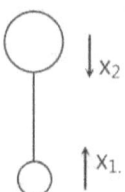

$x_1 \neq x_2$ = distance covered by man and balloon in ground frame, so

$x_1 + x_2 = L$

$\Rightarrow \left(\dfrac{M}{m} + 1\right) x_2 = L$

$\Rightarrow x_2 = \left(\dfrac{mL}{m+M}\right)$

in the downward direction

(c) No external force: initial velocity = final velocity = 0

Sol 10:

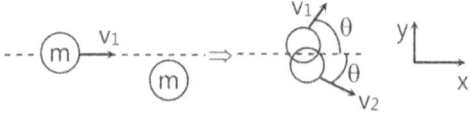

No generation of momentum would be there in the y-direction

Energy conservation $\Rightarrow$

$\dfrac{1}{2} mv^2 = \dfrac{1}{2} mv_1^2 + \dfrac{1}{2} mv_2^2$

$\Rightarrow v^2 = v_1^2 + v_2^2$ ...(i)

Also, $mv_1 \sin\theta = mv_2 \sin\theta$

$\Rightarrow v_1 = v_2$ ... (ii)

Now, using (i) and (ii), we get

$v^2 = 2v_1^2$

$\Rightarrow v = \sqrt{2}\, v_1$

and using momentum conservation in x-direction:

$mv = 2mv_1 \cos\theta$

$\Rightarrow \cos\theta = \dfrac{1}{\sqrt{2}}$

$\Rightarrow \theta = 45°$ so angle between them = 90°

Hence proved.

Sol 11:

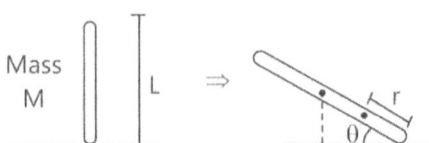

(a) Now as there is no horizontal force on rod, there would be no displacement of COM of the rod. Thus, the path followed will be a straight line.

(b) x comp. of $r = \dfrac{L}{2}\cos\theta - r\cos\theta$

$x = \left(\dfrac{L}{2} - r\right)\cos\theta$

and y comp. of $r = r\sin\theta = y$

$\Rightarrow \dfrac{x^2}{(L/2 - r)^2} + \dfrac{y^2}{r^2} = 1$

Sol 12:

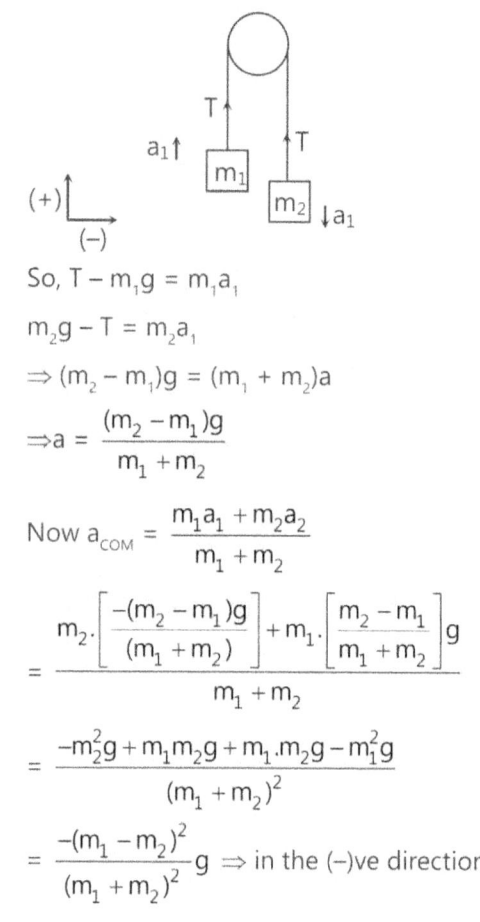

So, $T - m_1 g = m_1 a_1$

$m_2 g - T = m_2 a_1$

$\Rightarrow (m_2 - m_1)g = (m_1 + m_2)a$

$\Rightarrow a = \dfrac{(m_2 - m_1)g}{m_1 + m_2}$

Now $a_{COM} = \dfrac{m_1 a_1 + m_2 a_2}{m_1 + m_2}$

$= \dfrac{m_2 \cdot \left[\dfrac{-(m_2 - m_1)g}{(m_1 + m_2)}\right] + m_1 \cdot \left[\dfrac{m_2 - m_1}{m_1 + m_2}\right]g}{m_1 + m_2}$

$= \dfrac{-m_2^2 g + m_1 m_2 g + m_1 \cdot m_2 g - m_1^2 g}{(m_1 + m_2)^2}$

$= \dfrac{-(m_1 - m_2)^2}{(m_1 + m_2)^2} g \Rightarrow$ in the (–)ve direction.

Sol 13: There is no external force in horizontal direction, so x_{COM} is same after this relative horizontal distance

$= h\cot\theta$

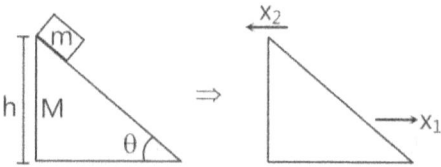

Let x_1 and x_2 be the distance travelled by the block and prism, respectively in ground frame.

so, $x_1 + x_2 = h \cos\theta$ ...(i)

and $m_1 x_1 - M x_2 = 0$

[taking x_{COM} as origin]

$\Rightarrow m_1 x_1 = M x_2$

$\Rightarrow x_1 = \dfrac{M}{m} x_2$, putting this in (i)

$\left(\dfrac{M}{m} + 1\right) x_2 = h \cot\theta$

$\Rightarrow \quad x_2 = \dfrac{m.h \cot\theta}{(m+M)}$

Sol 14:

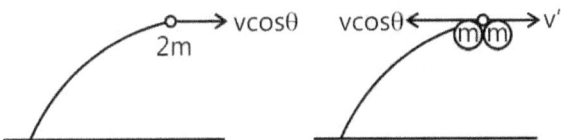

If the particle retraces its path, its velocity must be same as before i.e. $v\cos\theta$ in the opposite direction (independent of mass), so using the momentum conservation:

$2m.(v\cos\theta) = m.(-v\cos\theta) + m.v'$

$\Rightarrow mv' = 3mv\cos\theta$

$\Rightarrow \quad v' = 3v\cos\theta$

Sol 15:

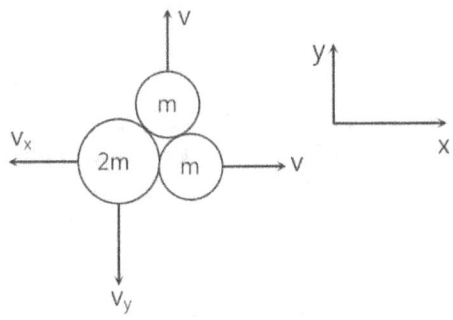

Momentum conservation in x-direction

$m.v + 2m.(-v_x) = 0$

$\Rightarrow v_x = \dfrac{v}{2}$

Similarly, $v_y = \dfrac{v}{2}$

so total energy

$= \dfrac{1}{2} \times m \times v^2 + \dfrac{1}{2} \times m \times v^2 + \dfrac{1}{2} \times (2m)(v_x^2 + v_y^2)$

$= mv^2 + m.\left[\dfrac{v^2}{4} + \dfrac{v^2}{4}\right] = \dfrac{3mv^2}{2}$

Sol 16:

Elastic collision $\Rightarrow \dfrac{v_2 - v_1}{v - 0} = 1$

$\Rightarrow v_2 - v_1 = v$ ...(i)

and using momentum conservation

$mv = mv_1 + 2mv_2$

$\Rightarrow mv = m(v_2 - v_1) + 2mv_2$

$\Rightarrow 2mv = 3mv_2$

$\Rightarrow v_2 = \dfrac{2v}{3}$ and thus $v_1 = \dfrac{-v}{3}$

so kinetic energy before collision:

$= \dfrac{1}{2} mv^2$

kinetic energy after collision

$= \dfrac{1}{2} \times m \times \dfrac{v^2}{9} = \dfrac{mv^2}{18}$

loss $= \dfrac{1}{2} mv^2 - \dfrac{mv^2}{18} = \dfrac{8mv^2}{18} = \dfrac{8}{9} \times \dfrac{1}{2} mv^2$

$= \dfrac{8}{9} \times (\text{initial K. E.})$

Sol 17:

v at first impact $= \sqrt{2gh}$

and time at first impact $= \sqrt{\dfrac{2h}{g}}$

(eq. of motion)

Now, u_n = velocity after nth impact

$= e^n. \sqrt{2hg}$

so total distance $= h + \displaystyle\sum_{n=1}^{\infty} \dfrac{u_n^2}{2g} \times 2$

$\left[\dfrac{u_n^2}{2g} \times 2 = \dfrac{u^2}{g} \Rightarrow \text{distance between two impacts}\right]$

$= h + 2e^2. h + 2e^4. h \ .....$

$= h. [1 + 2e^2(1 + e^2 + e^4 \ .....)]$

$$= h \cdot \left[1 + \frac{2e^2}{1-e^2}\right] = \frac{h.[1+e^2]}{[1-e^2]}$$

Similarly total time

$$= \sqrt{\frac{2h}{g}} + \sum_{n=1}^{\infty} \frac{2u_n}{g} \quad \left[\begin{array}{l} \frac{2 \times u_n}{8} = \text{time between} \\ \qquad \text{two impacts} \end{array}\right]$$

$$= \sqrt{\frac{2h}{g}} + \sqrt{\frac{2h}{g}} . 2e + \sqrt{\frac{2h}{g}} . 2e^2 + ...$$

$$= \sqrt{\frac{2h}{g}} [1 + 2e(1 + e + e^2 ....)]$$

$$= \sqrt{\frac{2h}{g}} \left[1 + \frac{2e}{1-e}\right] = \sqrt{\frac{2h}{g}} . \left[\frac{1+e}{1-e}\right]$$

Sol 18: If the block m_2 is moving with same velocity after wall collision $\Rightarrow$ the velocity of block m_1 and m_2 has same magnitude.

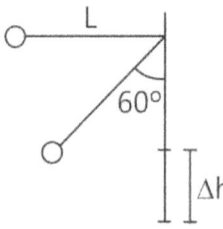

Now $\dfrac{u_2}{v_1 - (-v_1)} = 1$

$$\Rightarrow u_2 = 2v_1 \qquad\qquad\qquad .... \text{(i)}$$

and using momentum conservation

$$u_2 m_2 = m_1 v_1 - m_2 v_1$$

$$\Rightarrow 2m_2 v_1 = m_1 v_1 - m_2 v_1 \Rightarrow 3m_2 v_1 = m_1 v_1$$

$$\Rightarrow 3m_2 = m_1$$

$$\Rightarrow \quad m_2 = 50 \text{ kg}$$

Sol 19:

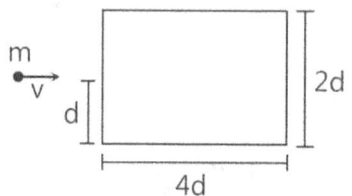

For the amplitude to be less than 60°.

$$\frac{1}{2} mv^2 < mg(\Delta h)$$

$$\Rightarrow v^2 < 2g\,(\Delta h)$$

$$v^2 < 2gL\,(1 - \cos 60°)$$

$$\Rightarrow v^2 < gL$$

v after n collisions = $e^n . \sqrt{2gL}$

$$\Rightarrow e^{2n} . (2gL) < gL$$

$$\Rightarrow e^{2n} < \frac{1}{2} \qquad \Rightarrow \left(\frac{4}{5}\right)^n < \frac{1}{2}$$

$$\Rightarrow n = 4 \text{ is the largest value satisfying.}$$

Sol 20: (a)

Now, using momentum conservation $\Rightarrow 4mv_2 + mv_1 = mv$

$$\Rightarrow 4v_2 + v_1 = v \qquad\qquad ... \text{(i)}$$

Elasticity $\dfrac{v_2 - v_1}{v} = 1$

$$\Rightarrow v_2 - v_1 = v \qquad\qquad ...\text{(ii)}$$

(i) + (ii) $\Rightarrow 5v_2 = 2v \Rightarrow v_2 = \dfrac{2v}{5}$ and

$$v_1 = \frac{-3v}{5}$$

To topple, the distance by 4m block should be 2d. (No horizontal force on 2m block, and hence that block remains stationary)

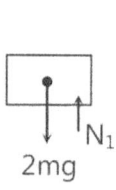 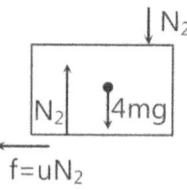

$$N_1 + 4mg = N_2,$$

Now $N_1 = 2mg$

$$\Rightarrow N_2 = 6 \text{ mg}$$

So f = 6μmg

thus acceleration (or deceleration)

$$a = \frac{f}{4m} = \frac{6\mu mg}{4m} = \frac{3\mu g}{2}$$

So now the velocity (or v_2) required

$\Rightarrow v^2 = u^2 + 2as$

$\Rightarrow 0 = v_2^2 + 2\left(\dfrac{-3\mu g}{2}\right) \times (2d) \Rightarrow v_2 = \sqrt{6\mu gd}$

so, $v_2 = \dfrac{2v}{5}$ from above,

$\Rightarrow v = \dfrac{5}{2}v_2 = \dfrac{5}{2}\sqrt{6\mu gd}$

(b) Now $v = 2v_0 = \dfrac{10}{2}\sqrt{6\mu gd}$

$v = 5\sqrt{6\mu gd}$ m/s

Now

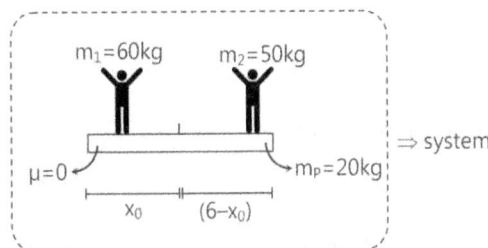

so $v_2 - v_1 = v$ (elasticity)

and $m \times v = m \times v_1 + 4m \times v_2$

[Momentum conservation]

$\Rightarrow v = v_1 + 4v_2$ ...(ii)

$\Rightarrow 2v = 5v_2$

$\Rightarrow v_2 = \dfrac{2v}{5}$ and $v_1 = \dfrac{-3v}{5}$

So $v_1 = -3\sqrt{6\mu gd}$

Now time $\Rightarrow \sqrt{\dfrac{2d}{g}}$

so distance $= v_1 \times t$

$= -3 \times \sqrt{6\mu gd} \times \sqrt{\dfrac{2d}{g}} = -6d.\sqrt{3\mu}$

Sol 21:

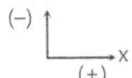

Now no external force thus $x_{COM} = $ constant ≈ 0.

$x_1, x_2, x_3 \Rightarrow$ displacement of man, woman and platform respectively w.r.t. ground frame then,

$60 \times x_1 + 50 \times x_2 + 20 \times x_3 = 0$

$\Rightarrow 6x_1 + 5x_2 + 2x_3 = 0$ ...(i)

and distance by man w.r.t. platform

$x_0 = x_1 - x_3$ ...(iii)

and displacement of woman w. r. t platform $= x_2 - x_3$

also $x_1 - x_3 - (x_2 - x_3) = 6$

$\Rightarrow x_1 - x_2 = 6$ ...(ii)

$x_2 = x_1 - 6$ and $x_1 = x_0 + x_3$

so thus putting these in (i)

6. $(x_0 + x_3) + 5(x_1 - 6) + 2x_3 = 0$

$6x_0 + 6x_3 + 5(x_0 + x_3) - 30 + 2x_3 = 0$

$11x_0 + 13x_3 - 30 = 0$

$\Rightarrow x_3 = \dfrac{30 - 11x_0}{13}$

so displacement $= \dfrac{30 - 11x_0}{13}$ in (+)ve direction.

Sol 22:

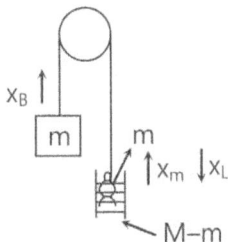

Now, we have

$x_B = x_L$ (constraint)

Now $x_m - (-x_L) = \Delta \vec{r}$

Now displacement of centre of mass

$\Rightarrow \Delta \vec{S} = \dfrac{M.x_B + m.x_m - (M-m)x_L}{M + M - m + m}$

$= \dfrac{M.x_B + m.x_m - (M-m)x_L}{2M}$

$= \dfrac{M(x_B - x_L) + m(x_m + x_L)}{2M}$

$[(x_B - x_L) = 0]$

$= \dfrac{m.\Delta \vec{r}}{2M}$

Sol 23: Let the total mass of both insects be 'm'

For the limiting case of toppling, normal reaction would pass through the corner.

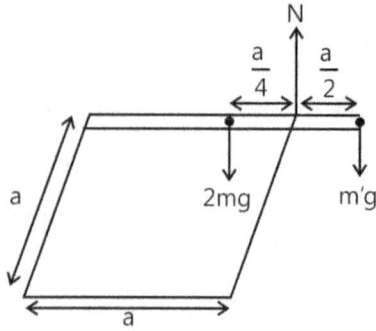

$N = 2mg + m'g$

$m'g\left(\dfrac{a}{2}\right) = 2mg\left(\dfrac{a}{4}\right)$

$\Rightarrow m' = m$

As the mass of first insect is $\dfrac{m}{2}$, the second insect would also have the same mass.

Hence, mass of the other insect $= m' - m = \dfrac{m}{2}$

Sol 24:

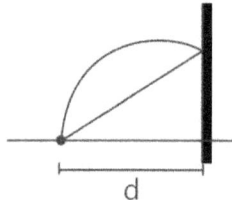

Horizontal velocity: $2\sqrt{ag}\cdot\cos\theta$

after collision: $\dfrac{2\sqrt{ag}\cos\theta}{(4\sin2\theta - 1)}$

so time (total)

$= \dfrac{d}{2\sqrt{ag}\cos\theta} + \dfrac{d(4\sin2\theta - 1)}{2\sqrt{ag}\cos\theta} = \dfrac{d\cdot(4\sin2\theta)}{2\sqrt{ag}\cos\theta}$

Now this time must be equal to time of flight

$= \dfrac{2u\sin\theta}{g} \Rightarrow \dfrac{2\cdot2\sqrt{ag}}{g} = \dfrac{d(4\sin2\theta)}{2\sqrt{ag}\cos\theta}$

$\Rightarrow \dfrac{a\cdot g\cdot\sin2\theta}{g} = d\cdot\sin2\theta \Rightarrow d = a$

also, $e < 1 \Rightarrow \dfrac{1}{4\sin2\theta - 1} < 1$

$\Rightarrow \sin2\theta > \dfrac{1}{2} \Rightarrow 2\theta > 30° \Rightarrow \theta > 15°$

Sol 25:

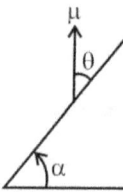 

Time for first collision, $T_1 = \dfrac{2u\sin\theta}{g\cos\alpha}$

Time for second collision, $T_2 = e\left(\dfrac{2u\sin\theta}{g\cos\alpha}\right)$

As velocity becomes perpendicular to the surface, its horizontal component (along the surface) must go to 'O'.

$O = \mu\cos\theta - g\sin\alpha(T_1 + T_2)$

$\Rightarrow \mu\cos\theta = g\sin\alpha\left(\dfrac{2\mu\sin\theta}{g\cos\alpha}\right)(1 + e)$

$\Rightarrow 1 + e = \dfrac{\cot\theta\cot\alpha}{2}$

$\Rightarrow e = \dfrac{\cot\theta\cot\alpha}{2} - 1$

Sol 26:

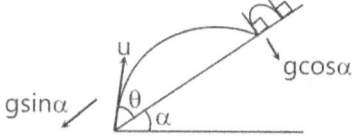

here time of flight $= \dfrac{2u\sin\theta}{g\cos\alpha}$

total time for second bounce

$= \dfrac{2u\sin\theta}{g\cos\alpha} + \dfrac{2ue\sin\theta}{g\cos\alpha}$

Also equation of motion along incline $= \dfrac{u\cos\theta}{g\sin\alpha}$

so $\dfrac{u\cos\theta}{g\sin\alpha} = \dfrac{2u\sin\theta}{g\cos\alpha} + \dfrac{2ue\sin\theta}{g\cos\alpha}$

$\Rightarrow \dfrac{\cos\theta}{\sin\alpha} - \dfrac{2\sin\theta}{\cos\alpha} = \dfrac{2e\sin\theta}{\cos\alpha}$

$\Rightarrow \dfrac{(\cot\theta\cdot\cot\alpha - 2)}{2} = e$

$\Rightarrow e = \dfrac{1}{2}\cot\alpha\cdot\cot\theta - 1$

Sol 27:

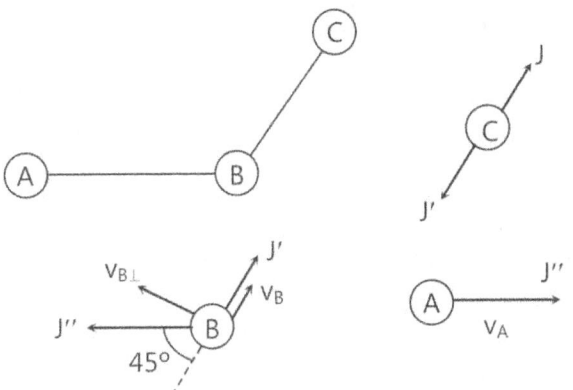

Now we have,

$$J - J' = mv_c \quad \text{... (i) (moment eq^{n})}$$

$$v_c = v_{B\parallel} \quad \text{...(ii) (constraint eq^{n})}$$

$$J' - J'' \cos 45° = mv_{B\parallel} \quad \text{...(iii)}$$

(impulse momentum eq^{n})

$$J'' \sin 45° = mv_{B\perp} \quad \text{...(iv)}$$

(impulse momentum eq^{n})

$$J'' = mv_A \quad \text{...(v)}$$

(impulse momentum eq^{n})

$$(v_{B\parallel} - v_{B\perp})\frac{1}{\sqrt{2}} = v_A \quad \text{...(vi)}$$

(constraint equation)

so we have 6 variables (J', J'', $v_{B\parallel}$, $v_{B\perp}$, v_A and v_c) and six equations,

on solving, we get

$$v_c = \frac{3J}{7m},$$

$$v_B = \frac{\sqrt{10}}{7m},$$

$$v_A = \frac{\sqrt{2}}{7m}$$

Sol 28:

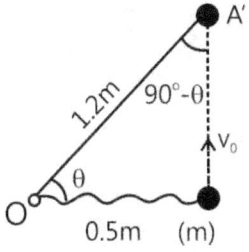

(a) The comp. $\perp$ to the string will only be there and the momentum along the thread will be lost.

So $mv_0 \cos(90° - \theta) < 3 \, N_s$

$$\Rightarrow 2 \times v_0 \times \sin\theta < 3$$

$$\Rightarrow v_0 < \frac{3}{2\sin\theta}$$

$$\Rightarrow v_0 < \frac{3 \times 1.2}{2 \times \sqrt{(1.2)^2 - (0.5)^2}}$$

$$\Rightarrow v_0 < 1.65 \, \text{m/s}$$

(b) The energy remaining

$$= \frac{1}{2} \times m(v_0 \cos\theta)^2 = \frac{1}{2}mv_0^2 \cos^2\theta$$

so loss $= \dfrac{1}{2}mv_0^2 - \dfrac{1}{2}mv_0^2 \cos^2\theta$

$$= \frac{1}{2}mv_0^2(1 - \cos^2\theta) = \frac{1}{2}mv_0^2 . \sin^2\theta$$

$$= \frac{1}{2} \times 2 \times (1.65)^2 \times \frac{(1.2)^2 - (0.5)^2}{(1.2)^2} = 2.25 \text{ Joules.}$$

Sol 29:

(a) (i)

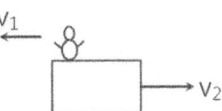

So $v_2 - v_1 = 5 \Rightarrow v_1 = v_2 - 5$

and

$(1225) \times 25 = 1150 \times v_2 + 75 \times v_1$

$1225 \times 25 = 1150 \times v_2 + 75 \times (v_2 - 5)$

$1240 \times 25 = 1225 \, v_2$

$$\Rightarrow v_2 = \frac{1240 \times 25}{1225} = 25.306 \text{ m/s}$$

(ii)

$v_1 \quad$

$1150 \rightarrow v_2$

$1150 \times 25.306 = 1075 \times v_2 + 75 \times v_1$

$\Rightarrow 1150 \times 25.306 = 1075v_2 + 75(v_2 - 5)$

$$\Rightarrow \frac{1150 \times 25.306 + 75 \times 5}{1075} = v_2$$

$\Rightarrow v_2 = 25.63 \text{ m/s}$

(iii)

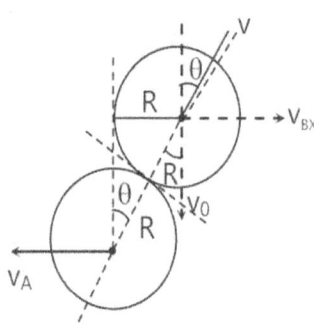

$$1075 \times 25.63 = 1000 \times v_2 + 75 \times v_1$$

$$\Rightarrow 1075 \times 25.63 = 1000v_2 + 75(v_2 - 5)$$

$$\Rightarrow \frac{1075 \times 25.63 + 75 \times 5}{1075} = v_2$$

$$\Rightarrow v_2 = 25.97 \text{ m/s}$$

(b) All together

$$\Rightarrow 1225 \times 25 = 1000 \times v_2 + 225 \times v_1$$

$$\Rightarrow 1225 \times 25 = 1000v_2 + 225(v_2 - 5)$$

$$\Rightarrow \frac{1225 \times 25 + 225 \times 5}{1225} = v_2 = 25.92 \text{ m/s}$$

Sol 30:

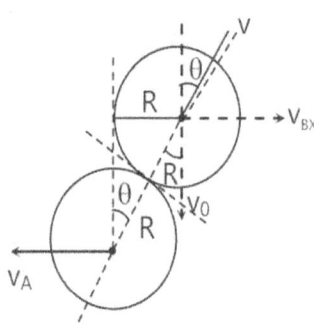

$$\sin \theta = \frac{R}{2R} = \frac{1}{2} \Rightarrow \theta = 30°$$

Now, using elasticity, (along line of impact)

$$\frac{v_A \cos 60° + v_{By} \cdot \frac{\sqrt{3}}{2} + v_{Bx} \cdot \frac{1}{2}}{v_0 \cdot \frac{\sqrt{3}}{2}} = 1$$

$$\Rightarrow \frac{v_A}{2} + \frac{\sqrt{3} v_{By}}{2} + \frac{v_{Bx}}{2} = \frac{v_0 \sqrt{3}}{2}$$

$$\Rightarrow v_A + v_{By} \cdot \sqrt{3} + v_{Bx} = v_0 \sqrt{3} \qquad \text{...(i)}$$

using energy balance,

$$\frac{1}{2} m v_0^2 = \frac{1}{2} m v_A^2 + \frac{1}{2} m (v_{Bx}^2 + v_{By}^2)$$

$$\Rightarrow v_0^2 = v_A^2 + v_{Bx}^2 + v_{By}^2 \qquad \text{...(ii)}$$

Momentum along x-direction

$$\Rightarrow v_A - v_{Bx} = 0 \Rightarrow v_A = v_{Bx} \qquad \text{...(iii)}$$

so on, solving $v_0 - \frac{2v_A}{\sqrt{3}} = v_{By}$

Putting this in (ii)

$$\Rightarrow v_0^2 = 2v_A^2 + \left(v_0 - \frac{2v_A}{\sqrt{3}} \right)^2$$

$$\Rightarrow v_A = \frac{2\sqrt{3} v_0}{\sqrt{3}} \Rightarrow 0.693 v_0$$

So, $v_{Bx} = 0.693 v_0$ and

$$v_{By} = v_0 - \frac{2v_A}{\sqrt{3}} = v_0 - \frac{4}{5} v_0 = \frac{v_0}{5} = 0.2 v_0$$

So, $v_B = \sqrt{(0.693)^2 + (0.2)^2} \cdot v_0$

$$\Rightarrow v_B = 0.721 v_0$$

Sol 31: $\frac{dP}{dt} = F = \frac{d(mu)}{dt} = \frac{u \, dm}{dt} + \frac{m \cdot du}{dt}$

$$-mg = \frac{u}{T} \cdot m_0 e^{t/T} + m \cdot \frac{du}{dt}$$

$$-g = \frac{u}{T} + \frac{du}{dt}$$

$$\Rightarrow \int_{u_0, 0}^{u, t} d(u.e^{t/T}) = \int_0^t -g.e^{t/T} .dt$$

$$u. e^{t/T} - u_0 = -gT. (e^{t/T} - 1)$$

$$\Rightarrow u(t) = u_0. e^{-t/T} - gT. (1 - e^{-t/T})$$

so here, u(t) = 0

$$\Rightarrow gT. e^{-t/T} - gT(1 - e^{-t/T}) = 0$$

$$\Rightarrow e^{-t/T} = \frac{1}{2}$$

$$\Rightarrow e^{t/T} = 2 \Rightarrow t = T \ln 2$$

So $m = m_0. e^{t/T}$

$$m = 2m_0$$

Sol 32:

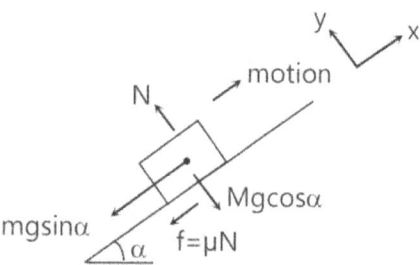

N = Mg cosα

So f = μN = μMg cosα

so acceleration of 2 kg block

$$= \frac{\mu mg\cos\alpha + mg\sin\alpha}{m} = \mu g\cos\alpha + g\sin\alpha$$

So for a distance of 6m, v just before impact would be:

$v^2 = u^2 - 2as$

$\Rightarrow v^2 = (10)^2 - 2 \times 6 \times [0.25 \times 10$

$\qquad \times 0.998 + 10 \times 0.05]$

$= 100 - 12 \times [3]$

$v^2 = 64$

$v_{in.} = 8$ m/s

Now, after collision:

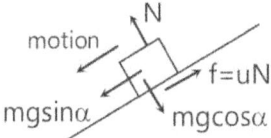

so acc. $= \mu g \cos\alpha - g \sin\alpha$

$= 2.5 - 0.5 = 2$ m/s^2

So $v^2 = u^2 - 2as$

$1^2 = u^2 - 2 \times 2 \times 6$

$\Rightarrow u^2 = 25 \Rightarrow u = 5$ m/s

so after collision v of 2 kg block = 5 m/s

For M block:

a_{cc} = same as in 1st part (ind. of mass,)

$= \mu g \cos\alpha + g \sin\alpha = 3$ m/s^2 (in (−)ve direction)

So $v^2 = u^2 + 2as$

$0 = u^2 + 2 \times (-3) \times (0.5)$

$\Rightarrow u = 1.732$ m/s

so coefficient of restitution

$= \dfrac{5+1.732}{8} = \dfrac{6.732}{8} = 0.84$

Using momentum conservation, just before impact

$m \times v_i = m \times v_f + M \times v_m$

$2 \times 8 = 2 \times (-5) + M \times 1.732$

$\Rightarrow \dfrac{26}{1.732} = M = 15.011$ kg

Exercise 2

Single Correct Choice Type

Sol 1: (B)

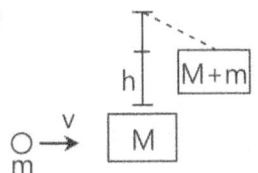

v of system after collision $= \sqrt{2gh}$

so using momentum conservation.

$(M + m) \sqrt{2gh} = mv$

$\Rightarrow v = \dfrac{(M+m)}{m} \sqrt{2gh}$

Sol 2: (C)

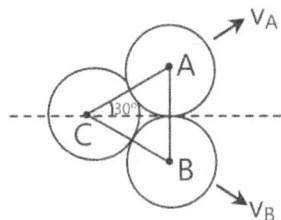

Let the initial velocity of

$C = v$

then momentum conservation in y gives

$V_A = V_B$

and using momentum conservation in x,

$mv = 2mv_A \cos 30°$

$\Rightarrow v = \sqrt{3}\, v_A \qquad \Rightarrow v_A = \dfrac{v}{\sqrt{3}}$

so coefficient of restitution

$= \dfrac{\text{final relative velocity}}{\text{initial relative velocity}} = \dfrac{v/\sqrt{3}}{v\sqrt{3}/2} = \dfrac{2}{3}$

Sol 3: (A)

$\underset{m}{\bigcirc} \xrightarrow{u} \underset{m}{\bigcirc} \Rightarrow \underset{m}{\bigcirc} \longrightarrow v_1 \quad \underset{m}{\bigcirc} \longrightarrow v_2$

$e = \dfrac{v_2 - v_1}{u} \Rightarrow v_2 - v_1 = ue \qquad\qquad …(i)$

and momentum conservation $\Rightarrow$

$mu = mv_1 + mv_2$

$\Rightarrow v_1 + v_2 = u \qquad\qquad …(ii)$

So $2v_2 = u(e + 1)$

$\Rightarrow v_2 = \dfrac{u(e+1)}{2}$ and thus,

$v_1 = \dfrac{(1-e)u}{2}$

Sol 4: (B)

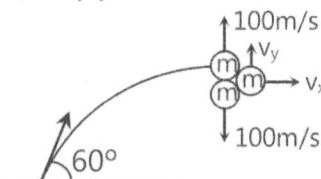

We have

Momentum cons.: in y-direction

$m \times 100 + m(-100) + mv_y = 0$

$\Rightarrow v_y = 0$

x-direction: $3m \cdot u \cos60° = mv_x$

$\Rightarrow v_x = 3 \times 200 \times \dfrac{1}{2} = 300$ m/s

Sol 5: (C)

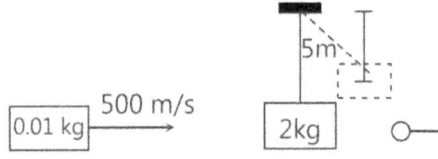

v of block after the bullet emerges can be found using energy conservation

$mgh = \dfrac{1}{2}mv^2$

$\Rightarrow v = \sqrt{2gh} = \sqrt{2 \times 9.8 \times 0.1} = 1.4$ m/s

so now using momentum conservation

$500 \times 0.01 = 2 \times 1.4 + 0.01 \times v'$

$\Rightarrow 5 - 2.8 = 0.01 \times v' \qquad \Rightarrow 2.2 = 0.01 \times v'$

$v' = 220$ m/s

Sol 6: (C)

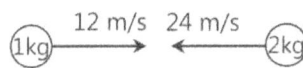

So $e = \dfrac{v_2 - v_1}{36} = \dfrac{2}{3}$

$\Rightarrow v_2 - v_1 = 24 \qquad\qquad\qquad …(i)$

Using momentum conservation

$12 \times 1 - 2 \times (24) = 1 \times v_1 + 2 \times v_2$

$\Rightarrow -36 = v_1 + 2v_2 \qquad\qquad …(ii)$

$\Rightarrow 3v_2 = -12$

$\Rightarrow v_2 = -4$ m/s and $v_1 = -28$ m/s

$E_{initial} = \dfrac{1}{2} \times (1) \times 12^2 + \dfrac{1}{2} \times 2 \times 24^2$

$= 72 + 576 = 648$ J

$E_{final} = \dfrac{1}{2} \times 1 \times (28)^2 + \dfrac{1}{2} \times 2 \times 4^2$

$= 16 + 392 = 408$ J

So $\Delta E = 240$ J

Sol 7: (B)

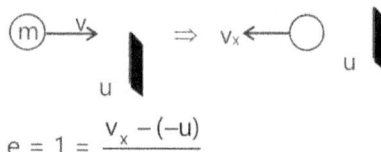

$e = 1 = \dfrac{v'_2 - v'_1}{v_1} \Rightarrow v'_2 - v'_1 = v_1 \qquad …(i)$

and $m_1 v_1 = m_1 v'_1 + m_2 v'_2$

$\Rightarrow v_1 = v'_1 + \dfrac{m_2}{m_1} \cdot v'_2 \qquad\qquad …(ii)$

so (1) + (2) $\Rightarrow \left(1 + \dfrac{m_2}{m_1}\right)v'_2 = 2v_1$

$\Rightarrow v'_2 = \dfrac{2v_1 \cdot m_1}{(m_1 + m_2)}$

and $v'_1 = v'_2 - v_1 = \dfrac{2v_1 \cdot m_1}{(m_1 + m_2)} - v_1$

$= \dfrac{m_1 v_1 - m_2 v_1}{m_1 + m_2} = \left(\dfrac{m_1 - m_2}{m_1 + m_2}\right)v_1$

Sol 8: (C) $v'_2 = \dfrac{2v_1 m_1}{(m_1 + m_2)}$

Sol 9: (B)

$e = 1 = \dfrac{v_x - (-u)}{-v - u}$

$v_x = -u - v - u = -v - 2u$

$2u + v \Rightarrow$ away from wall.

Sol 10: (C)

$e = \dfrac{v_2 - v_1}{u} = 1 \Rightarrow v_2 - v_1 = u \qquad …(i)$

Momentum conservation

$mu = Amv_2 + mv_1$

$\Rightarrow Av_2 + v_1 = u$

$\Rightarrow (A + 1)v_2 = 2u$ or $v_2 = \dfrac{2u}{(A+1)}$

$v_1 = v_2 - u = \dfrac{2u}{(A+1)} - u = \dfrac{(1-A)u}{(1+A)}$

$E = \dfrac{1}{2} \times m \times \dfrac{(1-A)^2 u^2}{(1+A)^2} = \dfrac{E.(1-A)^2}{(1+A)^2}$

Sol 11: (A) v at first imp. $= \sqrt{2gh}$

v at after 1st imp. $= e\sqrt{2gh}$

v at after nth imp. $= e^n\sqrt{2gh}$

$h = \dfrac{v^2}{2g} = e^{2n} . \dfrac{2gh}{2g} = e^{2n}. h$

Sol 12: (A)

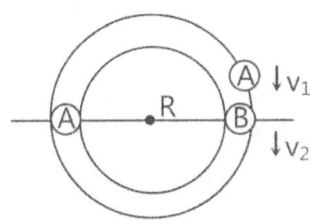

$v_A = \dfrac{\pi R}{t}$,

$v_2 - v_1 = ev_A$ (elasticity)

and $v_2 + v_1 = v_A$ (mom. cons.)

$\Rightarrow v_2 = \dfrac{(e+1)v_A}{2}$ and $v_1 = \dfrac{(1-e)v_A}{2}$

so total time $= \dfrac{D_{rel}}{v_{rel}} = \dfrac{2\pi r}{\left[\dfrac{e+1}{2} - \left(\dfrac{1-e}{2}\right)\right]v_A}$

$= \dfrac{2\pi r \times t}{e.\pi r} = \dfrac{2t}{e}$

Sol 13: (C)

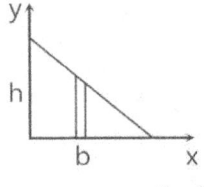

Equation $\Rightarrow \dfrac{y}{h} + \dfrac{x}{b} = 1$,

$\rho = \dfrac{M}{hb/2} = \dfrac{2M}{hb}$ So $x_{COM} = \int \dfrac{x.dm}{M}$

$x_{COM} = \int_0^b \dfrac{x.\rho.h.\left(1 - \dfrac{x}{b}\right)dx}{M} = \dfrac{2}{b}\left[\dfrac{x^2}{2} - \dfrac{x^3}{3b}\right]_0^b = \dfrac{b}{3}$

Similarly, $y_{COM} = \dfrac{h}{3}$

Sol 14: (B)

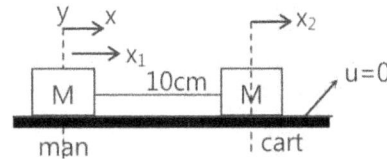

At x = 5 m

$M_1 \ddot{x}_1 + M_2 \ddot{x}_2 = 0$

$\Rightarrow \ddot{x}_1 + \ddot{x}_2 = 0$ and $\ddot{x}_1 - \ddot{x}_2 = 10$

$\Rightarrow \ddot{x}_1 = 5$ m & $\ddot{x}_2 = -5m$

Multiple Correct Choice Type

Sol 15: (C, D)

Elastic collision $\Rightarrow$ 100 % energy transfer

The relative velocity along tangent is zero but in oblique collision the tangent direction is not the one perpendicular to the line joining centres.

Assertion Reasoning Type

Sol 16: (D) $a_{COM} \neq 0$ as $F_{ext} \neq 0$. (in COM frame it is zero)

Sol 17: (A) Proper exp.

Sol 18: (B) It is average and it may be outside body.

Sol.19: (A) In explosion only internal forces are involved.

Sol 20: (D) Disk may be non-uniform.

Sol 21: (A) (A) true, (R) true $\Rightarrow$ correct reason

Sol 22: (B) A $\rightarrow$ true, R $\rightarrow$ true. But R not explanation of A.

Comprehension Type

Sol 23: (B) Obvious (inelastic)

Sol 24: (B) Inelastic collision leads to loss of energy.

Sol 25: (D) Basic concept.

Previous Years' Questions

Sol 1: (a) From conservation of linear momentum, momentum of composite body

$$\vec{p} = (\vec{p}_i)_1 + (\vec{p}_i)_2 = (mv)\hat{i} + (MV)\hat{j}$$

$$\therefore \quad |\vec{p}| = \sqrt{(mv)^2 + (MV)^2}$$

Let it makes an angle α with positive x-axis, then

$$\alpha = \tan^{-1}\left(\frac{p_y}{p_x}\right) = \tan^{-1}\left(\frac{MV}{mv}\right)$$

(b) Fraction of initial kinetic energy transformed into heat during collision

$$= \frac{K_f - K_i}{K_f} = \frac{K_f}{K_i} - 1$$

$$= \frac{p^2 / 2(M+m)}{\frac{1}{2}mv^2 + \frac{1}{2}MV^2} - 1$$

$$= \frac{(mv)^2 + (MV)^2}{(M+m)(mv^2 + MV^2)} - 1$$

$$= \frac{Mm(v^2 + V^2)}{(M+m)(mv^2 + MV^2)}$$

Sol 2: Applying conservation of linear momentum twice. We have

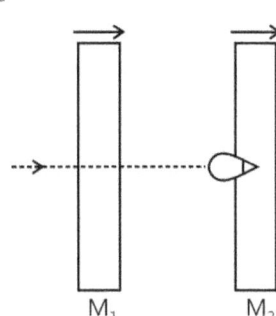

$$mv = M_1v_1 + mv_2 \qquad \text{....(i)}$$

$$mv_2 = (M_2 + m)v_1 \qquad \text{....(ii)}$$

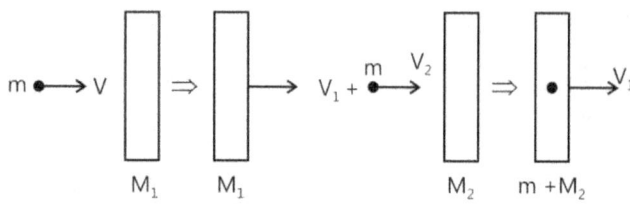

Solving Eqs. (i) and (ii), we get

$$\frac{v_2}{v} = \frac{M_2 + m}{M_1 + M_2 + m}$$

Substituting the values of $m : M_1$ and M_2 we get, percentage of velocity retained by bullet.

$$\frac{v_2}{v} \times 100 = \left(\frac{2.98 + 0.02}{1 + 2.98 + 0.02}\right) \times 100 = 75\%$$

$$\therefore \ \%loss = 25\%$$

Sol 3: Suppose r_1 be the distance of centre of mass of the remaining portion from centre of the bigger circle, then

$$A_1r_1 = A_2r_2$$

$$r_1 = \left(\frac{A_2}{A_1}\right)r_2$$

$$r_1 = \frac{\pi(42)^2}{\pi[(56)^2 - (42)^2]} \times 7 = 9 \text{ cm}$$

Sol 4: Before collision net momentum of the system was zero. No external force is acting on the system. Hence, momentum after collision should also be zero. A has come to rest. Therefore, B and C should have equal and opposite momenta or velocity of C should be V in opposite direction of velocity of B.

Sol 5: Collision between A and C is elastic and mass of both the blocks is same. Therefore, they will exchange their velocities i.e., C will come to rest and A will be moving will velocity v_0. Let v be the common velocity of A and B, then from conservation of linear momentum, we have

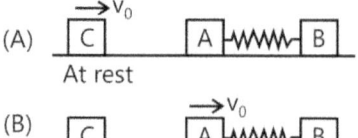

$$m_A v_0 = (m_A + m_B)v$$

or $\quad mv_0 = (m+2m)v$ or $v = \dfrac{v_0}{3}$

(b) From conservation of energy, we have

$$\frac{1}{2}m_A v_0^2 = \frac{1}{2}(m_A + m_B)v^2 + \frac{1}{2}kx_0^2$$

or $\quad \dfrac{1}{2}mv_0^2 = \dfrac{1}{2}(3m)\left(\dfrac{v_0}{3}\right)^2 + \dfrac{1}{2}kx_0^2$

or $\quad \dfrac{1}{2}kx_0^2 = \dfrac{1}{3}mv_0^2$ or $k = \dfrac{2mv_0^2}{3x_0^2}$

Sol 6: As shown in figure initially when the bob is at A, its potential energy is mgl. When the bob is released and it strikes the wall at B, its potential energy mgl is converted into its kinetic energy. If v be the velocity with which the bob strikes the wall, then

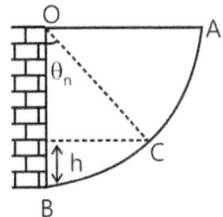

$$mgl = \frac{1}{2}mv^2 \text{ or } v = \sqrt{(2gl)} \qquad .....(i)$$

Speed of the bob after rebounding (first time)

$$v_1 = e\sqrt{(2gl)} \qquad .....(ii)$$

The speed after second rebound is $v_2 = e^2\sqrt{(2gl)}$

In general after n rebounds, the speed of the bob is

$$v_n = e^n\sqrt{(2gl)} \qquad .....(iii)$$

Let the bob rises to a height h after n rebounds. Applying the law of conservatioin of energy, we have

$$\frac{1}{2}mv_n^2 = mgh$$

$$\therefore \quad h = \frac{v_n^2}{2g} = \frac{e^{2n}.2gl}{2g} = e^{2n}.l$$

$$= \left(\frac{2}{\sqrt5}\right)^{2n}.l = \left(\frac{4}{5}\right)^n l \qquad .....(iv)$$

If θ_n be the angle after n collisions, then

$h = l - l\cos\theta_n = l(1-\cos\theta_n)$

From Eqs. (iv) and (v), we have

$$\left(\frac{4}{5}\right)^n l = l(1-\cos\theta_n) \text{ or } \left(\frac{4}{5}\right)^n = (1-\cos\theta_n)$$

For θ_n to be less than 60°, i.e., $\cos\theta_n$ is greater than $\dfrac{1}{2}$, i.e., $(1-\cos\theta_n)$ is less than $\dfrac{1}{2}$, we have

$$\left(\frac{4}{5}\right)^n < \left(\frac{1}{2}\right)$$

The condition is satisfied for n = 4.

$\therefore \quad$ Required number of collisions = 4.

Sol 7: (a) Since, only two forces are acting on the rod, its weight Mg (vertically downwards) and a normal reaction N at point of contact B (vertically upwards).

No horizontal force is acting on the rod (surface is smooth).

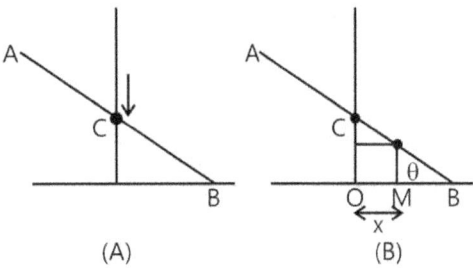

(A)   (B)

Therefore, CM will fall vertically downwards towards negative y-axis i.e., the path of CM is a straight line.

(b) Refer figure (B). We have to find the trajectory of a point P(x, y) at a distance r from end B.

CB = L/2

$\therefore \quad$ OB = (L/2) cosθ

MB = r cosθ

$\therefore \quad$ x = OB – MB = cosθ {(L/2 – r)}

or $\quad \cos\theta = \dfrac{x}{\{(L/2)-r\}} \qquad ....(i)$

Similarly, y = r sin θ

or $\quad \sin\theta = \dfrac{y}{r} \qquad ....(ii)$

Squaring and adding Eqs. (i) and (ii), we get

$$\sin^2\theta + \cos^2\theta = \frac{x^2}{\{(L/2)-r\}^2} + \frac{y^2}{r^2}$$

or $\quad \dfrac{x^2}{\{(L/2)-r\}^2} + \dfrac{y^2}{r^2} = 1 \qquad ....(iii)$

This is an equation of an ellipse. Hence, path of point P is an ellipse whose equation is given by (iii).

Sol 8: (a) Since, the collision is elastic, the wedge will return with velocity $v\hat{i}$

(i)

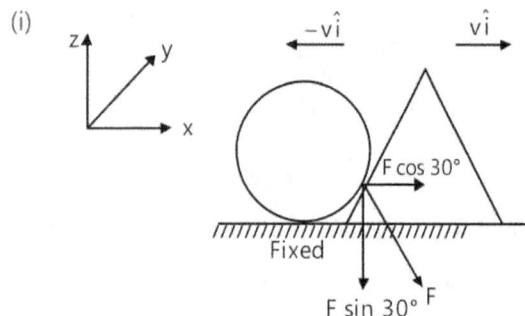

Now, linear impulse in x-direction

= change in momentum in x-direction.

$\therefore \quad (F\cos30°)\ \Delta t = mv - (-mv) = 2mv$

$\therefore \quad F = \dfrac{2mv}{\Delta t\cos30°} = \dfrac{4mv}{\sqrt{3}\,\Delta t}$

$F = \dfrac{4mv}{\sqrt{3}\,\Delta t}$

$\therefore \quad \vec{F} = (F\cos30°)\hat{i} - (F\sin30°)\hat{k}$

or $\quad \vec{F} = \left(\dfrac{2mv}{\Delta t}\right)\hat{i} - \left(\dfrac{2mv}{\sqrt{3}\,\Delta t}\right)\hat{k}$

(ii) Taking the equilibrium of wedge in vertical z-direction during collision.

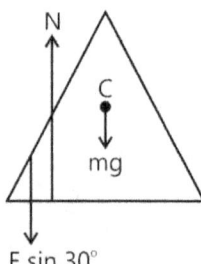

$N = mg + F\sin30°$

$N = mg + \dfrac{2mv}{\sqrt{3}\,\Delta t}$

or in vector form

$\vec{N} = \left(mg + \dfrac{2mv}{\sqrt{3}\,\Delta t}\right)\hat{k}$

(b) For rotational equilibrium of wedge [about (CM) anticlockwise torque of F = clockwise torque due to N.

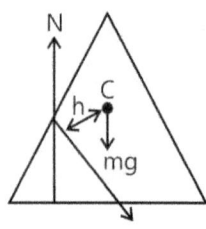

$\therefore$ Magnitude of torque of N about CM = magnitude of torque of F about CM

$= F.h$

$|\vec{\tau}_N| = \left(\dfrac{4mv}{\sqrt{3}\,\Delta t}\right)h$

Sol 9: After elastic collision,

$v'_A = \left(\dfrac{m-2m}{m+2m}\right)(9) + \dfrac{2(2m)}{m+2m}\,(0) = -3\,ms^{-1}$

Now from conservation of linear momentum after all collisions are complete,

$m(+9\,ms^{-1}) = m(-3\ ms^{-1}) + 3m\ (v_C)$

or $\quad v_C = 4\,ms^{-1}$

Sol 10: (A) Velocity of particle performing projectile motion at highest point

$= v_1 = v_0\cos\alpha$

Velocity of particle thrown vertically upwards at the position of collision

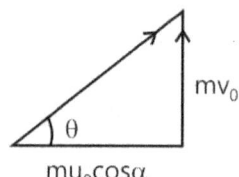

$= v_2^2 = u_0^2 - 2g\dfrac{u^2\sin^2\alpha}{2g} = v_0\cos\alpha$

So, from conservation of momentum

$\tan\theta = \dfrac{mv_0\cos\alpha}{mu_0\cos\alpha} = 1$

$\Rightarrow \theta = \pi/4$

Sol 11: The initial speed of 1st bob (suspended by a string of length l_1) is $\sqrt{5gl_1}$.

The speed of this bob at highest point will be $\sqrt{gl_1}$.

When this bob collides with the other bob there speeds will be interchanged.

$\sqrt{gl_1} = \sqrt{5gl_1} \Rightarrow \dfrac{l_1}{l_2} = 5$.

3. ROTATIONAL MECHANICS

1. INTRODUCTION

In this chapter we will be studying the kinematics and dynamics of a solid body in two kinds of motion. The first kind of motion of a solid body is rotation about a stationary axis, also called pure rotation. The second kind of motion of a solid is the plane motion wherein the center of mass of the solid body moves in a certain stationary plane while the angular velocity of the body remains permanently perpendicular to that plane. Here the body executes pure rotation about an axis passing through the center of mass and the center of mass itself translates in a stationary plane in the given reference frame. The axis through the center of mass is always perpendicular to the stationary plane. We will also learn about the inertia property in rotational motion, and the quantities torque and angular momentum which are rotational analogue of force and linear momentum respectively. The law of conservation of angular momentum is an important tool in the study of motion of solid bodies.

2. BASIC CONCEPT OF A RIGID BODY

A solid is considered to have structural rigidity and resists change in shape, size and density. A rigid body is a solid body which has no deformation, i.e. the shape and size of the body remains constant during its motion and interaction with other bodies. This means that the separation between any two points of a rigid body remains constant in time regardless of the kind of motion it executes and the forces exerted on it by surrounding bodies or a field of force.

A metal cylinder rolling on a surface is an example of a rigid body as shown in Fig. 7.1.

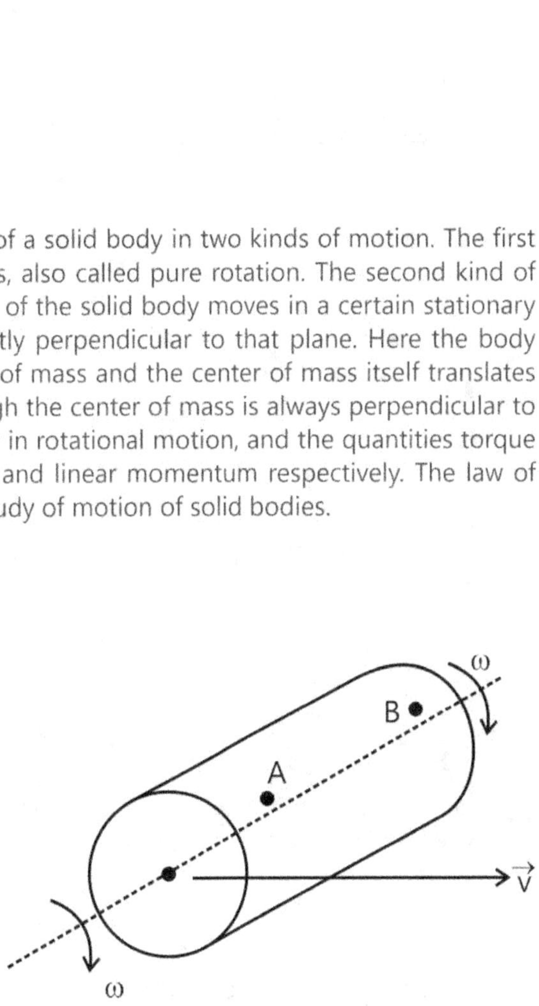

Figure 7.1: Metal cylinder rolling on a surface is a rigid body system. Relative distance between points A and B do not change.

Let velocities of points P and Q of a rigid body with respect to a reference frame be V_P and V_Q as shown in the Fig. 7.2.

As the body is rigid, the length PQ should not change during the motion of the body, i.e. the relative velocity between P and Q along the line joining P and Q should be zero i.e. velocity of approach or separation is zero. Let x-axis be along PQ, then

$\vec{V}_{QP}$ = relative velocity of Q with respect to P

$\vec{V}_{QP} = (V_Q \cos\theta_2 \, \hat{i} + V_Q \sin\theta_2 \, \hat{j}) - (V_P \cos\theta_1 \, \hat{i} - V_P \sin\theta_1 \, \hat{j})$

$\vec{V}_{QP} = (V_Q \cos\theta_2 - V_P \cos\theta_1) \, \hat{i} + (V_P \sin\theta_1 + V_Q \sin\theta_2) \, \hat{j}$

Now $V_P \cos\theta_1 = V_Q \cos\theta_2$

(Since velocity of separation is 0)

$\vec{V}_{QP} = (V_P \sin\theta_1 + V_Q \sin\theta_2)\hat{j}$ (which is perpendicular to line PQ).

Hence, we can conclude that for each and every pair of particles in a rigid body, relative motion between the two points in the pair will be perpendicular to the line joining the two points.

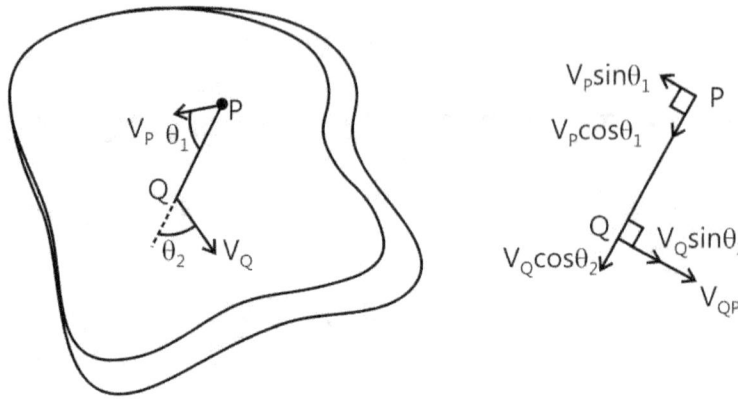

Figure 7.2: Relative velocity between two points of a rigid body

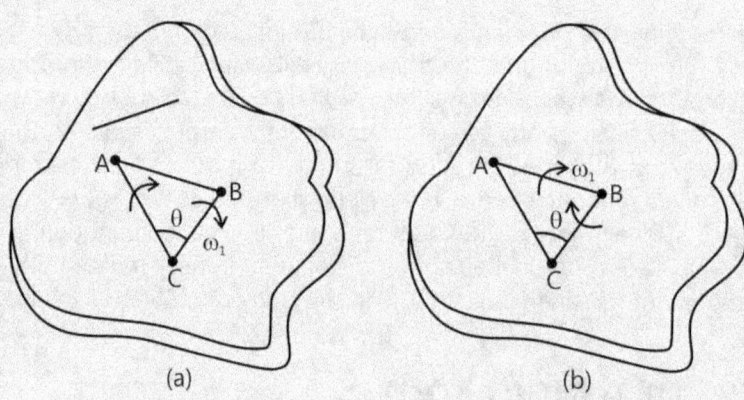

Figure 7.3: (a) Angular velocity of A and B w.r.t. C is ω_1 (b) Angular velocity of A and C w.r.t. B is ω_1

Suppose A, B, C are points of a rigid system hence during any motion the lengths of sides AB, BC, and CA will not change, and thus the angle between them will not change, and so they all must rotate through the same angle. Hence all the sides rotate by the same rate. Or we can say that each point is having the same angular velocity with respect to any other point on the rigid body.

3. MOTION OF A RIGID BODY

We will study the dynamics of three kinds of motion of a rigid body.

(a) Pure Translational motion

(b) Pure Rotational Motion

(c) Combined Translational and Rotational motion

Let us briefly discuss the characteristics of these three types of motion of a rigid body.

3.1 Pure Translational motion

A rigid body is said to be in pure translational motion if any straight

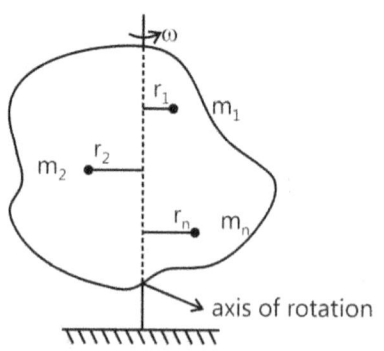

Figure 7.4: Body in pure rotational motion.

line fixed to it remains parallel to its initial orientation all the time. E.g. a car moving along a straight horizontal stretch of a road. In this kind of motion, the displacement of each and every particle of the rigid body is the same during any time interval. All the points of the rigid body have the same velocity and acceleration at any instant. Thus to study the translational motion of a rigid body, it is enough to study the motion of an individual point belonging to that rigid body i.e. the dynamics of a point.

3.2 Pure Rotational Motion

Suppose a rigid body of any arbitrary shape rotates about an axis which is stationary in a given reference frame. In this kind of motion every point of the body moves in a circle whose center lies on the axis of rotation at the foot of the perpendicular from the particle to this axis, and radius of the circle is equal to the perpendicular distance of the point from this axis. Every point of the rigid body moves through the same angle during a particular time interval. Such a motion is called pure rotational motion. Each particle has same instantaneous angular velocity (since the body is rigid) and different particles move in circles of different radii, the planes of all these circles are parallel to each other. Particles moving in smaller circles have less linear velocity and those moving in bigger circles have large linear velocity at the same instant.

In the Fig. 7.4 particles of mass m_1, m_2, m_3..... have linear velocities v_1, v_2, v_3....

If ω is the instantaneous angular velocity of the rigid body, then

$$v_1 = \omega r_1, \ v_2 = \omega r_2, \ v_3 = \omega r_3 \ ......, \ v_n = \omega r_n$$

Figure 7.5: Body in pure translational motion.

3.3 Combined Translational and Rotational Motion

A rigid body is said to be in combined translational and rotational motion if the body performs pure rotation about an axis and at the same time the axis translates with respect to a reference frame. In other words there is a reference frame K′ which is rigidly fixed to the axis of rotation, such that the body performs pure rotation in the K′ frame. The K′ frame in turn is in pure translational motion with respect to a reference frame K. So to describe the motion of the rigid body in the K frame, the translational motion of K′ frame is super-imposed on the pure rotational motion of the body in the K′ frame.

Illustration 1: A body is moving down into a well through a rope passing over a fixed pulley of radius 10 cm. Assume that there is no slipping between rope and pulley. Calculate the angular velocity and angular acceleration of the pulley at an instant when the body is going down at a speed of 20 cm s⁻¹ and has an acceleration of 4.0 m s⁻². **(JEE MAIN)**

Sol: Since the rope does not slip on the pulley, the linear speed and linear acceleration of the rim of the pulley will be equal to the speed and acceleration of the body respectively.

Therefore, the angular velocity of the pulley is

$$\omega = \frac{\text{linear velocity of rim}}{\text{radius of rim}} = \frac{20 \text{ cm s}^{-1}}{10 \text{ cm}} = 2 \text{ rad s}^{-1}$$

And the angular acceleration of the pulley is

$$\alpha = \frac{\text{linear acceleration of rim}}{\text{radius of rim}} = \frac{4.0 \text{ ms}^{-2}}{10 \text{cm}} = 40 \text{ rad s}^{-2}$$

4. ROTATIONAL KINEMATICS

Suppose a rigid body performing pure rotational motion about an axis of rotation rotates by an angle $\Delta\theta$ in a time interval Δt. The instantaneous angular velocity ω, is defined as,

$$\omega = \lim_{\Delta t \to 0} \frac{\Delta\theta}{\Delta t} = \frac{d\theta}{dt} \qquad \qquad ...(i)$$

Similarly, the instantaneous angular acceleration α is defined as,

$$\alpha = \frac{d\omega}{dt} = \frac{d^2\theta}{dt^2} \qquad \qquad ...(ii)$$

The relations between linear distance s, linear velocity v and linear acceleration a, and the corresponding angular variables describing circular motion θ, ω, and α respectively are given as:

$$s = r\theta; \quad v = r\omega; \qquad a_t = r\alpha \qquad \qquad ...(iii)$$

Here the subscript t along with a in the expression for acceleration signifies that this is the tangential component of linear acceleration.

If a body rotates with uniform angular acceleration,

$$\omega = \omega_0 + \alpha t \; ; \; \theta = \omega_0 t + \frac{1}{2}\alpha t^2 ; \; \omega^2 = \omega_0^2 + 2\alpha\theta \qquad \qquad ...(iv)$$

where ω_0 is initial angular velocity.

The equations for angular displacement, angular velocity and angular acceleration are similar to the corresponding equations of linear motion.

Illustration 2: A disc starts rotating with constant angular acceleration of $\pi/2$ rad s^{-2} about a fixed axis perpendicular to its plane and through its center. Calculate

(a) The angular velocity of the disc after 4 s

(b) The angular displacement of the disc after 4s and

(c) Number of turns accomplished by the disc in 4 s. **(JEE MAIN)**

Sol: Use the first and second equations of angular motion with constant angular acceleration.

Here $\alpha = \dfrac{\pi}{2}$ rad s^{-2}; $\quad \omega_0 = 0$; $\quad t = 4$ s;

(a) $\omega(4\text{ s}) = 0 + \left(\dfrac{\pi}{2}\text{rad s}^{-2}\right) \times 4\text{ s} = 2\pi \text{ rad s}^{-1}$

(b) $\theta_{(4s)} = 0 + \dfrac{1}{2}\left(\dfrac{\pi}{2}\text{rad s}^2\right) \times (16\text{s}^2) = 4\pi \text{ rad}$

(c) $\Rightarrow n \times 2\pi \text{ rad} = 4\pi \text{ rad} \Rightarrow n = 2.$

NOMORECLASS CONCEPTS

For variable angular acceleration we should proceed with differential equation $\dfrac{d\omega}{dt} = \alpha$

5. MOMENT OF INERTIA

Before discussing the dynamics of rigid body motion let us study about an important property of a rigid body called Moment of Inertia which is indispensable in understanding its dynamics.

Physical Significance of Moment of Inertia: As the name suggests, moment of inertia is the measure of the rotational inertia property of a rigid body, the rotational analog of mass in translational motion. "It is the property of the rigid body by virtue of which it opposes any change in its state of uniform rotational motion." The moment of inertia of a rigid body depends on its mass, on the location and orientation of the axis of rotation and on the shape and size of the body or in other words on the distribution of the mass of the body with respect to the axis of rotation. SI units of moment of inertia is Kg-m². Moment of inertia about a particular axis of rotation is a scalar positive quantity.

Definition: Moment of inertia of a system of n particles about an axis is defined as:

$$I = m_1 r_1^2 + m_2 r_2^2 + \ldots\ldots\ldots + m_n r_n^2 \text{ i.e. } I = \sum_{i=1}^{n} m_i r_i^2 \qquad \text{...(i)}$$

where, r_i is the perpendicular distance of ith particle of mass m_i from the axis of rotation.

For a continuous rigid body, the moment of inertia can be calculated as:

$$I = \int r^2 (dm) \qquad \text{...(ii)}$$

where dm is the mass of an infinitesimal element of the body at a perpendicular distance r from the axis of rotation.

Moment of inertia depends on:

(a) Mass of the rigid body.

(b) Shape and size of the rigid body.

(c) Location and orientation of the axis of rotation.

NOMORECLASS CONCEPTS

Moment of inertia does not change if the mass:

(i) Is shifted parallel to the axis of rotation because r_i does not change.

(ii) Is rotated about the axis of rotation in a circular path because r_i does not change.

Illustration 3: Two particles having masses m_1 & m_2 are situated in a plane perpendicular to line AB at a distance of r_1 and r_2 respectively as shown.

(i) Find the moment of inertia of the system about axis AB?

(ii) Find the moment of inertia of the system about an axis passing though m_1 and perpendicular to the line joining m_1 and m_2.

(iii) Find the moment of inertia of the system about an axis passing through m_1 and m_2.

(iv) Find moment of inertia about an axis passing though center of mass and perpendicular to line joining m_1 and m_2.　　　**(JEE MAIN)**

Sol: Use the formula for moment of inertia of a system of n particles. Find the distance of center of mass from m_1.

(i) Moment of inertia of particle on right is $I_1 = m_1 r_1^2$

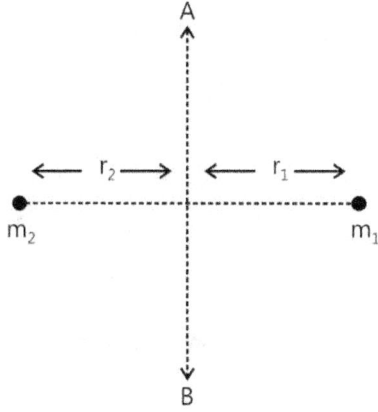

Figure 7.6

Moment of inertia of particle on left is $\qquad I_2 = m_2 r_2^2$

Moment of inertia of the system about AB is $\qquad I = I_1 + I_2 = m_1 r_1^2 + m_2 r_2^2$

(ii) Moment of inertia of particle on right is $\qquad I_1 = 0$

Moment of inertia of particle on left is $\qquad I_2 = m_2(r_1 + r_2)^2$

Moment of inertia of the system about axis is $\qquad I = I_1 + I_2 = 0 + m_2(r_1 + r_2)^2$

(iii) Moment of inertia of particle on right is $\qquad I_1 = 0$

Moment of inertia of particle of left is $\qquad I_2 = 0$

Moment of inertia of the system about axis is $\qquad I = I_1 + I_2 = 0 + 0$

(iv) of system $r_{cm} = m_2 \left(\dfrac{r_1 + r_2}{m_1 + m_2} \right) =$ Distance of center mass from mass m_1

Distance of center of mass from mass $m_2 = m_1 \left(\dfrac{r_1 + r_2}{m_1 + m_2} \right)$

So moment of inertia about center of mass $= I_{cm} = m_1 \left(m_2 \dfrac{r_1 + r_2}{m_1 + m_2} \right)^2 + m_2 \left(m_1 \dfrac{r_1 + r_2}{m_1 + m_2} \right)^2$

$$I_{cm} = \dfrac{m_1 m_2}{m_1 + m_2}(r_1 + r_2)^2.$$

Illustration 4: Three particles each of mass m, are situated at the vertices of an equilateral triangle PQR of side a as shown in the Fig 7.7. Calculate the moment of inertia of the system about

(i) The line PX perpendicular to PQ in the plane of PQR.

(ii) One of the sides of the triangle PQR

(iii) About an axis passing through the centroid and perpendicular to plane of the triangle PQR. **(JEE MAIN)**

Figure 7.7

Sol: Use the formula for moment of inertia of a system of n particles.

(i) Perpendicular distance of P from PX = 0; perpendicular distance of Q from PX = a perpendicular distance of R from PX = a/2. Thus, the moment of inertia of the particle at P is 0, that of particle Q is ma^2, and of the particle at R is $m(a/2)^2$.

The moment of inertia of the three particle system about PX is $I = 0 + ma^2 + m(a/2)^2 = \dfrac{5ma^2}{4}$

Note that the particles on the axis do not contribute to the moment of inertia.

(ii) Moment of inertia about the side PR = mass of particle Q × square of perpendicular distance of Q from side PR,

$$I_{PR} = m \left(\dfrac{\sqrt{3}}{2} a \right)^2 = \dfrac{3ma^2}{4}$$

(iii) Distance of centroid from each of the particles is $\dfrac{a}{\sqrt{3}}$, so moment of inertia about an axis passing through the

centroid and perpendicular to the plane of triangle PQR = $I_C = 3m \left(\dfrac{a}{\sqrt{3}} \right)^2 = ma^2$

Table 7.1: Formulae of MOI of symmetric bodies

S. No.	Body, mass M	Axis	Figure	I	K(Radius of Gyration)
1.	Ring or loop of radius R	Through its center and perpendicular to its plane		MR^2	R
2.	Disc, radius R	Perpendicular to its plane through its center		$\dfrac{MR^2}{2}$	$\dfrac{R}{\sqrt{2}}$
3.	Hollow cylinder, radius R	Axis of cylinder		MR^2	R
4.	Solid cylinder, radius R	Axis of cylinder		$\dfrac{MR^2}{2}$	$\dfrac{R}{\sqrt{2}}$
5.	Thick walled cylinder,	Axis of cylinder		$\dfrac{M\left(R_1^2 + R_2^2\right)}{2}$	$\dfrac{\sqrt{R_1^2 + R_2^2}}{2}$
6.	Solid sphere, radius R	Diameter		$\dfrac{2}{5}MR^2$	$\sqrt{\dfrac{2}{5}}R$
7.	Spherical shell radius, R	Diameter		$\dfrac{2MR^2}{3}$	$\sqrt{\dfrac{2}{3}}R$
8.	Thin rod, length L	Perpendicular to rod at middle point		$\dfrac{ML^2}{12}$	$\dfrac{L}{2\sqrt{3}}$

9.	Thin rod, length L	Perpendicular to rod at one end		$\dfrac{ML^2}{3}$	$\dfrac{L}{\sqrt{3}}$
10.	Solid cylinder, length l	Through center and perpendicular to length	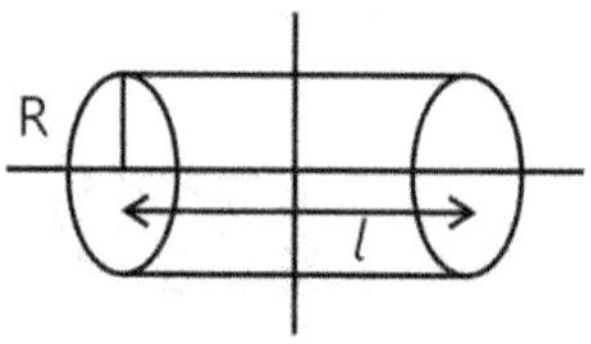	$\dfrac{MR^2}{4} + \dfrac{Ml^2}{12}$	$\sqrt{\dfrac{R^2}{4} + \dfrac{l^2}{12}}$
11.	Rectangular sheet, length l and breadth b	Through center and perpendicular to plane	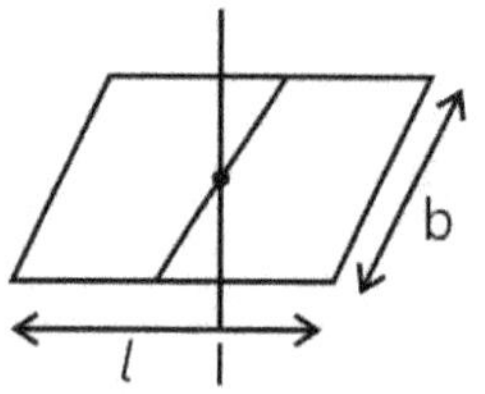	$\dfrac{M(l^2 + b^2)}{12}$	$\sqrt{\dfrac{l^2 + b^2}{12}}$

NOMORECLASS CONCEPTS

While deriving the MOI *of* any rigid body the element chosen should be such that:

Either perpendicular distance of axis from each point of the element is same or the moment of inertia of the element about the axis of rotation is known.

5.1 Theorems on Moment of Inertia

1. Theorem of Parallel Axes: This theorem is very useful in cases when the moment of inertia about an axis z_c passing through the center of mass (C.O.M) of the rigid body is known, and we sought to find the moment of inertia about any other axis z which is parallel to the axis z_c as shown in Fig. 7.8. The moment of inertia of the rigid body about axis z is equal to the sum of the moment of inertia about axis z_c and the product of the mass m of the body by the square of perpendicular distance between the two axes. If the moment of inertia of the rigid body about axis z_c is I_c, then the moment of inertia I of this body about any parallel axis z , is given by $I = I_c + Md^2$ ...(i)

where d is the perpendicular distance between the two axes.

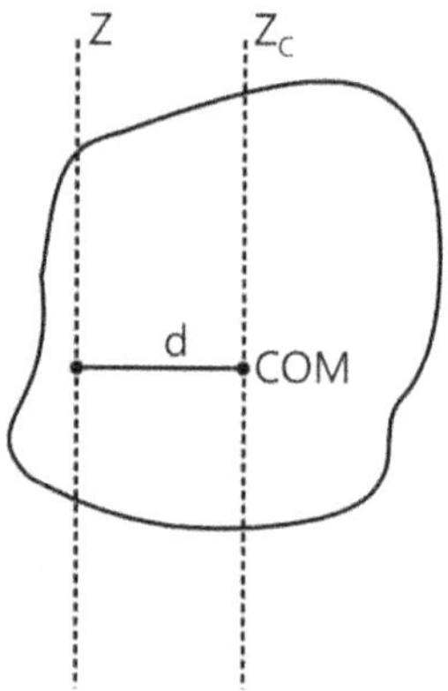

Figure 7.8: Parallel axes

Illustration 5: Find the moment of inertia of a uniform sphere of mass m and radius R about a tangent if the sphere is (i) solid (ii) hollow **(JEE MAIN)**

Sol: We know the formula for moment of inertia of sphere about an axis passing through its center. Use the parallel axes theorem to find the moment of inertia about the tangent.

(i) Using parallel axis theorem

$I = I_C + md^2$

For solid sphere

$I_C = \dfrac{2}{5}mR^2$, $d = R$; $I = \dfrac{7}{5}mR^2$

(ii) Using parallel axis theorem

$I = I_C + md^2$

For hollow sphere

$I_C = \dfrac{2}{3}mR^2$, $d = R$; $I = \dfrac{5}{3}mR^2$

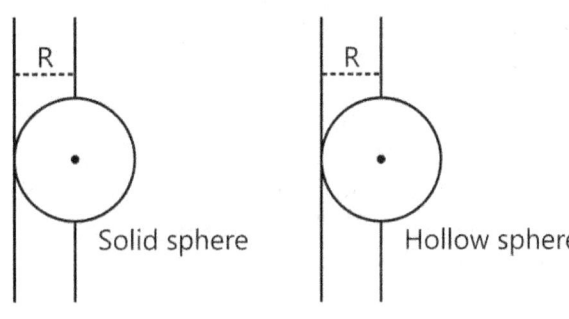

Figure 7.9

Illustration 6: Find the moment of inertia of the two uniform joint roads having mass m each about point P as shown in Fig 7.10. Use parallel axis theorem.   **(JEE MAIN)**

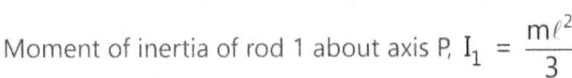

Sol: We know the formulae for moment of inertia of rod about the axes passing through its center and through one of its ends and perpendicular to it. Use the parallel axes theorem to find the moment of inertia about the point P.

Figure 7.10

Moment of inertia of rod 1 about axis P, $I_1 = \dfrac{m\ell^2}{3}$

Moment of inertia of rod 2 about axis p, $I_2 = \dfrac{m\ell^2}{12} + m\left(\sqrt{5}\dfrac{\ell}{2} \right)^2$

So moment of inertia of a system about axis p; $I = \dfrac{5m\ell^2}{3}$

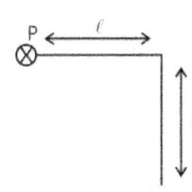

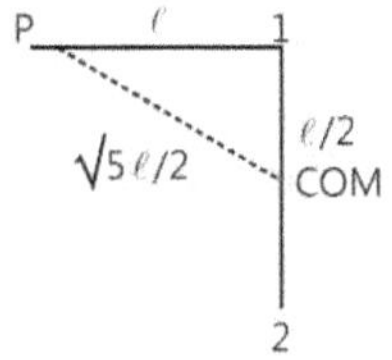

Figure 7.11

2. Theorem of Perpendicular Axes: This theorem is applicable only in case of two dimensional rigid body or planar lamina as shown in Fig. 7.12. Let the lamina lie in the x-y plane and I_x and I_y be the moment of inertia of the lamina about x and y axes respectively then the moment of inertia about z-axis perpendicular to the plane of the lamina and passing through the point of intersection of x and y axes is given as:

$I_z = I_x + I_y$ ...(ii)

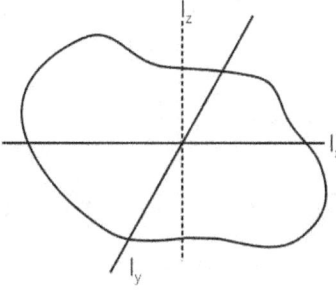

Figure 7.12: Perpendicular axes

Illustration 7: Find the moment of inertia of a half-disc about an axis perpendicular to the plane and passing through its center of mass. Mass of this disc is M and radius is R.   **(JEE MAIN)**

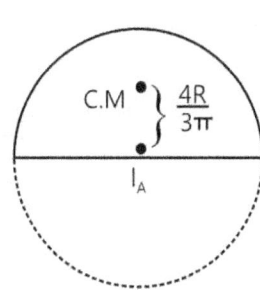

Sol: We know the formula for the moment of inertia of the half disc about a perpendicular axis through the center A. Use the parallel axes theorem to find the moment of inertia about a perpendicular axis through the center of mass.

The COM of half disc will be at distance $4R/3\pi$ from the center A. Let moment of inertia of half disc about a perpendicular axis passing through A be I_A.

First we fill the remaining half with same density to get a full disc of mass 2M.

Figure 7.13

The moment of inertia about center A of full disc will be $2I_A$,

So, $I_A = \dfrac{2MR^2}{2 \times 2} = \dfrac{MR^2}{2}$; $I_A = I_{CM} + M \times \left(\dfrac{4R}{3\pi} \right)^2$; $I_{CM} = \dfrac{MR^2}{2} - M \times \left(\dfrac{4R}{3\pi} \right)^2$

3.9

Illustration 8: Calculate the moment of inertia of a uniform disc of mass M and radius R about a diameter.

(JEE MAIN)

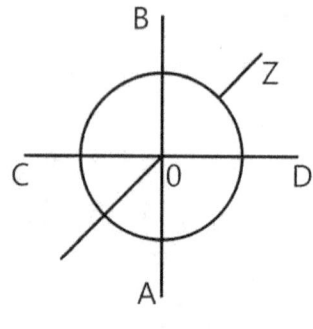

Sol: For a uniform disc all diameters are equivalent, i.e. moment of inertia about any diameter will be equal to that about any other diameter. We know the formula for moment of inertia of disc about axis perpendicular to its plane and passing through its center. Use the perpendicular axes theorem to find the moment of inertia about a diameter.

Let AB and CD be two mutually perpendicular diameters of the disc. Take them as x and y axes and the line perpendicular to the plane of the disc through the center as

Figure 7.14

the Z – axis. The moment of inertia of the ring about the Z – axis is $I = \frac{1}{2} MR^2$. As the disc is uniform, all of its diameters are equivalent and so $I_x = I_y$

From perpendicular an axis theorem $I_z = I_x + I_y$; hence $I_x = \frac{I_z}{2} = \frac{MR^2}{4}$

Illustration 9: In the Fig 7.15 shown find the moment of inertia of square plate having mass m and sides a about axis 2 passing through point C (center of mass) and in the plane of plate.

(JEE MAIN)

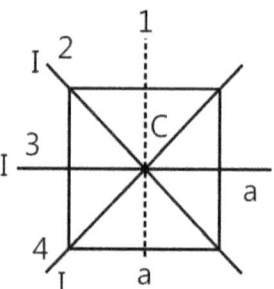

Sol: For uniform square plate axes 2 and 4 along diagonals are equivalent and axes 1 and 3 are equivalent. Suppose I_c is the moment of inertia about the axis perpendicular to the plane of plate and passing through the center C. Use perpendicular axes theorem to prove that the axes 1 and 2 are also equivalent.

Using perpendicular axes theorem $I_C = I_4 + I_2 = I' + I' = 2I'$ .... (i)

Using perpendicular axes theorem $I_C = I_3 + I_1 = I + I = 2I$ ....(ii)

Figure 7.15

From (i) and (ii) we get $I' = I$

$I_c = 2I = \frac{ma^2}{6} \Rightarrow I' = \frac{ma^2}{12}$

5.2 Radius of Gyration

The radius of gyration of a rigid body about an axis z is equal to the radius of a ring whose mass is equal to the mass of the rigid body, and the moment of inertia of the ring about an axis passing through its center and perpendicular to its plane is equal to the moment of inertia of the rigid body about the axis z. Radius of gyration can also be defined as the perpendicular distance from the axis of rotation where all mass of the rigid body can be assumed to be concentrated when the rigid body is performing pure rotation to get the equation of motion of the body. Thus, the radius of gyration is the "equivalent distance" of the rigid body from the axis of rotation.

$$I = MK^2$$

I = Moment of inertia of the rigid body about an axis

M = Mass of the rigid body

K = Radius of gyration about the same axis

or $\quad K = \sqrt{\dfrac{I}{M}}$ ...(iii)

Length K is the property of the rigid body which depends upon the shape and size of the body and on the orientation and location of the axis of rotation. S.I. Unit of K is meter.

Illustration 10: Find the radius of gyration of a hollow uniform sphere of radius R about its tangent. **(JEE MAIN)**

Sol: Use the formula for radius of gyration.

Moment of inertia of a hollow sphere about a tangent = $\dfrac{5}{3}MR^2$

$$MK^2 = \frac{5}{3}MR^2 \Rightarrow K = \sqrt{\frac{5}{3}}R$$

5.3 Moment of Inertia of a Body Having a Cavity

If we know the moments of inertia of different parts of a rigid body about the same axis, then the moment of inertia of the entire body can be calculated by simply adding the moments of inertia of the different parts (about the same axis) i.e. moment of inertia is an additive quantity. This principle can be used to calculate the moment of inertia of a body having hollow spaces by first assuming the hollow spaces to be filled with same density as that of the body and evaluating the moment of inertia of the whole body about the given axis and then add the moments on inertia of the hollow spaces about the same axis considering them to have negative mass.

Illustration 11: A uniform disc of radius R has a round disc of radius R/3 cut as shown in Fig 7.16. The mass of the disc equals M. Find the moment of inertia of such a disc relative to the axis passing through geometrical center of original disc and perpendicular to the plane of the disc. **(JEE ADVANCED)**

Sol: Consider the whole disc without the cavity. The cavity can be thought of as a negative mass of same density as disc. We know the formula for moment of inertia of uniform disc about axis perpendicular to its plane and passing through its center. Find the moment of inertia of cavity (negative mass) about the perpendicular axis passing through center of whole disc. The moment of inertia of disc with cavity is the sum of the moment of inertia of whole disc and the moment of inertia of cavity (negative).

Let the mass per unit area of the material of disc be σ. Now the empty space can be considered as having density $-\sigma$

Now $I_0 = I_\sigma + I_{-\sigma}$

$$I_\sigma = (\sigma\pi R^2)R^2/2 = \text{MI of } \sigma \text{ about } O = MR^2/2$$

$$I_{-\sigma} = \frac{-\sigma\pi(R/3)^2(R/3)^2}{2} + [-\sigma\pi(R/3)^2](2R/3)^2$$

$$= \text{M.I of } -\sigma \text{ About } O = -MR^2/18$$

$$I_0 = MR^2/2 - MR^2/18$$

$$I_0 = \frac{4}{9}MR^2$$

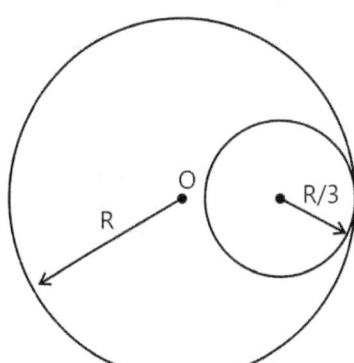

Figure 7.16

6 TORQUE

6.1 Torque About a Point

Torque of force F relative to a point O is defined as

$$\vec{\tau} = \vec{r} \times \vec{F} \qquad \qquad ...(x)$$

where $\vec{F}$ = force applied to a point on a body

$\vec{r}$ = position vector of the point of application of force relative to the point O in a chosen reference frame about which we want to determine the torque (see Fig. 7.17).

Torque is a vector quantity and its direction is given by the right hand rule for cross product of vectors.

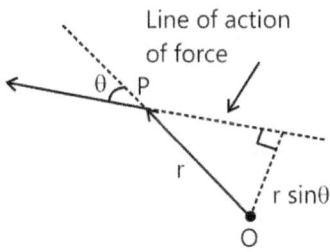

Figure 7.17: Torque of a force

Magnitude of torque $\quad |\vec{\tau}| = rF\sin\theta = r_\perp F = rF_\perp$

where θ is the angle between the force $\vec{F}$ and the position vector $\vec{r}$ of point of application.

$r_\perp = r\sin\theta$ = perpendicular distance of line of action of force from point O.

$F_\perp = F\sin\theta$ = component of $\vec{F}$ perpendicular to $\vec{r}$

SI unit of torque is N-m.

Illustration 12: Find the torque about point O and A.　　　　**(JEE MAIN)**

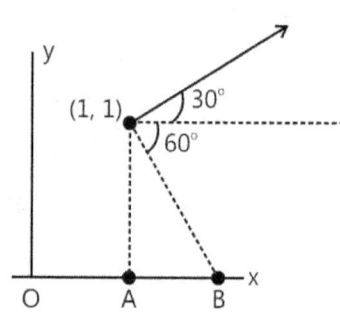

Sol: Express the position vector of A relative to O in terms of unit vectors $\hat{i}$ and $\hat{j}$. Force is given in terms of unit vectors $\hat{i}$ and $\hat{j}$.

Torque about point O, $\vec{\tau} = \vec{r_0} \times \vec{F}$, $\vec{r_0} = \hat{i} + \hat{j}$, $\vec{F} = 5\sqrt{3}\,\hat{i} + 5\hat{j}$

$\vec{\tau} = (\hat{i} + \hat{j}) \times (5\sqrt{3}\hat{i} + 5\hat{j}) = 5(1 - \sqrt{3})\hat{k}$

Torque about point A, $\vec{\tau} = \vec{r_a} \times \vec{F}$, $\vec{r_a} = \hat{j}$, $\vec{F} = 5\sqrt{3}\,\hat{i} + 5\hat{j}$

$\vec{\tau} = \hat{j} \times (5\sqrt{3}\hat{i} + 5\hat{j}) = -5\sqrt{3}\hat{k}$

Figure 7.18

Illustration 13: A particle of mass m is released in vertical plane from a point on the x – axis, it falls vertically along the y – axis. Find the torque τ about origin?

(JEE MAIN)

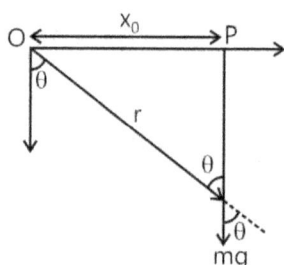

Sol: Torque is produced by the force of gravity. This will be equal to the product of force of gravity and the perpendicular distance between the line of action of force of gravity and the origin O.

$\vec{\tau} = rF\sin\theta\,\hat{k} \qquad$ Or $\qquad \tau = r_\perp F = x_0 mg$

$= r\,mg\dfrac{x_0}{r} = mgx_0\hat{k}$

Figure 7.19

6.2 Torque About An Axis

The torque of a force $\vec{F}$ about an axis AB is the component of the torque of $\vec{F}$ about point A along the axis AB.

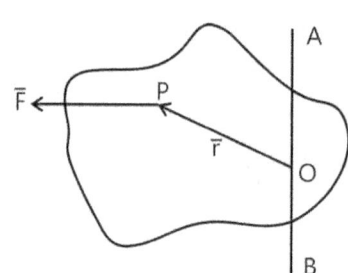

Alternatively to find torque of force $\vec{F}$ about axis AB we choose any point O on the axis AB and find the torque of $\vec{F}$ about O as $\vec{\tau}_0 = \vec{r} \times \vec{F}$. Then we calculate the component of $\vec{\tau}_0$ along AB to get $\vec{\tau}_{AB}$ (see Fig. 7.20).

There are a few special cases of torque of a force about an axis:

Case I: Applied force is parallel to the axis of rotation, i.e. $\vec{F} \parallel \overrightarrow{AB}$

Figure 7.20: Torque about an axis

Therefore torque $\vec{r} \times \vec{F}$ about any point on the axis will be perpendicular to $\vec{F}$ and hence perpendicular to $\overrightarrow{AB}$. Therefore the component of $\vec{r} \times \vec{F}$ along $\overrightarrow{AB}$ will be zero.

Case II: The line of action of the applied force intersects the axis of rotation

($\vec{F}$ intersects $\overrightarrow{AB}$)

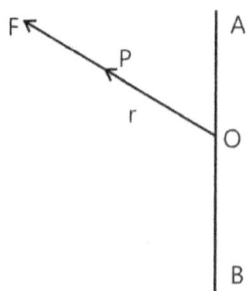

If we choose the point of intersection of line of action of $\vec{F}$ and $\overrightarrow{AB}$ as the origin O then the position vector $\vec{r}$ and applied force $\vec{F}$ will be collinear (see Fig. 7.21). Therefore the torque about O is $\vec{r} \times \vec{F} = 0$ and thus the component of this torque along line AB will also be zero.

Figure 7.21: Force intersects axis

Case III: Line of action of $\vec{F}$ and axis AB are skew and $\vec{F} \perp \overline{AB}$

Let O be the origin on the axis AB and P be the point of application of force $\vec{F}$ such that OP is perpendicular to the axis AB (see Fig. 7.22). Then torque $\vec{\tau} = \overrightarrow{OP} \times \vec{F}$ will be parallel to axis AB and the component of $\vec{\tau}$ along AB will be equal to its magnitude i.e.

$$\tau_{AB} = F \times (OP)\sin\theta = F \times l$$

where $l = (OP)\sin\theta$ is the length of the common perpendicular to the line of action of force and the axis called the lever arm or moment arm of this force.

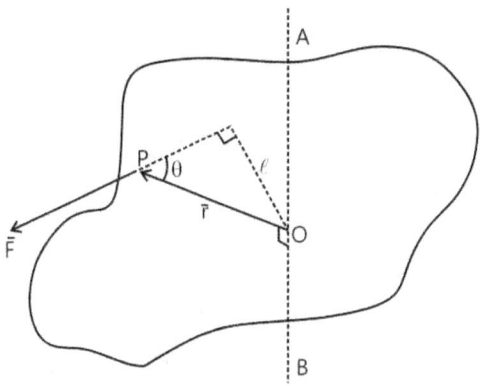

Figure 7.22: Force and axis are skew

Illustration 14: Find the torque of weight about the axis passing through point P. **(JEE MAIN)**

Sol: Required torque is equal to the product of force of gravity and the perpendicular distance between the line of action of force of gravity and the point P.

$\vec{F} = mg$ Downwards

$\vec{\tau} = \vec{F} \times \vec{r} \qquad = F.r \sin\theta$

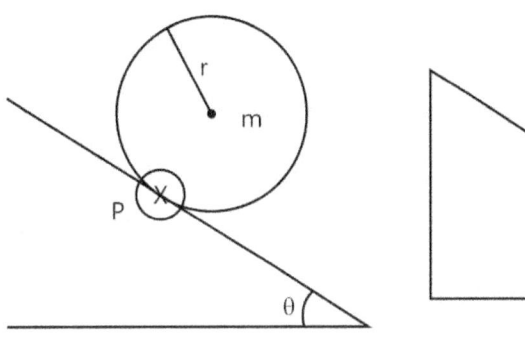

Figure 7.23

Illustration 15: A bob of mass m is suspended at point O by string of length l. Bob is moving in a horizontal circle find out. (i) Torque of gravity and tension about point O and O' (ii) Net torque about axis OO'. **(JEE ADVANCED)**

Sol: Torque of a force about an origin is equal to the product of force and the perpendicular distance between the line of action of force and the origin.

(i) Torque about point O

Torque of tension (T), $\tau_{net} = 0$ (tension is passing through point O)

Torque of gravity $\tau_{mg} = mg\,l \sin\theta$ (along negative $\hat{j}$)

Torque about point O'

Torque of gravity $\qquad \tau_{mg} = mgr = mg\,l \sin\theta$ (along negative $\hat{j}$)

Torque of tension $\qquad \tau_T = Tr \sin(90 + \theta)$ (T $\cos\theta = mg$)

$\tau_T = Tr\cos\theta = \dfrac{mg}{\cos\theta}(l\sin\theta)\cos\theta = mg\,l\sin\theta$ (along positive $\hat{j}$)

(ii) Torque about axis OO'

Torque of gravity about axis OO' $\tau_{mg} = 0$ (force mg is parallel to axis OO')

Torque of tension about axis OO' $\tau_T = 0$ (force T intersects the axis OO')

Net torque about axis OO' $\qquad \tau_{net} = 0$

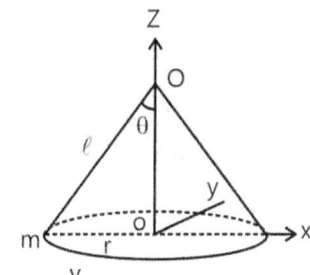

Figure 7.24

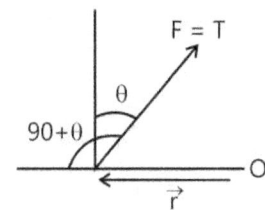

Figure 7.25

6.3 Force Couple

A pair of forces each of same magnitude and acting in opposite directions is called a force couple.

Torque due to couple = magnitude of one force x distance between their lines of action.

Magnitude of torque = τ = F (d)

A couple does not exert a net force on an object even though it exerts a torque.

Net torque due to a force couple is the same about any point (see Fig. 7.26).

Total torque about A = $x_1F + x_2F = F(x_1 + x_2) = Fd$

Total torque about B = $y_1F - y_2F = F(y_1 - y_2) = Fd$

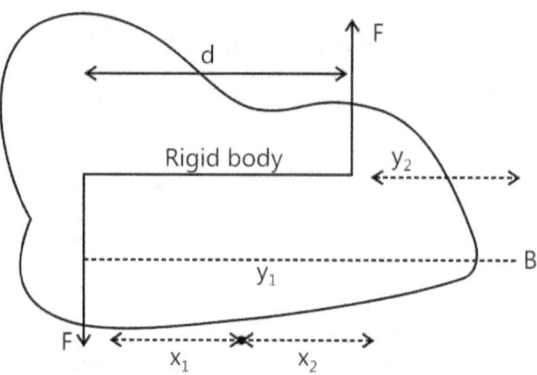

Figure 7.26: Force couple

6.4 Torque on a Rigid Body Executing Pure Rotation

Suppose I is the moment of inertia of a rigid body about the axis of rotation which is stationary in a given reference frame. The body is executing pure rotational motion about this fixed axis.

τ_{ext} = resultant torque about the axis of rotation due to all the external forces acting on the body

α = instantaneous angular acceleration of the body.

ω = instantaneous angular velocity of the body.

Consider one particle of the body say i^{th} particle of mass m_i at perpendicular distance r_i from the axis.

Radial force on the particle $F_r = m\omega^2 r$ towards the center of its circular path.

Tangential force on the particle $F_t = m_i a_t = m_i \alpha r_i$

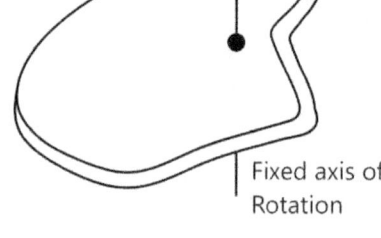

Figure 7.27: Rigid body executing pure rotation

Torque of the radial force about the axis of rotation is zero as it intersects the axis. Torque of tangential force about the axis will be,

$$\tau_i = r_i F_t = m_i r_i^2 \alpha$$

To find the total torque on the rigid body about the axis we take summation of torques acting on all the particles of the body. The total torque comes out to be equal to the resultant torque due to external forces only as the torques due to internal forces cancel each other in pairs when summation is taken on all the particles of the body (By Newton's third law of motion internal forces form pairs of equal and opposite collinear forces. So the lever arms of the forces of a pair with respect to the axis will be equal so their torques will have equal magnitude but opposite directions and cancel each other in the summation). So

$$\tau_{ext} = \sum_i \tau_i = \left(\sum_i m_i r_i^2\right) \alpha = I \alpha \qquad \qquad ...(i)$$

Remember: This formula is applicable only for pure rotational motion of a rigid body about a fixed axis.

7. KINETIC ENERGY OF BODY IN PURE ROTATION

When a rigid body performs pure rotational motion about a given axis, all of its constituent particles move in circular paths with centers on the axis and radii r_1, r_2 ...... and r_n (say), and with linear velocities $v_1 = \omega r_1$, $v_2 = \omega r_2$,.......... and $v_n = \omega r_n$. If m_1, m_2 ,...... and m_n are the masses of the particles then the total kinetic energy of the rigid body is given by

$$K = \frac{1}{2} m_1 v_1^2 + \frac{1}{2} m_2 v_2^2 + ... + \frac{1}{2} m_n v_n^2 \qquad \qquad ...(i)$$

$$= \frac{1}{2} m_1 \omega^2 r_1^2 + \frac{1}{2} m_2 \omega^2 r_2^2 + \dots + \frac{1}{2} m_n \omega^2 r_n^2$$

$$= \frac{1}{2}(m_1 r_1^2 + m_2 r_2^2 + \dots + m_n r_n^2)\omega^2$$

Now as we have learnt the term $m_1 r_1^2 + m_2 r_2^2 + \dots + m_n r_n^2$ is the moment of inertia of the rigid body.

Hence, the rotational kinetic energy of a body is given by

$$K = \frac{1}{2} I \omega^2 \qquad \qquad \dots(ii)$$

NOMORECLASS CONCEPTS

Most of the problems involving incline and a rigid body, can be solved by using the conservation of energy. Care has to be taken in writing down the total Kinetic energy. Rotational Kinetic Energy term has to be taken into consideration along with translational kinetic energy. And while writing the rotational energy, the axis about which the moment of inertia is taken should be carefully chosen.

The point about which the conservation is done should be inertial to avoid calculating the work done by pseudo forces or the point itself should be the COM so that the work done by the torque of pseudo forces would be 0.

Illustration 16: A uniform circular disc has radius R and mass m. A particle, also of mass m, is fixed at a point A on the edge of the disc as shown in Fig 7.28. The disc can freely rotate about a fixed horizontal chord PQ that is at a distance R/4 from the center C of the disc. The line CA is ⊥ to PQ. Initially the disc is held vertical with point A at its highest point. It is then allowed to fall so that it starts rotating about PQ. Find the linear speed of the particle as it reaches lowest point. **(JEE ADVANCED)**

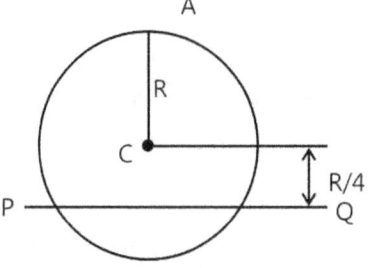

Figure 7.28

Sol: Find the moment of inertia of circular disc and the particle at point A about the chord PQ. The loss in potential energy of the system comprising the disc and the particle will be equal to the gain in its rotational kinetic energy.

$$I = \frac{1}{2} \times \frac{mR^2}{2} + m\left(\frac{R}{4}\right)^2 + m\left(\frac{5R}{4}\right)^2 = \frac{15mR^2}{8}$$

Energy equation

$$mg \frac{5R}{4} + \frac{mgR}{4} = \frac{1}{2}I\omega^2 - mg \frac{5R}{4} - \frac{mgR}{4}$$

$$\omega = 4\sqrt{\frac{g}{5R}}$$

$$V = \frac{5R\omega}{4} = \sqrt{5gR}$$

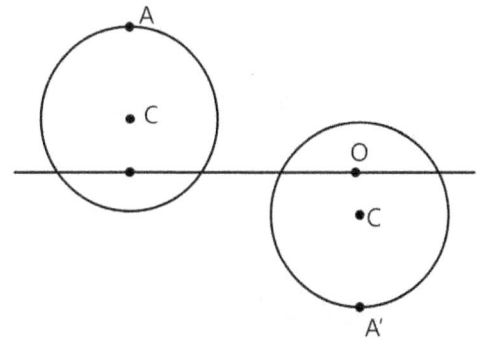

Figure 7.29

Illustrations 17: A pulley having radius r and moment of inertia I about its axis is fixed at the top of an inclined plane of inclination θ as shown in Fig 7.30. A string is wrapped round the pulley and its free end supports a block of mass m which can slide on the plane initially. The pulley is rotated at a speed ω_0 in a direction such that the block slides up the plane. Calculate the distance moved by the block before stopping? **(JEE ADVANCED)**

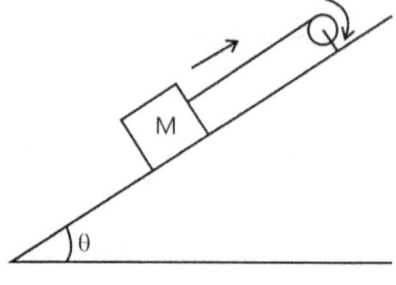

Sol: Apply Newton's second law of motion for block M along the inclined plane. Find the torque (about its axis) of force of tension acting on pulley. This will be equal to the product of moment of inertia I and the angular acceleration of pulley.

Suppose the deceleration of the block is a. The linear deceleration of the rim of the pulley is also a. The angular deceleration of the pulley is $\alpha = a/r$. If the tension in the string is T, the equations of motion are as follows:

Figure 7.30

$$mg \sin\theta - T = ma \quad \text{and} \quad Tr = I\alpha = Ia/r.$$

Eliminating T from these equations,

$$mg \sin\theta - I\frac{a}{r^2} = ma; \text{ Giving, } a = \frac{mgr^2 \sin\theta}{I + mr^2}$$

The initial velocity of the block up the incline is $v = \omega_0 r$ Thus, the distance moved by the block before stopping is

$$x = \frac{v^2}{2a} = \frac{(I + mr^2)\omega_0^2}{2mg \sin\theta}$$

Illustration 18: A uniform rod of mass m and length ℓ can rotate in vertical plane about a smooth horizontal axis hinged at point H.

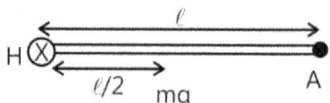

(i) Find angular acceleration α of the rod just after it is released from initial horizontal position from rest?

Figure 7.31

(ii) Calculate the acceleration (tangential and radial) of point A at this moment.

(iii) Calculate net hinge force acting at this moment.

(iv) Find α and ω when rod becomes vertical.

(v) Find hinge force when rod become vertical. **(JEE ADVANCED)**

Sol: The axis of rotation passing through H is fixed. So the torque of force of gravity about axis through H is equal to the product of moment of inertia about axis through H and angular acceleration of rod. Angular acceleration at an instant can be found if the torque of force of gravity at the instant is known.

(i) $\tau_H = I_H \alpha$

$$mg.\frac{\ell}{2} = \frac{m\ell^2}{3}\alpha \Rightarrow \alpha = \frac{3g}{2\ell}$$

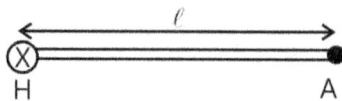

(ii) $a_{tA} = \alpha\ell = \frac{3g}{2\ell}.\ell = \frac{3g}{2}$

Figure 7.32

$a_{CA} = \omega^2 r = 0.\ell = 0$ ($\because \omega = 0$ just after release)

(iii) Suppose hinge exerts normal reaction in component form as shown

In vertical direction

$F_{ext} = ma_{CM}$

$$\Rightarrow mg - N_1 = m.\frac{3g}{4}$$

(We get the value of a_{CM} from previous example)

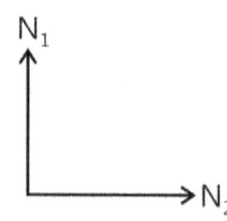

$$\Rightarrow N_1 = \frac{mg}{4}$$

Figure 7.33

3.16

In horizontal direction

$F_{ext} = ma_{CM} \Rightarrow N_2 = 0$ ($\because a_{CM}$ in horizontal = 0 as ω = 0 just after release)

(iv) Torque = 0 when rod becomes vertical so α = 0

Using energy conservation $\dfrac{mg\ell}{2} = \dfrac{1}{2}I\omega^2$ $\left(I=\dfrac{m\ell^2}{3}\right)$

(Work done by gravity when COM moves down by (½) ℓ = change in K.E.)

$$\omega = \sqrt{\dfrac{3g}{\ell}}$$

(v) When rod becomes vertical

α = 0, $\omega = \sqrt{\dfrac{3g}{\ell}}$ (Using $F_{net} = Ma_{CM}$)

$F_H - mg = \dfrac{m\omega^2\ell}{2}$ (a_{CM} = centripetal acceleration of COM)

Ans. $F_H = \dfrac{5mg}{2}$

Illustration 19: A bar of mass m is held as shown between 4 disks each of mass m' and radius r = 75 mm. Determine the acceleration of the bar immediately after it has been released from rest, knowing that the normal forces exerted on the disks are sufficient to prevent any slipping and assuming that.   (a) m = 5 kg and m' = 2 kg.

(b) The mass m' of the disks is negligible.

(c) The mass m of the bar is negligible   **(JEE ADVANCED)**

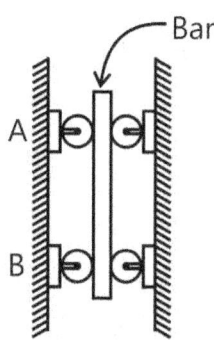

Figure 7.34

Sol: Apply Newton's second law of motion in vertical direction for the motion of center of mass of bar. Write the equation of torque due to force of friction acting on disc, for rotational motion about fixed axis through center of disk. Acceleration of rod will be equal to the tangential acceleration of the disc in the case of no slipping.

(a) Equation of center of mass of rod,

mg – 4f = ma   ....(i)

(where f is frictional force from one disk)

Torque acting on each disk due to frictional force is

$fr = \dfrac{m'r^2}{2}\dfrac{a}{r}$ ....(ii)

From (i) and (ii) we get

mg – 2m' a = ma   ....(iii)

$5g = (5 + 2 \times 2)a;\quad a = \dfrac{5g}{9}$

(b) Putting m'=0 in eqn. (iii) we get a = g

(c) Putting m = 0 in eqn. (iii) we get a = 0

(a) $\dfrac{5g}{9}$ ↓

(b) g↓   c) 0

7.1 Work Done and Power Delivered by Torque

If a torque τ rotates a body through an angle $d\theta$, the work, dW done by it is given by

$$dW = \tau . d\theta$$

The total work done W in rotating a body from the initial angle θ_1 to the final angle θ_2, is

$$W = \int_{\theta_1}^{\theta_2} \tau . d\theta = \int_{\omega_1}^{\omega_2} I \frac{d\omega}{dt} \omega dt = \int_{\omega_1}^{\omega_2} I \omega \, d\omega = \frac{1}{2} I\omega_2^2 - \frac{1}{2} I\omega_1^2$$

So the work done by torque is equal to the change in the rotational kinetic energy.

$$W = \Delta K_{rot} = \Delta\left(\frac{1}{2} I\omega^2\right) \qquad \text{...(i)}$$

This is called the **Work-Energy Theorem for rotation of rigid body**.

The rate at which work is done is called power P, given by

$$P = \frac{dW}{dt} = \tau \frac{d\theta}{dt} = \tau\omega \qquad \text{...(ii)}$$

Also, the power P delivered by the torque on the rigid body is equal to the rate of change of kinetic energy

$$K = \frac{1}{2} I\omega^2 \qquad \therefore P = \frac{dK}{dt} = \frac{d}{dt}\left(\frac{1}{2} I\omega^2\right)$$

$$\therefore P = \frac{1}{2} \times I \times 2\omega \frac{d\omega}{dt} = I \frac{d\omega}{dt} \omega = \tau \omega$$

8. EQUILIBRIUM OF RIGID BODIES

A rigid body can be in linear equilibrium as well as in rotational equilibrium. If a rigid body is in linear equilibrium, then the vector sum of all the forces acting on it should be zero.

i.e. $\Sigma \vec{F}_{ext} = 0$

Taking scalar components along the three axes x, y and z we get $\Sigma F_x = 0, \Sigma F_y = 0, \Sigma F_z = 0$

If a rigid body is in rotational equilibrium then the vector sum of all the external torques acting on it with respect to an axis in a given reference frame must be zero.

$$\Sigma \tau_{ext} = 0 \Rightarrow \Sigma \tau_x = 0, \Sigma \tau_y = 0, \Sigma \tau_z = 0$$

Illustrations 20: Two boys weighing 20 kg and 25 kg are trying to balance a seesaw of total length 4 m, with the fulcrum at the center. If one of the boys is sitting at an end, where should the other sit? **(JEE MAIN)**

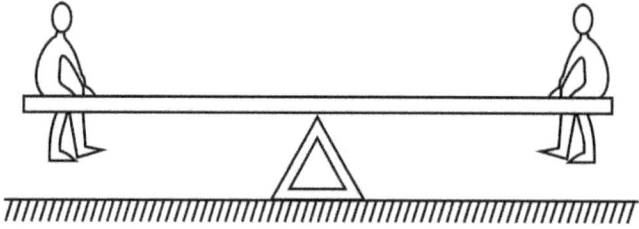

Figure 7.35

Sol: For rotational equilibrium, the net torque about the fulcrum of all the forces acting on the boys and the seesaw should be zero.

It is clear that the 20 kg kid should sit at the end and the 25 kg kid should sit closer to the center. Suppose his distance from the center is x. As the boys are in equilibrium, the normal force between a boy and the seesaw equals the weight of that boy. Considering the rotational equilibrium of the seesaw, the torque of the forces acting on it should add to zero. The forces are

(a) (25kg) g downward by the 25 kg boy

(b) (20kg) g downward by the 20 kg boy

(c) Weight of the seesaw and

(d) The normal force by the fulcrum.

Taking torques about the fulcrum.

(25 kg) g x = (20 kg) g (2 m) or x = 1.6 m

9. ANGULAR MOMENTUM

9.1 Angular Momentum of a Particle About a Point

If $\vec{p}$ is the linear momentum of a particle in a given reference frame, then angular momentum of the particle about an origin O in this reference frame is defined as

$$\vec{L} = \vec{r} \times \vec{p} \qquad \text{...(i)}$$

where $\vec{r}$ is the position vector of the particle with respect to origin O (see Fig. 7.36).

Magnitude of angular momentum is $L = rp\sin\theta$

or $L = r_\perp p$ or $L = p_\perp r$

θ = angle between vectors $\vec{r}$ and $\vec{p}$

$r_\perp$ = component of position vector $\vec{r}$ perpendicular to vector $\vec{p}$.

$p_\perp$ = component of vector $\vec{p}$ perpendicular to position vector $\vec{r}$.

SI unit angular momentum is $kg\,m^2\,s^{-1}$.

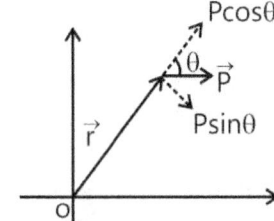

Figure 7.36: Angular momentum about a point

Relation between Torque and Angular Momentum

$\because \quad \vec{L} = \vec{r} \times \vec{p}$

Differentiating with respect to time we get

$$\frac{d\vec{L}}{dt} = \frac{d\vec{r}}{dt} \times \vec{p} + \vec{r} \times \frac{d\vec{p}}{dt} = \vec{v} \times (m\vec{v}) + \vec{r} \times \vec{F} = 0 + \vec{r} \times \vec{F} = \vec{\tau}$$

$$\Rightarrow \frac{d\vec{L}}{dt} = \vec{\tau} \qquad \text{...(ii)}$$

For a single particle moving in a circle of radius r with angular velocity ω we have

$v = \omega r$ and $p = m\omega r$

So angular momentum comes out to be $L = r\,p = mr^2\omega$

Illustration 21: A particle of mass m is projected at time t = 0 from a point O with a speed u at an angle of 45° to the horizontal. Calculate the magnitude and direction of the angular momentum of the particle about the point O at time t = u/g. **(JEE ADVANCED)**

Sol: Express the position and velocity of particle in Cartesian coordinates in terms of unit vectors $\hat{i}$ and $\hat{j}$ and then calculate the cross product in

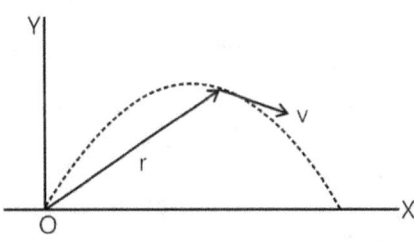

Figure 7.37

Cartesian coordinates.

Let us take the origin at O, X –horizontal axis and

Y – Axis along the vertical upward direction as shown in Fig 7.37 for horizontal during the time 0 to t.

$v_x = u\cos 45° = u/\sqrt{2}$ and $x = v_x t = \dfrac{u}{\sqrt{2}} \cdot \dfrac{u}{g} = \dfrac{u^2}{\sqrt{2}g}$

For vertical motion,

$v_y = u\sin 45° - gt = \dfrac{u}{\sqrt{2}} - u = \dfrac{(1-\sqrt{2})}{\sqrt{2}}u$

and $\quad y = (u\sin 45°)\, t - \dfrac{1}{2}gt^2 = \dfrac{u^2}{\sqrt{2}g} - \dfrac{u^2}{2g} = \dfrac{u^2}{2g}(\sqrt{2}-1)$

The angular momentum of the particle at time t about the origin is

$\vec{L} = \vec{r} \times \vec{P} = m\vec{r} \times \vec{v} = m(\hat{i}x + \hat{j}y) \times (\hat{i}v_x + \hat{j}v_y) = m(\hat{k}xv_y - \hat{k}yv_x)$

$= m\,\hat{k}\left[\left(\dfrac{u^2}{\sqrt{2}g}\right)\dfrac{u}{\sqrt{2}}(1-\sqrt{2}) - \dfrac{u^2}{2g}(\sqrt{2}-1)\dfrac{u^2}{\sqrt{2}}\right] = -\hat{k}\dfrac{mu^3}{2\sqrt{2}g}$

Thus, the angular momentum of the particle is $\dfrac{mu^3}{2\sqrt{2}g}$ in the negative z – direction i.e., perpendicular to the plane of motion, going into the plane.

Illustration 22: A cylinder is given angular velocity ω_0 and kept on a horizontal rough surface the initial velocity is zero. Find out distance travelled by the cylinder before it performs pure rolling and work by frictional force.

(JEE ADVANCED)

Sol: Due to backward slipping force of friction will act forwards. The cylinder is accelerated forwards. The torque due to friction and hence the angular acceleration is opposite to the initial angular velocity. So the angular velocity will decrease and the linear velocity of center of mass of cylinder will increase in the forward direction, till the slipping stops and pure rolling starts. The work done by frictional force is equal to change in the kinetic energy of the cylinder. The kinetic energy includes both rotational kinetic energy and translational kinetic energy.

$\mu Mg\, R = \dfrac{MR^2\alpha}{2}$

$\alpha = \dfrac{2\mu g}{R}$ ... (i)

Initial velocity u = 0

$v^2 = u^2 + 2as$

$v^2 = 2as$ ... (ii)

$f_k = ma; \quad \mu Mg = Ma; \quad a = \mu g$ ... (iii)

$\omega = \omega_0 - \alpha t$

From equation (i) $\omega = \omega_0 - \dfrac{2\mu g}{R}t; \quad V = u + at$

From equation (iii) $v = \mu g t$

$\omega = \omega_0 - \dfrac{2v}{R}; \quad \omega = \omega_0 - 2\omega; \quad \omega = \dfrac{\omega_0}{3}$

From equation (ii)

$\left(\dfrac{\omega_0 R}{3}\right)^2 = (2as) = 2\mu\, gs; \quad S = \left(\dfrac{\omega_0^2 R^2}{18\mu g}\right)$

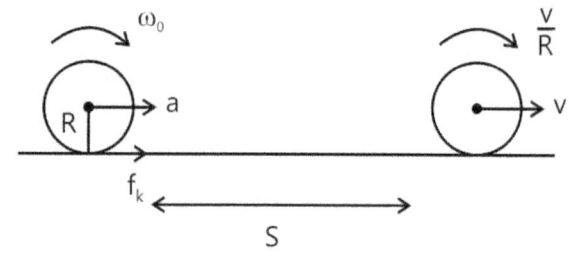

Figure 7.38

Work done by the frictional force

$$W = (-f_k R d\theta + f_k \Delta s) = -\mu mg R \Delta\theta + \frac{\mu mg \times \omega_0^2 R^2}{18\mu g};$$

$$\Delta\theta = \omega_0 \times t - \frac{1}{2}\alpha t^2 = \left\{ \omega_0 \times \left(\frac{\omega_0 R}{3\mu g}\right) \frac{1}{2} \times \frac{2\mu g}{R}\left(\frac{\omega_0 R}{3\mu g}\right)^2 \right\} = \left(\frac{\omega_0^2 R}{3\mu g} - \frac{\omega_0^2 R}{9\mu g}\right) = \frac{2\omega_0^2 R}{9\mu g}$$

$$W = \left\{ \left(-\mu mg \times R \frac{2\omega_0^2 R}{9\mu g}\right) + \left(\mu mg \times \frac{\omega_0^2 R^2}{18\mu g}\right) \right\} = -\frac{m\omega_0^2 R^2}{6}$$

Illustration 23: A hollow sphere is projected horizontally along a rough surface with speed v and angular velocity ω_0. Find out the ratio $\dfrac{v}{\omega_0}$, so that the sphere stops moving after some time. **(JEE ADVANCED)**

Sol: For the sphere to stop after sometime, the acceleration should be opposite to velocity, i.e. the force of friction should be backwards (forward slipping). Also, the torque due to friction should be opposite to angular velocity, i.e. if the torque due to friction is clockwise (see Fig. 7.39), then the initial angular velocity should be anti-clockwise.

Torque about lowest point of sphere

$$f_k \times R = I\alpha; \quad \mu mg \times R = \frac{2}{3}mR^2\alpha; \quad \alpha = \frac{3\omega g}{2R} \text{ (Angular acceleration in opposite direction of angular velocity)}$$

$$\omega = \omega_0 - \alpha t \qquad \text{(Final angular velocity } \omega = 0)$$

$$\omega_0 = \frac{3\omega g}{2R} \times t; \qquad t = \frac{\omega_0 \times 2R}{3i\,g}$$

Acceleration $a = \mu g$

$$v_t = v - at \qquad \text{(Final velocity } v_t = 0);$$

$$v = \mu g \times t; \qquad t = \frac{v}{\mu g}$$

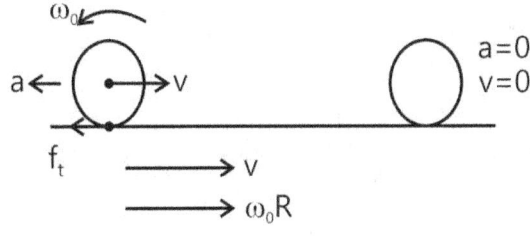

Figure 7.39

To stop the sphere, time at which v and ω are zero, should be same.

$$\frac{v}{\mu g} = \frac{2\omega_0 R}{3i\,g}; \quad \Rightarrow \frac{v}{\omega_0} = \frac{2R}{3}$$

Illustration 24: A rod AB of mass 2m and length ℓ is lying on a horizontal frictionless surface. A particle of mass m travelling along the surface hits the end of the rod with a velocity v_0 in a direction perpendicular to AB. The collision is elastic. After the collision the particle comes to rest. Find out after collision

(a) Velocity of center of mass of rod

(b) Angular velocity. **(JEE ADVANCED)**

Sol: Conserve linear momentum and angular momentum of the system constituting "the rod and the particle" before and after collision. Here the linear and angular momentum of the rod before collision is zero. Angular momenta of the rod and particle are calculated about the center of the rod.

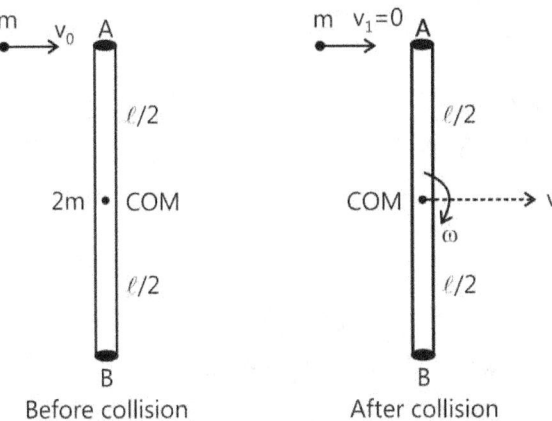

Figure 7.40

(a) Let just after collision the speed of COM of rod is v and angular velocity about COM is ω.

External force on the system (rod + mass) in horizontal plane is zero.

Apply conservation of linear momentum in x direction;

$$mv_0 = 2mv \hspace{6cm} \text{..... (i)}$$

(b) Net torque on the system about any point is zero

Apply conservation of angular momentum about COM of rod.

$$mv_0 \frac{\ell}{2} = I\omega \Rightarrow mv_0 \frac{\ell}{2} = \frac{2m\ell^2}{12}\omega \hspace{3cm} \text{...... (ii)}$$

From equation (i) velocity of center of mass $v = \dfrac{v_0}{2}$

From equation (ii) angular velocity $\omega = \dfrac{3v_0}{\ell}$.

9.2 Angular Momentum of a Rigid Body Rotating About Fixed Axis

For a system of particles the total angular momentum about an origin is the sum of the angular momenta of all the particles calculated about the same origin.

$$\vec{L} = \sum_i \vec{L}_i$$

Differentiating with respect to time we get,

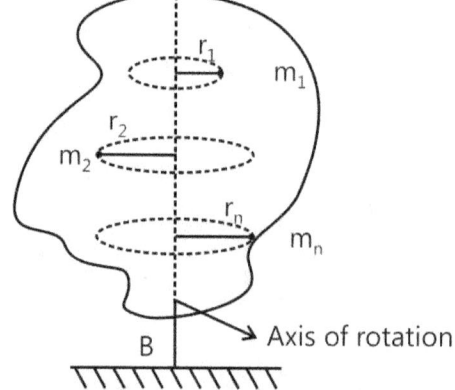

$$\frac{d\vec{L}}{dt} = \sum_i \frac{d\vec{L}_i}{dt} = \sum_i \left(\sum_k \vec{\tau}_{ik} + \vec{\tau}_i^{ext}\right) = \sum_i \sum_k \vec{\tau}_{ik} + \sum_i \vec{\tau}_i^{ext} = 0 + \vec{\tau}^{ext}$$

The double summation term corresponds to the sum of torques due to internal forces and as explained earlier, according to Newton's third law of motion these internal torques cancel out in pairs.

So for a system of particles

$$\frac{d\vec{L}}{dt} = \vec{\tau}^{ext} \hspace{4cm} \text{...(xvii)}$$

Impulse of a torque is defined as $J = \int d\vec{L} = \int \vec{\tau}^{ext} dt$

Figure 7.41: Angular momentum of rigid body

Angular momentum of a rigid body rotating about a fixed axis can be calculated as below:

Angular momenta of its individual particles about the axis are

$L_1 = m_1 r_1^2 \omega$, $L_2 = m_2 r_2^2 \omega$, $L_3 = m_3 r_3^2 \omega$, ... $L_n = m_n r_n^2 \omega$ where ω is the instantaneous angular velocity of the rigid body

Total angular momentum of the body

$$L = m_1 r_1^2 \omega + m_2 r_2^2 \omega + m_3 r_3^2 \omega \ldots\ldots\ldots + m_n r_n^2 \omega$$

$$L = \sum_i m_i (r_i)^2 \omega = I\omega$$

So $L = I\omega$

Remember: This formula is applicable only for rotation of the rigid body about a fixed axis.

Again differentiating this relation with respect to time we get,

$$\frac{dL}{dt} = I\frac{d\omega}{dt} = I\alpha = \tau_{ext}$$

Illustration 25: Two small balls of mass m each are attached to a light rod of length ℓ, one at its center and the other at its free end. The rod is fixed at the other end and is rotated in horizontal plane at an angular speed ω. Calculate the angular momentum of the ball at the end with respect to the ball at the center. **(JEE MAIN)**

Sol: Both the balls A and B have same angular velocity but different linear velocities.

The situation is shown in Fig 7.42. The velocity of the ball A with respect to the fixed end O is $v_A = \omega(\ell/2)$ and that of B with respect to O is $v_B = \omega\ell$. Hence the velocity of B with respect to A is $v_B - v_A = \omega(\ell/2)$. The angular momentum of B with respect to A is, therefore,

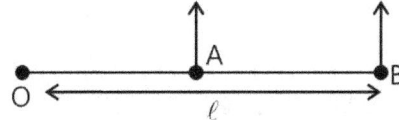

Figure 7.42

$$L = mvr = m\omega\left(\frac{\ell}{2}\right)\frac{\ell}{2} = \frac{1}{4}m\omega\ell^2$$

along the direction perpendicular to the plane of rotation.

9.3 Conservation of Angular Momentum

In the previous article we have proved the relation

$$\frac{d\vec{L}}{dt} = \vec{\tau}^{ext} \text{ where } \vec{L} \text{ and } \vec{\tau}^{ext} \text{ are evaluated about the same origin.}$$

From the above equation we see that if $\vec{\tau}^{ext} = 0$ then $\vec{L}$ of the system of particles remains constant.

In some situations the component of external torque about an axis is zero even if the net external torque is not zero. So in these cases the component of the total angular momentum, about the particular axis, remains constant.

Illustration 26: A uniform rod of mass m and length ℓ can rotate freely on a smooth horizontal plane about a vertical axis hinged at point H. A point mass having same mass m coming with an initial speed u perpendicular to the rod strikes the rod in-elastically at its free end. Find out the angular velocity of the rod just after collision?
(JEE MAIN)

Sol: After collision the rod and the particle execute pure rotational motion about vertical axis through fixed point H.

Angular momentum is conserved about H because no external force is present in horizontal plane which is producing torque about H.

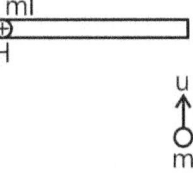

$$mul = \left(\frac{m\ell^2}{3} + m\ell^2\right)\omega \implies \omega = \frac{3u}{4\ell}$$

Figure 7.43

Illustration 27: A uniform rod of mass m_1 and length ℓ lies on a frictionless horizontal plane. A particle of mass m_2 moving at a speed v_0 perpendicular to the length of the rod strikes it at a distance $\ell/3$ from the center and stops after the collision. Calculate (a) the velocity of the center of the rod and (b) the angular velocity of the rod about its center just after the collision. **(JEE ADVANCED)**

Sol: Conserve the linear momentum of the system comprising "the rod and the particle" before and after the collision. Conserve the angular momentum, about the center of the rod, of the system comprising "the rod and the particle" before and after the collision.

The situation is shown in the Fig 7.44. Consider the rod and the particle together as the system. As there is no external resultant force, the linear momentum of the system will remain constant. Also there is no resultant external torque on the system and so the resultant external torque on the system and the angular momentum of the

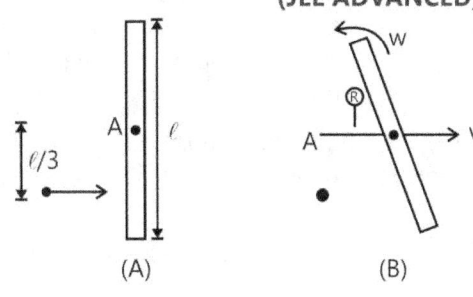

Figure 7.44

system about the line will remains constant. Suppose the velocity of the center of the rod is V and the angular velocity about the center is ω .

(a) The linear momentum before the collision is $m_2 v_0$ and that after the collision is $M_1 V$.

Thus $m_2 v_0 = m_1 V$, or $V = \left(\dfrac{m_2}{m_1}\right) v_0$

(b) Let A be the center of the rod when at rest. Let AB be the line perpendicular to the plane of the Fig 7.44. Consider the angular momentum of N "the rod plus the particle" system about AB.

Initially the rod is at rest. The angular momentum of the particle about AB is

$\quad L = m_2 v_0 (\ell/3)$

After collision the particle comes to rest. The angular momentum of the rod about a is

$\vec{L} = \vec{L}_{CM} + m_1 \vec{r_0} \times \vec{V}$

As $\vec{r_0} \parallel \vec{V}$, $\vec{r_0} \times \vec{V} = 0$ thus, $\vec{L} = \vec{L}_{CM}$

Hence the angular momentum of the rod about AB is

$L = I\omega = \dfrac{m_1 \ell^2}{12}\omega \quad$ Thus, $\dfrac{m_2 v \ell}{3} = \dfrac{m_1 \ell^2}{12}\omega \quad$ Or $\quad \omega = \dfrac{4 m_2 v_0}{m_1 \ell}$

10. RIGID BODY IN COMBINED TRANSLATIONAL AND ROTATIONAL MOTION

As discussed earlier, in this type of motion the rigid body is performing pure rotational motion about an axis and the axis itself is performing pure translational motion relative to a given reference frame.

Consider a car moving over a straight horizontal road with some instantaneous velocity v with respect to a reference frame K fixed to the road. Now let us observe the motion of a wheel of the car from the K frame. This motion of the wheel in K frame is an example of combined translational and rotational motion. Let us suppose a reference frame K' which is translating with respect to frame K with same instantaneous velocity v. In other words frame K' is rigidly fixed to the body of the car. In this frame the wheel of the car performs pure rotational motion. The body of the car itself is performing pure translational motion.

Take another example of motion of a fan fixed inside the car.

If the fan is switched off while the car is moving on the road, the motion of fan is **pure translational** with respect to K frame.

If the fan is switched on while the car is at rest, the motion of fan is **pure rotational** about its axis, as the axis is at rest in the K frame.

If the fan is switched on while the car is moving on the road, the motion of the fan with respect to K frame is neither pure translational nor pure rotational but a combination of both. Now if an observer A is sitting inside the car, as the car moves, the motion of fan will appear to him as pure rotational while the motion of the observer A with respect to K frame is pure translational. Hence in this case we can see that the motion of the fan can be resolved into two components, pure rotational motion relative to observer A and pure translational motion of observer A relative to K frame.

Such a resolution of motion of a rigid body into components of pure rotational and pure translational motion is an important tool used in the study of their dynamics.

10.1 Kinematics of a General Rigid Body Motion

For a rigid body the value of angular displacement θ, angular velocity ω, and angular acceleration α is same for all points on the rigid body. Also, if we choose any point of the rigid body as origin O and any other point, say A, of the body has a position vector $\vec{r}$ relative to O, and during any time interval the vector $\vec{r}$ rotates by an angle θ relative to its initial direction, then position vector of any other point, say B, relative to any other origin, say O', inside the rigid body will also rotate by the same angle θ. This means the angular variables θ, ω, and α do not depend on the choice of origin in the rigid body.

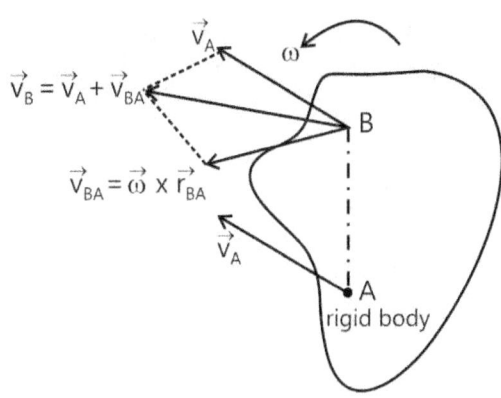

Figure 7.45: Kinematics of rigid body

The above concept is very important as it enables us to calculate the velocity of each point of the rigid body if we know the velocity of any one point (say A) in the rigid body with respect to a reference frame K and angular velocity of any point in the rigid body relative to any other point in the rigid body.

Suppose we want to calculate the velocity of a point B in the rigid body which has a position vector $\vec{r}_{BA}$ relative to A (see Fig. 7.45).

The velocity of point A is $\vec{v}_A$, so we have velocity of B as

$$\vec{v}_B = \vec{v}_A + \vec{v}_{BA} = \vec{v}_A + \vec{\omega} \times \vec{r}_{BA}$$

Direction of $\vec{u}$ is given by right hand thumb rule. If we curl the fingers of the right hand in the direction of rotation of the body, thumb gives the direction of $\vec{\omega}$.

Similarly the acceleration of point B is: $\vec{a}_B = \vec{a}_A + \vec{\alpha} \times \vec{r}_{BA}$

Illustration 28: Consider the general motion of a wheel (radius r) which can be viewed as pure translation of its center O (with the velocity v) and pure rotation about O (with angular velocity $\vec{u}$)

Find out $\vec{v}_{AO}, \vec{v}_{BO}, \vec{v}_{CO}, \vec{v}_{DO}$ and $\vec{v}_A, \vec{v}_B, \vec{v}_C, \vec{v}_D$ **(JEE MAIN)**

Sol: Express the angular velocity, linear velocity of point O and position vectors of points A, B and C relative to O in Cartesian coordinates.

$$\vec{v}_{AO} = (\vec{\omega} \times \vec{r}_{AO}) = \left(\omega(-\hat{k}) \times \vec{OA}\right)$$
$$= \left(\omega(-\hat{k}) \times r(-\hat{j})\right) = -\omega r \hat{i}$$

Similarly $\vec{v}_{BO} = \omega r (-\hat{j})$; $\vec{v}_{CO} = \omega r(\hat{i})$; $\vec{v}_{DO} = \omega r(\hat{j})$

$$\vec{v}_A = \vec{v}_O + \vec{v}_{AO} = v\hat{i} - \omega r\hat{i} ;$$
$$\vec{v}_B = \vec{v}_O + \vec{v}_{BO} = v\hat{i} + \omega r\hat{j}$$
$$\vec{v}_C = \vec{v}_O + \vec{v}_{CO} = v\hat{i} + \omega r\hat{i} ; \qquad \vec{v}_D = \vec{v}_O + \vec{v}_{DO} = v\hat{i} + \omega r\hat{j}$$

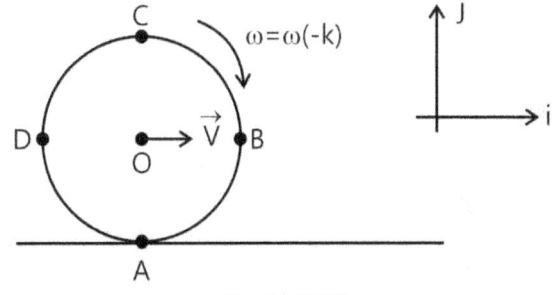

Figure 7.46

10.2 Dynamics of a General Rigid Body Motion

Combined rotation and translation of a rigid body is considered as combination of pure rotation in C frame about an axis passing through the center of mass and translation of center of mass in a reference frame K. Dynamics of combined rotational and translational motion of a rigid body in K frame is defined by two vector equations. One of them describes the dynamics of the center of mass of the rigid body in the K frame, and the other the equation of dynamics of pure rotation of the body about center of mass in the C frame.

So if the total mass of the rigid body is M and the net external force acting on it is $\vec{F}_{ext}$ then we have in the K frame,

$$M\frac{d\vec{V}_C}{dt} = \vec{F}_{ext} \qquad \qquad ...(i)$$

If I_C is the moment of inertia of the rigid body about the axis passing through center of mass and $\vec{\tau}_C$ is the net torque of all external forces about the axis passing through the center of mass, then we have in the C frame,

$$\vec{\tau}_C = I_C\frac{d\vec{\omega}}{dt} = I_C\vec{\alpha} \qquad \qquad ...(ii)$$

If $\vec{P}_{total}$ is the total linear momentum of the rigid body in the K frame, $\bar{L}_C$ is angular momentum of the body in C frame about center of mass and $\vec{r}_C$ is the position vector of center of mass relative to some origin in K frame, then we have,

$$\vec{P}_{total} = M\vec{V}_C$$

Total Kinetic energy

$$K = \frac{1}{2}MV_C^2 + \frac{1}{2}I_C\omega^2 \qquad \text{...(iii)}$$

$$\bar{L}_C = I_C\,\vec{\omega} \qquad \text{...(iv)}$$

Angular momentum in K frame = $\bar{L}_C$ about C.O.M + $\bar{L}$ of the C.O.M about some origin in K frame

$$\vec{L} = \bar{I}_C\,\vec{\omega} + \vec{r}_C \times M\vec{V}_C \qquad \text{...(v)}$$

10.3 Pure Rolling (Rolling Without Slipping)

Pure rolling is a special case of combined translational and rotational motion of a rigid body with circular cross section (e.g. wheel, disc, ring, cylinder, sphere etc.) moving on a surface. Here, there is no slipping between the rolling body and the surface at the point of contact.

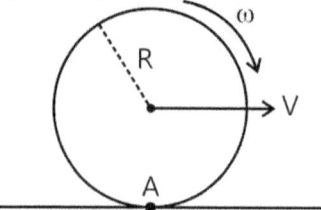

Suppose a sphere rolls on a stationary surface and the point of contact between the sphere and the surface is A (see Fig. 7.47). Let the velocity of the center of sphere be v, radius be R and its angular velocity be ω. For pure rolling the relative velocity between the point A of the sphere and the surface must be zero. As the surface is at rest, the velocity of point A is also zero.

Figure 7.47: Sphere rolling on a stationary surface.

$$\therefore v_A = v - \omega R = 0$$
$$\therefore v = \omega R$$

If sphere is rolling on a plank moving velocity v_0, then for pure rolling, $v_A = v - \omega R = v_0$ (see Fig. 7.48)

Same is true for the tangential acceleration of the point of contact in case of pure rolling.

Now let's discuss the case where a rolling cylinder of mass m moves forward on a rough plate of same mass with acceleration "a" and the rough plate moves forward with an acceleration "a_0" under action of force F on a smooth surface.

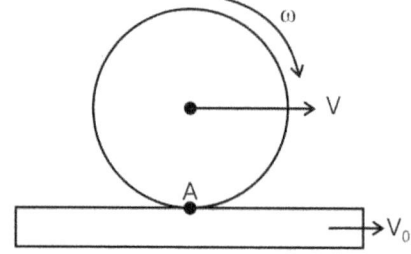

As the cylinder accelerates in the forward direction, so by Newton's second law, the friction on the cylinder at the point of contact will be in forward direction and on the plate in backward direction by Newton's third law (see Fig. 7.50).

Figure 7.48: Sphere rolling on a moving surface.

Equation of torque about center of cylinder:

$$\tau_C = fR = \frac{mR^2}{2}\alpha$$

$$\Rightarrow \alpha = \frac{2f}{mR} \qquad \text{....... (i)}$$

Equation of motion of center of cylinder:

$$f = ma \qquad \text{...... (ii)}$$

From (i) and (ii) we get

$$a = \frac{\alpha R}{2}$$

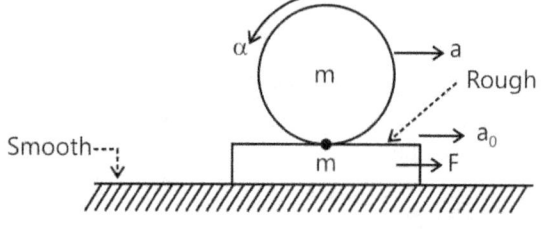

Figure 7.49: Cylinder rolling on an accelerating plate.

At contact point

$\alpha_0 = a + \alpha R = \dfrac{3\alpha R}{2} = 3a$ ...... (iii)

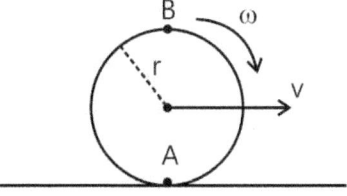

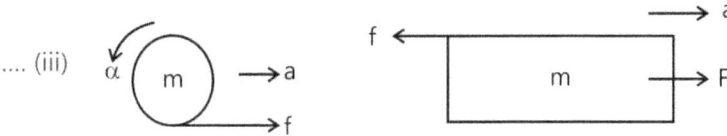

Equation of motion of plate:

$F - f = ma_0$

$F = m(a + a_0)$

Figure 7.50: (a) FBD of Cylinder. (b) FBD of Plate.

$F = 4ma$; $a = \dfrac{F}{4m}$; $a_0 = \dfrac{3F}{4m}$

Illustration 29: A wheel of radius r rolls (rolling without slipping) on a level road as shown in fig 7.51.

Find out velocity of point A and B. **(JEE MAIN)**

Sol: Linear velocity of any point on the rim of the wheel has magnitude ωr in the reference frame of center of wheel (C-frame). Velocity in ground frame is the vector sum of velocity in C-frame and the velocity of center of wheel.

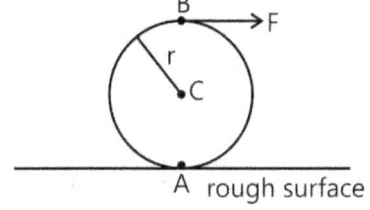

Figure 7.51

Contact point at surface is in rest for pure rolling

Velocity of point is A zero.

So $v = \omega r$

Velocity of point B $= v + \omega r = 2v$

Illustration 30: A uniform sphere of mass 200 g rolls without slipping on a plane surface so that its center moves at a speed of 2.00 cm s⁻¹. Find its kinetic energy. **(JEE MAIN)**

Sol: The kinetic energy of sphere is the sum of the translational kinetic energy and the rotational kinetic energy.

As the sphere rolls without slipping on the plane surface its angular speed about center is

$\omega = \dfrac{v_{cm}}{r}$. The kinetic energy is $K = \dfrac{1}{2} I_{cm}\omega^2 + \dfrac{1}{2}Mv_{cm}^2 = \dfrac{1}{2}\cdot\dfrac{2}{5}Mr^2\omega^2 + \dfrac{1}{2}Mv_{cm}^2$

$= \dfrac{1}{5}Mv_{cm}^2 + \dfrac{1}{2}Mv_{cm}^2 = \dfrac{7}{10} Mv_{cm}^2 = \dfrac{7}{10}(0.200\ kg)(0.02\ m\ s^{-1})^2 = 5.6 \times 10^{-5} J$

Illustration 31: A constant force F acts tangentially at the highest point of a uniform disc of mass m kept on a rough horizontal surface as shown in Fig 7.52. If the disc rolls without slipping, calculate the acceleration of the Center C and point A and B of the disc. **(JEE ADVANCED)**

Sol: Apply Newton's second law for the motion of center of mass of the disc. Find the torque of the force F and the force of friction acting on the disc at point A about the center of mass of the disc and thus obtain the equation relating the angular acceleration in the C-frame to the torques of all the external forces.

The situation is shown in Fig 7.52. As the force F rotates the disc, the point of contact has a tendency to slip towards left so that the static friction on the disc will act towards right. Let r be the radius of the disc and be the linear acceleration of the center of the disc. The angular acceleration about the center of the disc is

$\alpha = a/r$, as there is no slipping.

3.27

For the linear motion of the center,

$$F + f = ma \qquad \text{........ (i)}$$

And for the rotation motion about the center,

$$Fr - fr = I\alpha = Fr - fr = I = \left(\frac{1}{2}mr^2\right)\left(\frac{a}{r}\right) \quad \text{or} \quad F - f = \frac{1}{2}ma \qquad \text{........ (ii)}$$

From (i) and (ii),

$$2F = \frac{3}{2}ma \quad \text{or} \quad a = \frac{4F}{3m}$$

Acceleration of point A is zero

Acceleration of point B is $2a = 2\left(\frac{4F}{3m}\right) = \left(\frac{8F}{3m}\right)$.

Illustration 32: A circular rigid body of mass m, radius R and radius of gyration (k) rolls without slipping on an inclined plane of an inclination θ. Find the linear acceleration of the rigid body and force of friction on it. What must be the minimum value of coefficient of friction so that rigid body may roll without sliding?

(JEE ADVANCED)

Sol: Apply Newton's second law for the motion of center of mass of the rigid body. Find the torque of the force F and the force of friction acting on the rigid body about the center of mass of the disc and thus obtain the equation relating the angular acceleration in the C-frame to the torques of all the external forces.

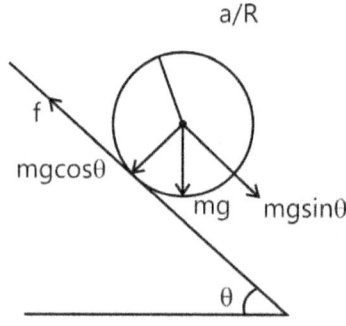

If a is the acceleration of the center of mass of the rigid body and f the force of friction between sphere and the plane, the equation of translational and rotational motion of the rigid body will be

$$Mg\sin\theta - f = ma \text{ (Translational motion)}$$

$$fR = I\alpha \text{ (Rotational motion)}$$

$$f = \frac{I\alpha}{R} \quad I = mk^2, \text{ due to pure rolling } a = \alpha R$$

$$mg\sin\theta - \frac{I\alpha}{R} = ma R = ma R + \frac{I\alpha}{R} = ma + \frac{mk^2\alpha}{R} = a\left[\frac{R^2 + k^2}{R^2}\right]$$

$$a = \frac{g\sin\theta}{\left(\frac{R^2 + k^2}{R^2}\right)} = \frac{g\sin\theta}{\left(1 + \frac{k^2}{R^2}\right)}; \qquad f = \frac{I\alpha}{R} = \frac{mk^2 a}{R^2} \Rightarrow \frac{mg\,k^2\sin\theta}{R^2 + k^2}$$

$$f \le \mu N; \qquad \frac{mk^2}{R^2}a \le \mu \le mg\cos\theta$$

$$R^2\frac{k^2}{R^2} \times \frac{g\sin\theta}{\left(k^2 + R^2\right)} \le \mu g\cos\theta; \quad \mu \ge \frac{\tan\theta}{\left[1 + \frac{R^2}{k^2}\right]}; \qquad \mu_{min} \ge \frac{\tan\theta}{\left[1 + \frac{R^2}{k^2}\right]}$$

Figure 7.53

- From above example if rigid bodies are solid cylinder, hollow cylinder, solid sphere and hollow sphere (having radius 'r' and mass 'm')

- Increasing order of acceleration

$$a_{\text{solid sphere}} > a_{\text{hollow sphere}} > a_{\text{solid cylinder}} > a_{\text{hollow cylinder}}$$

- Increasing order of required friction force for pure rolling

$$f_{\text{hollow cylinder}} > f_{\text{hollow sphere}} > f_{\text{solid cylinder}} > f_{\text{solid sphere}}$$

- Increasing order of required minimum friction coefficient for pure rolling

$$\mu_{\text{hollow cylinder}} > \mu_{\text{hollow sphere}} > \mu_{\text{solid cylinder}} > \mu_{\text{solid sphere}}$$

- I would advise you to derive these, verify and remember!

10.4 Instantaneous Axis of Rotation

The combined translational and rotational motion of a rigid body can be reduced to a purely rotational motion. When we know the velocity V_c of the center of mass and the instantaneous angular velocity ω of the body then we can find a point whose velocity comes out to be zero at a given moment of time. The axis passing through this point at the given moment is called instantaneous axis of rotation and the rigid body performs pure rotation about this axis with same angular velocity at that moment.

The position of the instantaneous axis of rotation changes with time. E.g. in pure rolling the point of contact with the surface is the instantaneous axis of rotation (see Fig. 7.54).

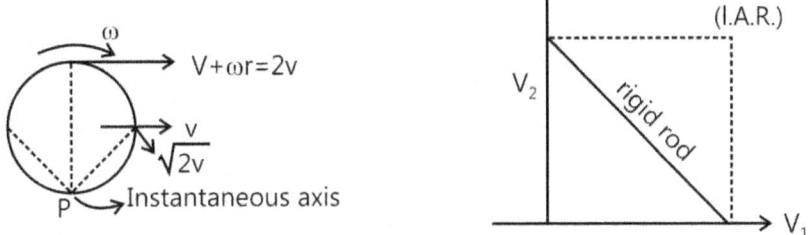

Figure 7.54: IAR (a) pure rolling; (b) Rod slipping down a wall

Geometrical construction of instantaneous axis of rotation (I.A.R.). If we know the velocity vectors of any two points in the rigid body then the I.A.R. is the axis passing through the point of intersection of the perpendiculars drawn to the velocity vectors at those points.

Once location of I.A.R is known, we find the moment of inertia of the body about this axis, and then the equations of rotation about fixed axis can be used for this axis.

Illustration 33: Prove that kinetic energy = $1/2 \ I_p\omega^2$ **(JEE MAIN)**

Sol: Kinetic energy is the sum of the translational kinetic energy and the rotational kinetic energy.

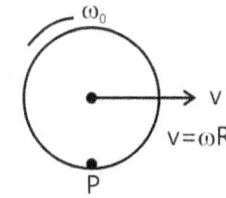

Figure 7.55

3.29

$$\text{K.E.} = \frac{1}{2} I_{cm}\omega^2 + \frac{1}{2} Mv_{cm}^2 = \frac{1}{2}I_{cm}\omega^2 + \frac{1}{2} M\omega^2 R^2 = \frac{1}{2}(I_{cm} + MR^2)\omega^2 = \frac{1}{2}\left(I_{\text{contact point}}\right)\omega^2$$

Notice that in pure rolling of uniform object, equation of torque can also be applied about the contact point.

Illustration 34: A uniform bar of length ℓ and mass m stands vertically touching a vertical wall (y – axis). When slightly displaced, its lower end begins to slide along the floor (x – axis). Obtain an expression for the angular velocity (ω) of the bar as a function of θ. Neglect friction everywhere.

(JEE ADVANCED)

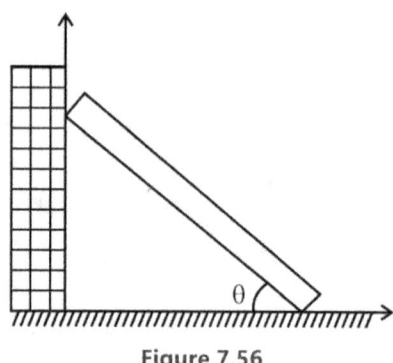

Figure 7.56

Sol: As the rod falls, it executes pure rotational motion about the instantaneous axis of rotation. The loss in gravitational potential energy is equal to the gain in the rotational kinetic energy.

The position of instantaneous axis of rotation (IAOR) is shown in Fig 7.57.

$$C = \left(\frac{\ell}{2}\cos\theta, \frac{\ell}{2}\sin\theta\right); \qquad r = \frac{\ell}{2} = \text{half of the diagonal}$$

All surfaces are smooth, therefore, mechanical energy will remain conserved.

$\therefore$ Decrease in gravitational potential energy of bar = increase in rotational kinetic energy of bar about IAOR.

$$mg\frac{\ell}{2} (1 - \sin\theta) = \frac{1}{2}I\omega^2 \qquad \text{... (i)}$$

Here, $I = \dfrac{m\ell^2}{12} + mr^2$ (about IAOR) or I

$$= \frac{m\ell^2}{12} + \frac{m\ell^2}{4} = \frac{m\ell^2}{3} \text{ Substituting in Eq. (i)}$$

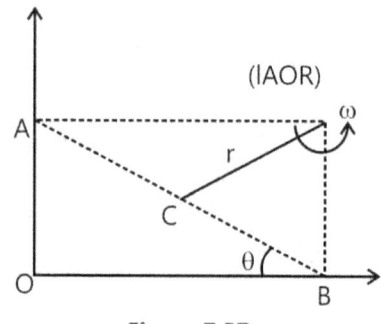

Figure 7.57

We have $mg\dfrac{\ell}{2} (1 - \sin\theta) = \dfrac{1}{2}\left(\dfrac{m\ell^2}{3}\right)\omega^2$ or $\omega = \sqrt{\dfrac{3g(1-\sin\theta)}{\ell}}$

NOMORECLASS CONCEPTS

Nature of friction for rigid bodies:

- A rigid body rolling with a speed of v and angular velocity of ω at an instant. Then it falls under one of the following cases.

Cases	Rough/Smooth	Diagram	Inference
$V < r\omega$	Rough Surface	ω, v, f_k Rough surface	1. There is relative motion at point of contact. With respect to the body the surface moves slower than itself. So the surface tries to decrease its angular velocity by a frictional force in forward direction. And this friction is kinetic friction.
			2. It increases v and decreases ω So, after sometime, v =rω and pure rolling will resume.

Cases	Rough/Smooth	Diagram	Inference
	Smooth Surface	Smooth surface	No friction is possible and it is not pure rolling.
$v > r\omega$	Rough surface	Rough surface	With respect to the COM of the cylinder, the surface moves at a higher speed than itself. So the surface tries to increase its angular velocity by exerting a frictional force in backward direction. And this friction would be kinetic friction. 2. The friction tries to reduce V and increase ω
$V > r\omega$	Smooth Surface	Smooth surface	No friction and no pure rolling.
$V = r\omega$	Rough Surface	Smooth surface	This is pure rolling. However there might be static friction acting on the body.
	Smooth Surface	Smooth surface	No friction is possible and it is pure rolling

Illustration 35: A rigid body of mass m and radius r rolls without slipping on a surface. A force is acting on the rigid body at x distance from the center as shown in Fig 7.58. Find the value of x so that static friction is zero.

(JEE MAIN)

Sol: For static friction to be zero, the linear and angular accelerations a and α caused by the force F should be related as a = α R, for rolling without slipping.

Torque about center of mass $Fx = I_{cm}\alpha$ ... (i)

$$F = ma \qquad \text{... (ii)}$$

From equation (i) & (ii) $max = I_{cm}\alpha$ (a = α R) ;

$$x = \frac{I_{cm}}{mR}$$

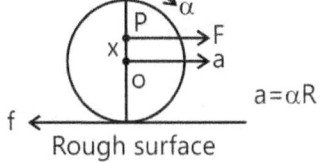

a=αR

Figure 7.58

Illustration 36: There are two cylinders of radii R_1 and R_2 having moments of inertia I_1 and I_2 about their respective axes as shown in Fig 7.59. Initially, the cylinders rotate about their axes with angular speed ω_1 and ω_2 as shown in the Fig 7.59. The cylinders are moved close to touch each other keeping the axes parallel. The cylinders first slip over each other at the contact but the slipping finally ceases due to the friction between them. Calculate the angular speeds of the cylinders after the slipping ceases. **(JEE ADVANCED)**

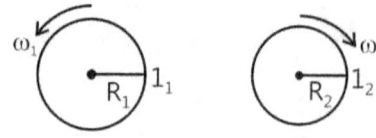

Figure 7.59

Sol: The force of friction acting on the cylinder moving faster will be such that its angular velocity decreases. The force of friction acting on the cylinder moving slower will be such that its angular velocity increases. When slipping ceases, the linear speeds of the points of contact of the two cylinders will be equal.

If ω'_1 and ω'_2 be the respective angular speeds at the instant slipping ceases, we have

$$\omega'_1 R_1 = \omega'_2 R_2 \qquad \qquad \qquad ...(i)$$

The change in the angular speed is brought about by the frictional force which acts as long as the slipping exists. If this force f acts for a time t. the torque on the first cylinder is fR_1 and that on the second is fR_2. Assuming $\omega_1 > \omega_2$. The corresponding angular impulses are $- fR_1 t$ and $fR_2 t$,

We therefore, have

$$- fR_1 t = I_1(\omega'_1 - \omega_1) \text{ and } fR_2 t = I_2(\omega'_2 - \omega_2)$$

$$\text{or } -\frac{I_1}{R_1}(\omega'_1 - \omega_1) = \frac{I_2}{R_2}(\omega'_2 - \omega_2) \qquad \qquad ...(ii)$$

Solving (i) and (ii) $\omega'_1 = \dfrac{I_1\omega_1 R_2 + I_2\omega_2 R_1}{I_2 R_1^2 + I_1 R_2^2} R_2$ and $\omega'_2 = \dfrac{I_1\omega_1 R_2 + I_2\omega_2 R_1}{I_2 R_1^2 + I_1 R_2^2} R_1$.

10.5 Energy Method in Solving Problems of Rolling Body

We can conserve energy in case of pure rolling of a rigid body because the point of contact between the surfaces remains at rest and so the frictional forces acting at the point of contact do not do any work. Thus only conservative force do work on the body.

Thus Potential energy + total K.E. = constant

As shown in the Fig. 7.60, a disc is rolling down on an inclined plane. Then we can conserve total mechanical energy. If the disc falls a height h then loss in potential energy is equal to gain in kinetic energy.

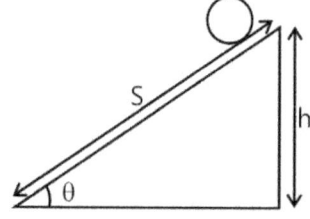

Figure 7.60

$$Mgh = \frac{1}{2}I\omega^2 + \frac{1}{2}MV_C^2 \qquad \qquad ... (i)$$

Its total kinetic energy $= \frac{1}{2}MV_C^2 + \frac{1}{2}I\omega^2 = \frac{1}{2}MV_C^2\left(1 + \frac{K^2}{R^2}\right) \qquad ...(ii)$

where K is the radius of gyration of the disc and V_C the velocity of center of mass.

$$\text{So } \frac{1}{2}MV_C^2\left(1 + \frac{K^2}{R^2}\right) = Mgh; \quad V_C^2 = \frac{2gh}{\left(1 + \dfrac{K^2}{R^2}\right)}$$

Thus the velocity of center of mass of a body rolling down an inclined plane is given by

$$V_c = \frac{\sqrt{2gh}}{\left(1 + \dfrac{K^2}{R^2}\right)^{1/2}}$$

If a_c is linear acceleration of center of mass down this plane, and distance covered on the plane is s, then if the body starts from rest we have

$$V_c^2 = 2a_c s \therefore a_c = \frac{V_c^2}{2s} = \frac{2gh}{\left(1 + \dfrac{K^2}{R^2}\right) \times 2 \times \dfrac{h}{\sin\theta}} \quad \text{or} \quad a_c = \frac{g\sin\theta}{1 + \dfrac{K^2}{R^2}}$$

NOMORECLASS CONCEPTS

Rather than going in a conventional way, using this method greatly simplifies our effort. But take care while writing the kinetic energy!

Illustration 37: A solid sphere is released from rest from the top of an incline of inclination θ and length ℓ. If the sphere rolls without slipping. What will be its speed when it reaches the bottom? **(JEE MAIN)**

Sol: The loss in the gravitational potential energy of the solid sphere is equal to the gain in the kinetic energy. The kinetic energy of the sphere comprises the rotational kinetic energy as well as the translational kinetic energy.

Let the mass of the sphere be m and its radius be r. Suppose the linear speed of the sphere when it reaches the bottom is v. As the sphere rolls without slipping, its angular speed ω about its axis is v/r. The kinetic energy at the bottom will be

$$K = \frac{1}{2}I\omega^2 + \frac{1}{2}mv^2 = \frac{1}{2}\left(\frac{2}{5}mr^2\right)\omega^2 + \frac{1}{2}mv^2 = \frac{1}{5}mv^2 + \frac{1}{2}mv^2 = \frac{7}{10}mv^2$$

This should be equal to the loss of potential energy $mg\,\ell\sin\theta$. Thus

$$\frac{7}{10}mv^2 = mg\ell\sin\theta \qquad \text{Or} \quad v = \sqrt{\frac{10}{7}g\ell\sin\theta}\,.$$

11. TOPPLING

When an external force is applied to the upper edge of a body with a flat base to cause it to slide along a surface, the body may topple before sliding starts. Toppling is more likely to happen when the width of the base of the body is small.

Toppling occurs due to the turning effect of torques of applied force at the upper edge and frictional force at the base.

Let the surface be quite rough and the force F is applied at height h above the base of the block as shown in Fig. 7.61. Width of the base is b. The static friction at the base is f = F. The normal reaction is N = mg. The couple of forces F and f try to topple the block about point S. To cancel the effect of this unbalanced torque the normal reaction N shifts towards S by a distance x so that torque of N counter balances torques of F and f.

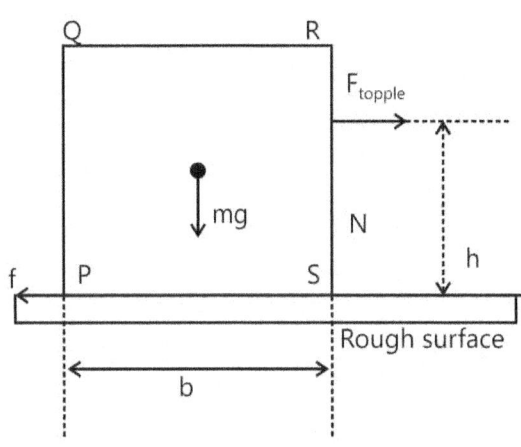

Figure 7.61: Block toppling on rough surface

3.33

$$Fh = (mg)x \qquad \text{or} \qquad x = \frac{Fh}{mg}$$

If F or h or both increase, distance x also increases, but it cannot go beyond the maximum value of $x_{max} = b/2$ i.e in extreme case N passes through edge S. If F is further increased block will topple.

So, $\quad F_{topple} = \dfrac{mgb}{2h}$

Here we assumed that the surface is sufficiently rough so that sliding starts only when

$F = f_{max} = \mu mg > F_{topple}$ or $\mu > \dfrac{b}{2h}$ (toppling before sliding)

If surface is not sufficiently rough, the body slides before F is increased to F_{topple} i.e. the body will slide before toppling. This is the case when

$F = f_{max} = \mu mg < F_{topple}$ or $\mu < \dfrac{b}{2h}$

Illustration 38: A uniform cube of side 'a' and mass m rests on a rough horizontal table. A horizontal force F is applied normal to one of the faces at a point directly below the center of the face, at a height $\dfrac{a}{4}$ above the base.

(i) What is the minimum value of F for which the cube begins to tip about an edge?

(ii) What is the minimum value of μ_s so that toppling occurs?

(iii) If $f_1 = \mu_{min}$, find minimum force for toppling.

(iv) Minimum μ_s so that F_{min} can cause toppling. **(JEE ADVANCED)**

Sol: For part (i) we consider toppling before sliding. The normal reaction will pass through the edge. In part (ii) it is not mentioned whether the toppling occurs before sliding or sliding occurs before toppling. So the toppling will occur for any value of μ_s, sliding or no sliding. Part (iii) is same as part (i). Part (iv) is the case of toppling before sliding.

(i) In the limiting case normal reaction will pass through O. The cube will tip about O if torque of F about O exceeds the torque of mg.

Hence, $F\left(\dfrac{a}{4}\right) > mg\left(\dfrac{a}{2}\right)$ or $F > 2\,mg$

Therefore, minimum value of F is 2 mg.

(ii) In this case since it is not acting at COM, toppling can occur even after body started sliding even if there is no friction by increasing the torque of F about COM. Hence $\mu_{min} = 0$.

(iii) Now body is sliding before toppling. O is not I.A.R., torque equation cannot be applied across it. It can be applied about COM.

$F \times \dfrac{a}{4} = N \times \dfrac{a}{2}$ ... (i)

$N = mg$ ... (ii)

From (i) and (ii) -> F = 2 mg

(iv) $F > 2\,mg$ .... (i) (From sol. (i))

$N = mg$ ... (ii)

$F = \mu_s N = \mu_s mg$ ... (iii)

From (i) and (iii) $\mu_s = 2$

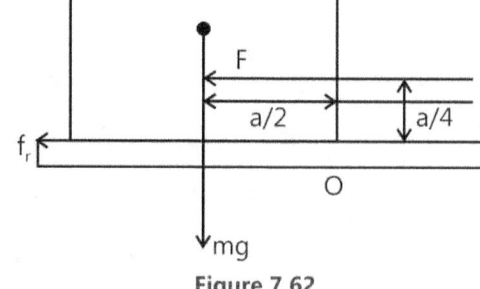

Figure 7.62

Illustration 39: Find minimum value of ℓ so that truck can avoid the dead end, without toppling the block kept on it.

(JEE ADVANCED)

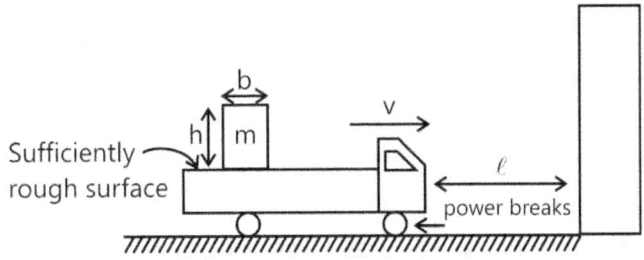

Figure 7.63

Sol: The block kept on truck will experience pseudo force in forward direction and friction force due to the floor of the truck in backward direction. We assume the case of toppling before sliding. In extreme case the normal reaction N = mg will pass through the edge.

$$ma\frac{h}{2} \le mg\frac{b}{2} \implies a \le \frac{b}{h}g$$

Final velocity of truck is zero. So that $0 = v^2 - 2(\frac{b}{h}g)\ell$

$$\ell = \frac{h}{2b}\frac{v^2}{g}$$

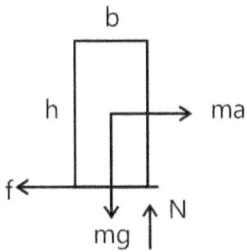

Figure 7.64

PROBLEM-SOLVING TACTICS

- Most of the problems involving incline and a rigid body can be solved by using conservation of energy during pure rolling. In case of non-conservative forces, work done by them also has to be taken into consideration in the equation. Care has to be taken in writing down the Kinetic energy. Rotational Kinetic Energy term has to be taken into consideration. And while writing the rotational energy, the axis about which the moment of inertia is taken should pass through the COM.

- The motion of a body in pure rolling can be viewed as pure rotation about the bottommost point of the body or the point of contact with the ground. Hence an axis passing through the point of contact and tangential to the point would be the Instantaneous axis of rotation. So problems on pure rolling can be solved easily by using the concept of instantaneous axis of rotation.

- Problems on toppling can be easily solved by writing the moments on the body and visualizing them as forces acting on the body. If the net moment is tending to stabilize the body, then the body doesn't topple. For any condition else it may get toppled.

- Problems which include the concept of sliding and rolling can be solved easily by using the concept of conservation of angular momentum. But care has to be taken in selecting the proper axis so that net moment about that axis vanishes.

FORMULAE SHEET

S. No	Term	Description	Linear Motion	Rotational motion & relation
1	Displacement	Displacement (linear or angular) is the physical change in the position of the body when a body moves linearly or angular in position. **(a)** The linear displacement Δs is difference between final and initial position measured in linear direction. **S.I. unit**: meter **m** **(b)** The angular displacement of the body while rotating about a fixed axis is the displacement $\Delta \theta$ it swept out with respect to its initial position in sense of rotation. It can be positive (anti clockwise) or negative (clockwise) **S.I. unit**: radians **rad**,	s	θ $(s = r\theta)$
2	Velocity	Velocity of any moving object is the time rate of change of position. The velocity is the vector quantity. Linear velocity is in the plane of motion. Angular velocity can be positive or negative & its direction is perpendicular to the plane of rotation Linear velocity is categorized as - Average velocity= $\Delta s / \Delta t$ - Instantaneous velocity= ds/dt. **S.I. unit: m/s** Angular velocity is categorized as - Average angular velocity $\Delta \theta / \Delta t$ - Instantaneous angular velocity $\omega = d\theta / dt$ **S.I. unit: rad/s**	$v = \dfrac{ds}{dt}$	$\omega = \dfrac{d\theta}{dt} \ (v = r\omega)$
3	Acceleration	Acceleration is the time rate change of velocity of a body. It's a vector quantity. Linear acceleration can be positive or negative and related to direction of motion. Linear acceleration is categorized as - Average acceleration= $\Delta v / \Delta t$ - Instantaneous acceleration = dv/dt.	$a = \dfrac{dv}{dt}$	$\alpha = \dfrac{d\omega}{dt} \ (a = r\alpha)$

S. No	Term	Description	Linear Motion	Rotational motion & relation
		S.I. unit: m/s^{-2} Angular acceleration is categorized as - Average angular acceleration $\Delta\omega / \Delta t$ - Instantaneous angular acceleration $\alpha = d\omega / dt$ S.I. unit: rad/s^{-2}		
4	Mass	Mass is the basic entity of any body by virtue of which the body gains weight. In linear kinematics the mass of whole body is constant. **S.I. unit:** kilogram **kg** In angular kinematics mass of body is distributed among various tiny rigid points so mass is measured about inertia of rotating body- moment of inertia **I**	M	I (I = $\sum mr^2$)
5	Momentum	Momentum of body is product of mass and its velocity of motion. It's a vector quantity. Linear momentum= mv **S.I. unit: kg m/s** Angular momentum of body is a vector in direction perpendicular to plane of rotation given by $\vec{L}$ **S.I. unit: kg m²/s**	p = mv	$\vec{L} = I$ $\vec{L} = \vec{r} \times \vec{p}$
6	Impulse	Impulse is the product of force and time period And it is categorized as -Linear impulse -Angular impulse	$\int F\, dt$	$\int \tau dt$
7	Force (Newton's second law of motion)	From the newton second law of motion, force is time rate of change of momentum. It's a vector quantity. Linear force F= $\dfrac{dp}{dt}$ = ma **S.I. unit:** Newton **N** Angular force $\vec{\tau} = I \times \vec{\alpha}$ **Laws of conservation of momentum** - Linear momentum is said to be conserved if $\dfrac{dp}{dt}$ = 0, than P remains constant - Angular momentum is said to be conserved if $\dfrac{d\vec{L}}{dt}$ = 0 than L remains constant	F = ma If = 0 the body is in equilibrium with its surrounding	$\vec{\tau} = r \times \vec{F} = I \times \vec{\alpha}$ $= \dfrac{d\vec{L}}{dt}$ If = 0 the body is in equilibrium with its surrounding

S. No	Term	Description	Linear Motion	Rotational motion & relation
8	Work	Work is the product of displacement of body under action of external applied force.	$W = \int F\,ds$	$W = \int \tau d\theta$
9	Power	Power is the time rate change of work done	$P = F$	$P = \tau\,\omega$
10	Kinetic energy	The phenomenon associated with the moving bodies	$K.E._{tran} = \frac{1}{2}mv^2$	$K.E._{rot} = \frac{1}{2}I\omega^2$
11	Kinematics of Motion	Kinematical equation are the interrelation of displacement, velocity, acceleration and time and are categorized as follows: -Linear kinematical equation -Angular kinematical equation	$v = u + at$ $s = ut + \frac{1}{2}at^2$ $v^2 = u^2 + 2as$	$\omega = \omega_0 + \alpha t$ $\theta = \omega_0 t + \frac{1}{2}\alpha t^2$ $\omega^2 = \omega_0^2 + 2\alpha\theta$
12	Parallel Axis Theorem	$I_{xx} = I_{cc} + Md^2$ where I_{cc} is the moment of inertia about the center of mass		
13	Perpendicular Axis Theorem	$I_{xx} + I_{yy} = I_{zz}$ It is valid for plane laminas only.		
14	Work energy principle	Work energy principle is used to determine the change in the kinetic energy of moving body	$W = \frac{1}{2}mv^2 - \frac{1}{2}mu^2$	$W = \frac{1}{2}I\omega^2 - \frac{1}{2}I\omega_0^2$

Solved Examples

JEE Main/Boards

The first five Examples discussed below show us the strategy to tackle down any problem in the rigid body motion. Hence follow them up properly! They may be lengthy but are very learner friendly!!

Example 1: A person of mass M is standing on a railroad car, which is rounding an unbanked turn of radius at speed v. His center of mass is at a height of L above the car midway between his feet, which are separated by a distance of d. The man is facing the direction of motion. What is the magnitude of the normal force on each foot?

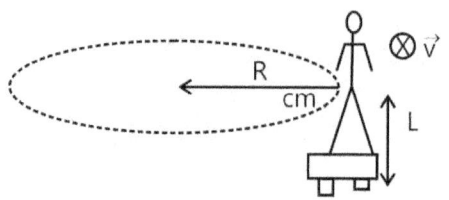

Sol: The frictional forces acting on the feet of man will provide the necessary centripetal acceleration to move in a circular path. Apply the Newton's second law of motion at the center of mass of the man to get the equation of motion along the circular path. In the vertical plane the man is in rotational and translational equilibrium under the action of its weight acting vertically downwards and the normal reactions at its feet acting vertically upwards. Get one equation each for rotational and translational equilibrium in vertical plane.

$$\tau_{cm2} = -\frac{d}{2}N_2 + Lf_2 \quad \text{(clockwise)}$$

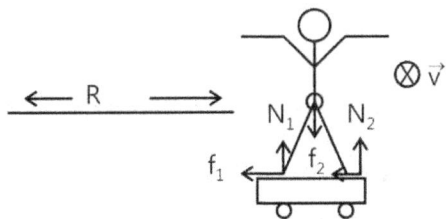

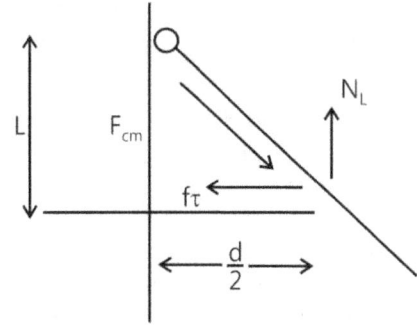

We draw the free body diagram of the man, as shown in figure.

Static friction $\vec{f}_1$ and a normal reaction $\vec{N}_1$ is acting on the inner foot. Static friction $\vec{f}_2$ and normal reaction $\vec{N}_2$ is acting on the outer foot. We do not assume the limiting value of frictional forces. The weight of the man acts at its center of mass.

As the man is moving in a circular path with speed, by Newton's Second Law the forces of friction should act towards the center of the circular path.

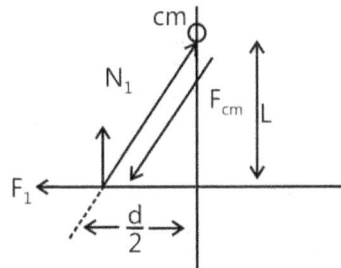

$$f_1 + f_2 = m\frac{v^2}{R} \qquad \text{....(i)}$$

For vertical equilibrium we should have

$$N_1 + N_2 - mg = 0$$

or $\quad N_1 + N_2 = mg$ \qquad ....(ii)

For rotational equilibrium of the man about its center of mass we have $\quad \vec{\tau}_{cm}^{total} = 0$

The gravitational force does not contribute to the torque about center of mass because it is acting at the center of mass itself. We draw a torque diagram in the figure showing the line of action of the forces at the inner foot.

The torque on the inner foot about COM is given by

$$\tau_{cm1} = \frac{d}{2}N_1 + Lf_1 \quad \text{(clockwise)}$$

We draw a similar torque diagram for the forces at the outer foot.

The torque on the outer foot about COM is given by

Both these torques about the center of mass must add up to zero.

Therefore

$$(\frac{d}{2}N_1 + Lf_1) + (-\frac{d}{2}N_2 + Lf_2) = 0$$

$$\frac{d}{2}(N_1 - N_2) + L(f_1 + f_2) = 0 \qquad \text{....(iii)}$$

Putting (i) in (iii) we get,

$$\frac{d}{2}(N_1 - N_2) + Lm\frac{v^2}{R} = 0$$

or $\quad N_2 - N_1 = \dfrac{2Lmv^2}{Rd}$ \qquad ....(iv)

Solving (ii) and (iv) we get

$$N_1 = \frac{1}{2}\left(mg - \frac{2Lmv^2}{Rd}\right) \qquad \text{....(v)}$$

$$N_2 = \frac{1}{2}\left(\frac{2Lmv^2}{Rd} + mg\right) \qquad \text{.... (vi)}$$

Example 2: A Yo-Yo of mass m has an axle of radius b and a spool of radius R. Moment of inertia about the center can be taken to be $I_{cm} = (1/2)MR^2$ and the thickness of the string can be neglected. The Yo-Yo is released from rest. You will need to assume that the center of mass of the Yo-Yo descends vertically, and that the string is vertical as it unwinds.

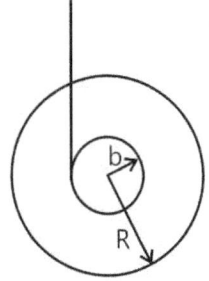

(a) What is the tension in the cord as the Yo-Yo descends?

(b) What is the magnitude of the angular acceleration as the Yo-Yo descends and the magnitude of the linear acceleration?

(c) Find the angular velocity of the Yo-Yo when it reaches the bottom of the string when a length L of the string has unwound.

Sol: Apply the Newton's second law of motion at the center of mass of Yo-Yo to get the equation of motion along the vertical direction. Get the relation between net torque, of all the external forces acting on Yo-Yo, and its moment of inertia, both these quantities calculated about the axis passing through the center of mass of Yo-Yo. As the Yo-Yo descends, the loss in the gravitational potential energy is equal to the gain in the translational and rotational kinetic energy.

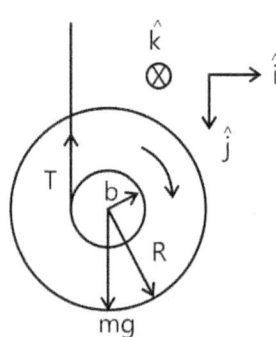

(a) The torque of tension in the cord about the center of mass of the Yo-Yo is in the clockwise direction. So as the Yo-Yo descends with linear acceleration a_{cm}, it rotates in the clockwise direction with angular acceleration α.

$$\tau_{cm} = bT = I_{cm}\alpha \text{ (clockwise)} \qquad ....(i)$$

Applying Newton's Second Law for the motion of COM in the vertical direction,

$$Mg - T = Ma_{cm} \qquad ....(ii)$$

As the string is stationary, and the Yo-Yo does not slip on the string, the angular acceleration and the linear acceleration of COM are related by the constraint condition,

$$a_{cm} - b\alpha = 0 \Rightarrow a_{cm} = b\alpha \qquad ....(iii)$$

From (ii) and (iii) we get,

$$Mg - T = Mb\alpha \qquad ....(iv)$$

Eliminating α from (i) and (iv) we get

$$Mg - T = \frac{Mb^2T}{I_{cm}}$$

or

$$T = \frac{Mg}{\left(1 + \frac{Mb^2}{I_{cm}}\right)} = \frac{Mg}{\left(1 + \frac{Mb^2}{(1/2)MR^2}\right)} = \frac{Mg}{\left(1 + \frac{2b^2}{R^2}\right)} \qquad ....(v)$$

(b) Substitute Eq. (v) into Eq. (i) to determine the angular acceleration

$$\alpha = \frac{bT}{I_{cm}} = \frac{2bg}{(R^2 + 2b^2)} \qquad ....(vi)$$

From (iii) and (vi) we get

$$a_{cm} = b\alpha = \frac{2b^2g}{(R^2 + 2b^2)} = \frac{g}{1 + (R^2/2b^2)} \qquad ....(vii)$$

For a typical Yo-Yo, the acceleration is much less than that of an object in free fall.

(c) Use conservation of energy to determine the angular velocity of the Yo-Yo when it reaches the bottom of the string (Tension force does not perform any work because point of contact between string and Yo-Yo is always at rest).

Loss in gravitational potential energy = Gain in kinetic energy MgL

$$= MgL = \frac{1}{2}Mv_{cm}^2 + \frac{1}{2}I_{cm}\omega^2 = \frac{1}{2}M(v_{cm}^2 + \frac{1}{2}R^2\omega^2) \qquad ....(viii)$$

Linear velocity of COM and angular velocity are related by the constraint condition,

$$v_{cm} - b\omega = 0 \Rightarrow v_{cm} = b\omega \qquad ....(ix)$$

Solving (viii) and (ix) for ω, we get $\quad \omega = \sqrt{\dfrac{4gL}{(2b^2 + R^2)}}$

Example 3: A uniform cylinder of radius R and mass M with moment of inertia about the center of mass $I_{cm} = (1/2)MR^2$ starts rolling due to the mass of the cylinder, and has dropped a vertical distance h when it reaches the bottom of the incline. Let g denote the gravitational constant. The coefficient of static friction between the cylinder and the surface is μ. The cylinder rolls without slipping down the incline. The goal of this problem is to find the magnitude of the velocity of the center of mass of the cylinder when it reaches the bottom of the incline.

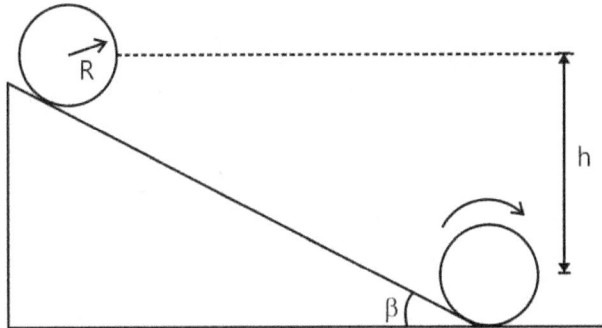

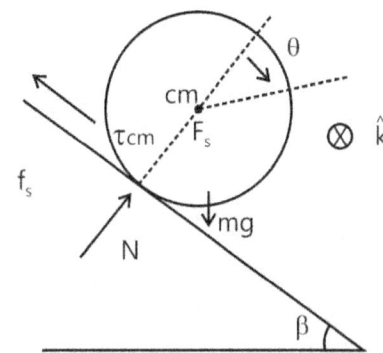

Sol: This problem can be solved either by applying law of conservation of mechanical energy, or by applying Newton's laws of motion.

We shall solve this problem in three different ways.

1. Applying the torque equation about the center of mass and the force equation for the center of mass motion.

2. Applying the energy equation.

3. Using torque about a fixed point that lies along the line of contact between the cylinder and the surface.

Applying the torque equation about the center of mass and the force equation for the center of mass motion.

We will find the acceleration and hence the speed at the bottom of the incline using kinematics. A figure showing the force is shown below.

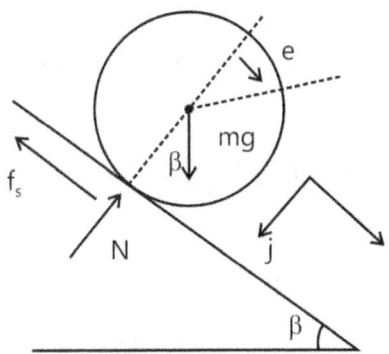

Choose x = 0 as the point where the cylinder just starts to roll. With the unit vectors shown in the figure above, Newton's second Law, applied in the x – and y – directions in turn, yields

$$Mg \sin \beta - f_s = Ma_x \qquad \text{...(i)}$$

$$-N + Mg \cos\beta = 0 \qquad \text{...(ii)}$$

Choose the center of the cylinder to compute the torque about (see figure below).

Then the only force exerting a torque about the center of mass is the friction force, and so we have $f_s R = I_{cm}\alpha_z$

$$\qquad \text{...(iii)}$$

Use $I_{cm} = (1/2) M R^2$ and the kinematic constraint for the no-slipping condition $\alpha_z = a_x/R$ in Eq. (xxxiv) to solve for the magnitude of the static friction force yielding

$$f_x = (1/2)Ma_x \qquad \text{...(iv)}$$

Substituting Eq. (iv) into Eq. (v)

$$Mg \sin \theta = (1/2) M a_x = M a_x \qquad \text{...(v)}$$

Which we can solve for the acceleration

$$a_x = \frac{2}{3} \sin \beta \qquad \text{...(vi)}$$

The displacement of the cylinder is $x_f = h/\sin \beta$ in the time it takes the bottom, t_f. The x – component of the velocity v_x at the bottom is $v_{x.f} = a_{x.f}$. The displacement in the time interval t_f satisfies $x_f = (1/2) a_x t_f^2$. After eliminating t_f, we have $x_f = v_{x.f}^2/2a_x$, so the magnitude of the velocity when the cylinder reaches the bottom of the inclined plane is

$$\begin{aligned} v_{x.f} &= \sqrt{1a_x x_f} \\ &= \sqrt{2((2/3)g \sin \beta)(h/\sin\beta)} = \sqrt{(4/3)gh} \end{aligned}$$

$$\qquad \text{...(viii)}$$

Note that if we substitute Eq. (vi) into Eq. (iv) the magnitude of the friction force is

$$fs = (1/3) Mg \sin\beta \qquad \text{...(ix)}$$

In order for the cylinder to roll without slipping.

$$f_s \leq \mu_s Mg \cos \beta \qquad \text{...(x)}$$

So combining Eq. (ix) and Eq. (x) we have the condition that

$$(1/3) Mg \sin \beta \leq \mu_s Mg \cos\beta \qquad \text{...(xi)}$$

Thus in order to roll without slipping the coefficient of static friction must satisfy

$$\mu_s \geq \frac{1}{3}\tan\beta \qquad \text{...(xii)}$$

Applying the energy equation

We shall use the fact that the energy of the cylinder-earth system is constant since the static frictional force does no work. Choose a zero reference point for potential energy at the center of mass when the cylinder reaches the bottom of the incline plane.

Then the initial potential energy is $\quad U_t = Mgh \quad$ ...(xiii)

$Mg - N = 0$

For the given moment of inertia, the final kinetic energy is

$$K_f = \frac{1}{2}Mv_{x.f}^2 + \frac{1}{2}I_{cm}\omega_{z.f}^2$$

$$= \frac{1}{2}Mv_{x.f}^2 + \frac{1}{2}(1/2)MR^2(v_{x.f}/R)^2 \qquad \text{... (xiv)}$$

$$= \frac{3}{4}Mv_{x.f}^2$$

Setting the final kinetic energy equal to the initial gravitational potential energy leads to

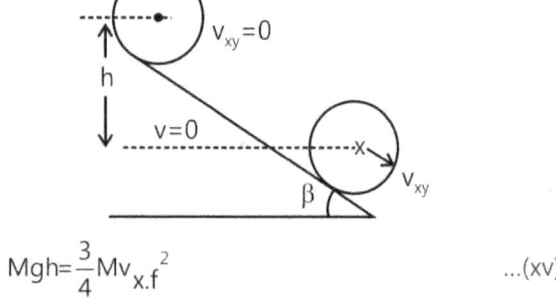

$$Mgh = \frac{3}{4}Mv_{x.f}^2 \qquad \text{...(xv)}$$

The magnitude of the velocity of the center of mass of the cylinder when it reaches the bottom of the incline is

$$v_{x.f} = \sqrt{(4/3)gh} \qquad \text{...(xvi)}$$

In agreement with Eq. (viii)

Using torque about a fixed point that lies along the line of contact between the cylinder and the surface

Choose a fixed point that lies along the line of contact between the cylinder and the surface. Then the torque diagram, is shown below.

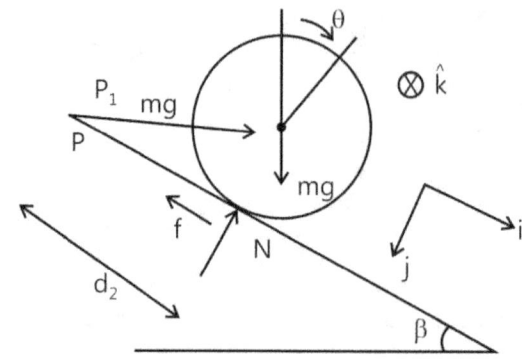

The gravitational force $M\vec{g} = Mg\sin\beta\,\hat{j}$ acts at the center of mass. The vector from the point P to the center of mass is given by $\vec{r}_{p.mg} = d_p\,\hat{i} - R\hat{j}$, so the torque due to the gravitational force about the point P is given by

$$\vec{\tau}_{p.mg} = \vec{r}_{p.mg} \times M\vec{g} = (d_p\hat{i} - R\hat{j}) \times (Mg\sin\beta\,\hat{i} + Mg\cos\beta\hat{j})$$

$$= (d_p Mg\cos\beta + RMg\sin\beta)\hat{k} \qquad \text{...(xvii)}$$

The normal force acts at the point of contact between the cylinder and the surface and is given by $\vec{N} = -N\hat{j}$. The vector from the point P to the point of contact between the cylinder and the surface is $\vec{\tau}_{P.N} = d_p\hat{i}$. So the torque due to the normal force about the point P is given by

$$\vec{\tau}_{p.N} = \vec{\tau}_{p.N} \times \vec{N} = (d_p\hat{i}) \times (-N\hat{j}) = -d_p\hat{k} \qquad \text{...(xviii)}$$

Substituting Eq. (xxxiii) for the normal force into Eq. (xviii) yields

$$\vec{\tau}_{p.N} = -d_p Mg\cos\beta\hat{k} \qquad \text{...(xix)}$$

Therefore the sum of the torques about the point P is

$$\vec{\tau}_p = \vec{\tau}_{pMg} + \vec{\tau}_{p.N} = (Mg\cos\beta + RMg\sin\beta)\,\hat{k} - d_p Mg\cos\beta\,\hat{k} = Rmg\sin\beta\,\hat{k} \qquad \text{...(xx)}$$

The angular momentum about the point P is given by

$$\vec{L}_P = \vec{L}_{cm} + \vec{r}_{p.cm} \times M\vec{V}_{cm}$$

$$= I_{cm}\omega_z\hat{k} + (d_p\hat{i} - R\hat{j}) \times (Mv_x)\hat{i}$$

$$= (I_{cm}\omega_z + RMv_x)\hat{k} \qquad \text{...(xxi)}$$

The time derivative of the angular momentum about the point P is then

$$\frac{d\vec{L}_P}{dt} = I_{cm}\alpha_z + RMa_x\hat{k} \qquad \text{...(xxii)}$$

Therefore the torque equation

$$\vec{\tau} = \frac{d\vec{L}_P}{dt} \qquad \text{...(xxiii)}$$

3.42

Becomes $RMg \sin\beta \hat{k} = (I_{cm}\alpha_z + RMa_x)\hat{k}$ ...(xxiv)

Using the fact that $I_{cm} = (1/2)MR^2$ and $\alpha_x = a_x/R$, we can conclude that $RM\,a_x = (3/2)MR\,a_x$ ...(xxv)

We can now solve Eq. (xxv) for the x – component of the acceleration

$$a_x = (2/3)g\sin\beta \qquad ...(xxvi)$$

In agreement with Eq. (vi).

Example 4: A bowling ball of mass m and radius R is initially thrown down an alley with an initial speed v_0 and it slides without rolling but due to friction it begins to roll. The moment of inertia of the ball about its center of mass is $I_{cm} = (2/5)mR^2$. Using conservation of angular momentum about a point (you need to find that point), find the speed v_f of the bowling ball when it just start to roll without slipping?

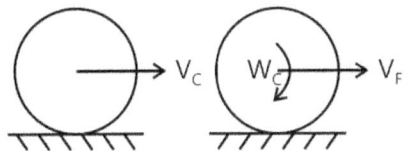

Sol: The angular momentum of any rigid body about a fixed point in ground reference frame is the sum of the angular momentum in the C-frame and the angular momentum corresponding to the translation of the center of mass relative to the fixed point in ground frame.

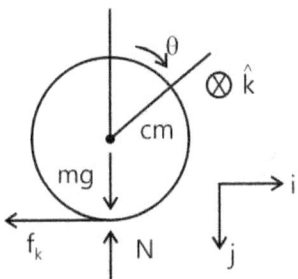

At t = 0, when the ball is released $v_{cm,i} = v_0$ towards right and $\omega_i = 0$, so the ball slips towards right on the surface and hence the frictional force on the ball, will be towards left.

The frictional force will pass through the point of contact S with the surface.

The weight of the ball as well as the normal reaction from the surface are equal in magnitude and opposite

in direction and have same line of action and will also pass through the point of contact S. (Point of contact is vertically below the COM of the ball).

Thus we choose the initial point of contact S as the origin and the net torque of all the forces about the origin S comes out to be zero at all times. So we can conserve the angular momentum of the ball about the initial point of contact (origin S).

The initial angular momentum about the origin S is only due to the translation of the center of mass.

$$L_i = mv_0 R$$

The final angular momentum about the origin S has both translational and rotational contribution.

$$L_f = m\,v_{cm,f}\,R + I_{cm}\omega$$

Finally the ball rolls without slipping, so we have $v_{cm,f} = R\omega$

Now $I_{cm} = (2/5)mR^2$

Therefore the final angular momentum about the origin S is

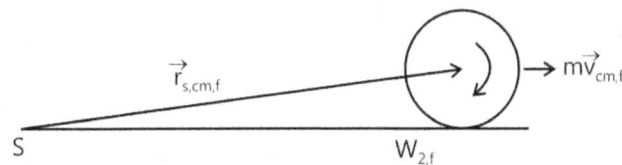

$$L_f = (mR + (2/5)mR)v_{cm,f} = (7/5)mRv_{cm,f}$$

Now equating $L_i = L_f$ we get

$$mR\,v_0 = (7/5)\,m\,R\,v_{cm,f}$$

or $\quad v_{cm,f} = (5/7)v_0$

Example 5: A long narrow uniform stick of length ℓ and mass m lies motionless on ice (assume the ice provides a frictionless surface). The center of mass of the stick is the same as the geometric center (at the midpoint of the stick). The moment of inertia of the stick about its center of mass is I_{cm}. A puck (with putty on one side) has same mass m as the stick. The puck slides without spinning on the ice with a speed of v_0 towards the stick, hits one end of the stick, and attaches to it. You may assume that the radius of the puck is much less than the length of the stick so that moment of inertia of the puck about its center of mass is negligible compared to I_{cm}.

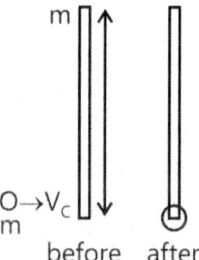

before   after

(a) How far from the midpoint of the stick is the center of mass of the stick–puck combination after the collision?

(b) What is the linear velocity of the stick plus puck after the collision?

(c) Is mechanical energy conserved during the collision? Explain your reasoning.

(d) What is the angular velocity of the stick plus puck after the collision?

(e) How far does the stick's center of mass move during one rotation of the stick?

Sol: Apply the law of conservation of linear momentum and law of conservation of angular momentum before and after collision. The angular momentum is to be calculated about the center of mass of "stick-puck system".

(a) From the midpoint of the stick the center of mass of the stick–puck combination after the collision is at a distance d_{cm}.

(b) $d_{cm} = \dfrac{m_{stick} d_{stick} + m_{puck} d_{puck}}{m_{stick} + m_{puck}} = \dfrac{m \times 0 + m(\ell / 2)}{m + m} = \dfrac{\ell}{4}$

There are no external forces acting on this system comprising "stick and puck" so the momentum of the system before and after the collision is conserved.

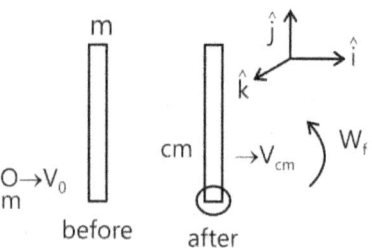

before   after

After the collision, suppose the center of mass of the system is moving with speed v_f

Equating initial and final linear momentum we get,

$$mv_0 = (2m)v_f \Rightarrow v_f = \dfrac{v_0}{2}$$

The direction of the velocity is the same as the initial direction of the puck's velocity.

(c) As the collision is perfectly inelastic the mechanical energy of the system is not conserved.

(d) Choose the center of mass of the stick-puck combination, as found in part (a) as the point about which we find the angular momentum before and after the collision. This choice is advantageous as there will be no angular momentum due to the translation of the center of mass just after the collision about the center of mass itself. Before the collision, the angular momentum was entirely due to the motion of the puck,

$$L_0 = (\ell / 4)(mv_0)$$

After the collision, the angular momentum is $L_f = I_{cm}\omega_f$

where I_{cm} is the moment of inertia about the center of mass of the stick-puck combination.

This moment of inertia of the stick about the new center of mass is found from the parallel axis theorem, and the moment of inertia of the puck is $m(\ell / 4)^2$, and so

$$I_{cm} = m(\ell^2 / 12) + m(\ell / 4)^2 + m(\ell / 4)^2 = 5m\ell^2 / 24$$

Equating $L_0 = L_f$ we get $(\ell / 4)(mv_0) = 5m\ell^2\omega_f / 24$

This gives $\omega_f = \dfrac{6v_0}{5\ell}$

(e) Time taken by stick-puck system for one rotation is $T = 2\pi/\omega_f$

Distance travelled by COM during this time is

$$x_{cm} = v_{cm}T = \dfrac{2\pi}{\omega_f}v_{cm} = \dfrac{2\pi}{6v_0 / 5\ell}(v_0/2)$$

This gives $x_{cm} = \dfrac{5\pi\ell}{6}$

Example 6: Two small kids of masses 10 kg and 15 kg are trying to balance a seesaw of total length 5.0 m with the fulcrum at the center. If one of the kids is sitting at ends, where should the other sit?

Sol: For rotational equilibrium, the net torque about the fulcrum of all the forces acting on the boys and the seesaw should be zero.

It is clear that the 10 kg kid should sit at the end and the 15 kg kid should sit closer to the center. Suppose this distance from the center is X. As the kids are in equilibrium, the normal force between a kid and the see-saw equals the weight of that kid. Considering the rotational equilibrium of the seesaw, the torques of the forces acting on it should add to zero. The forces are.

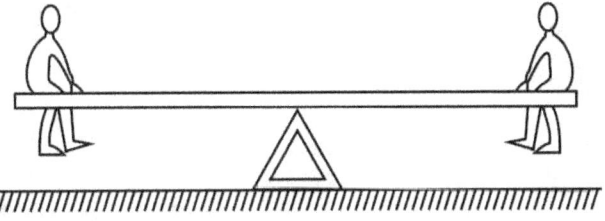

$$T = N\frac{2}{\sqrt{3}} = (392N) \times \frac{2}{\sqrt{3}} = 450 \text{ N}.$$

(a) (15kg) g downward by the 15kg kid,

(b) (10 kg) g downward by the 10kg kid,

(c) Weight of the seesaw and

(d) The normal force by the fulcrum.

Taking torques about by the fulcrum

(15kg) g x = (10 kg) g (2.5m) or x = 1.7 m.

Example 7: The ladder shown in figure has negligible mass and rests on a frictionless floor. The crossbar connects the two legs of the ladder at the middle. The angle between the two legs is 60°.

The fat person sitting on the ladder has a mass of 80 kg. Find the contact force exerted by the floor on each leg and the tension in the crossbar.

Sol: The forces of normal reaction at the feet of the ladders balance the weight of the person. For rotational equilibrium of ladder the toque due to normal reaction at the foot is balanced by the torque due to tension in the crossbar. Both the torques are calculated about the upper end of the ladder.

The forces acting on the different parts are shown in figure. Consider the vertical equilibrium of "the ladder plus the person" system. The forces acting on this system are its weight (80kg)g and the contact force N + N = 2N due to the floor. Thus

2N = (80kg) g or   N = (40 kg) (9.8m/s²) = 392 N

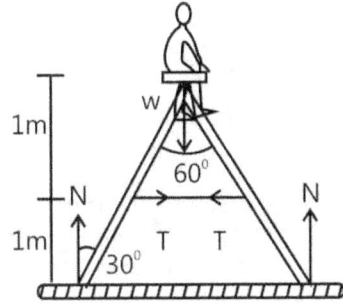

Next consider the equilibrium of the left leg of the ladder. Taking torque of the forces acting on it about the upper end.

N (2 m) tan 30°= T (1 m) or

Example 8: A solid cylinder of mass m and radius r starts rolling down an inclined plane of inclination θ. Friction is enough to prevent slipping. Find the speed of its center of mass when, its center of mass has fallen a height h.

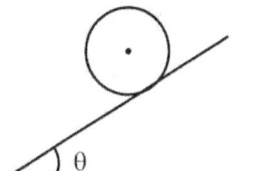

Sol: Loss in the gravitational potential energy of the cylinder is equal to the gain in rotational plus translational kinetic energy.

Consider the two shown positions of the cylinder. As it does not slip, total mechanical energy will be conserved.

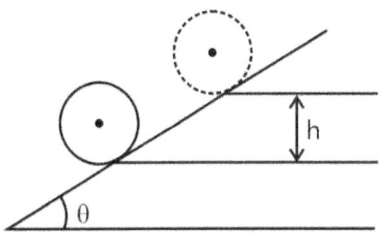

Energy at position 1 is $E_1 = mgh$

Energy at position 2 is $E_2 = \frac{1}{2}mv_{cm}^2 + \frac{1}{2}I_{cm}\omega^2$

$\therefore \frac{v_{cm}}{r} = \omega$ and $I_{c.m} = \frac{mr^2}{2} \Rightarrow E_2 = \frac{3}{4}mv_{c.m}^2$

From COE, $E_1 = E_2$; $v_{c.m} = \sqrt{\frac{4}{3}gh}$

Example 9: A uniform disc of radius R and mass M is rotated to an angular speed ω_0 in its own plane about its center and then placed on a rough horizontal surface such that plane of the disc is parallel to the horizontal surface. If co-efficient of friction between the disc and the surface is μ then how long will it take for the disc to come to stop.

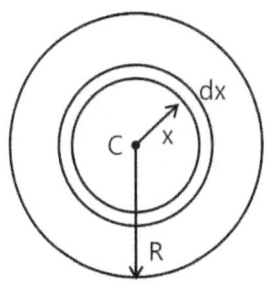

Sol: The disc can be thought of made-up of elementary rings of infinitesimal thickness. The torque about the center of disk due to friction force on each ring will be different from the other rings in the disc as the radii of rings are different, varying from 0 to R. So use the method of integration to find the torque on the entire disc.

Consider a differential circular strip of the disc of radius x and thickness dx. Mass of this strip is $dm = 2\rho\,\pi x\,dx$, where $\rho = \dfrac{M}{\pi R^2}$. Frictional force on this strip is along the tangent and is equal to $dF = 2\mu\rho\pi g\,x\,dx$

Torque on the strip due to frictional force is equal to $d\tau = \mu\rho g 2\pi x^2 dx$

The disc is supposed to be the combination of Number of such strips hence torque on the disc is given by

$$\tau = \int d\tau = \mu\rho\, g2\pi \int_0^R x^2 dx = \mu\rho g2\pi\frac{R^3}{3}$$

$$\Rightarrow \tau = \mu Mg(2/3)R$$

$$\Rightarrow \alpha = \frac{2\mu MgR}{3\left(\dfrac{MR^2}{2}\right)} = \frac{4}{3}\frac{\mu g}{R}$$

The α is opposite to the $\dot{\omega}$

$$\because\ \omega(t) - \omega_0 + \alpha t \qquad \Rightarrow 0 = \omega_0\frac{4\mu g}{3R}t$$

$$\Rightarrow t = \frac{3\omega_0 R}{4\mu g}\ .$$

Example 10: A sphere of mass M and radius r shown in figure slips on a rough horizontal plane. At some instant it has translational velocity v_0 and rotational velocity about the center is $\dfrac{v_0}{2r}$. Find the translational velocity after the sphere starts pure rolling.

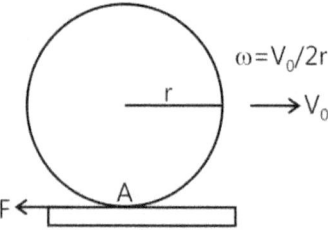

Sol: Due to forward slipping the friction will act backwards. So the sphere will decelerate. The torque due to friction will be in the direction of initial angular velocity. So the angular velocity will increase.

The slipping will stop when the condition of pure rolling is satisfied.

Velocity about the center $= \dfrac{v_0}{2r}$. Thus $v_0 > \omega_0 r$. The sphere slips forward and thus the friction by the plane on the sphere will act backward. As the friction is kinetic its value of N is given by $\mu N = \mu Mg$ and sphere will be decelerated by $a_{cm} = f/M$. Hence.

This friction will also have a torque $T = fr$ about the center. This torque is clockwise and in the direction of ω_0. Hence the angular acceleration about the center will be

$$\alpha = f\frac{r}{(2/5)Mr^2} = \frac{5f}{2Mr}$$

and the clockwise angular velocity at time t will be

$$\omega(t) = \omega_0 + \frac{5f}{2Mr}t = \frac{v_0}{2r} + \frac{5f}{2Mr}t$$

Pure rolling starts when

Eliminating t from (i) and (ii)

$$\frac{5}{2}v(t) + v(t) = \frac{5}{2}v_0 + \frac{v_0}{2} \quad \text{Or}$$

Thus the sphere rolls with translational velocity $6v_0/7$ in the forward direction.

JEE Advanced/Boards

Example 1: A carpet of mass M made of inextensible material is rolled along its length in the form of a cylinder of radius R and is kept on a rough floor. The carpet starts unrolling without sliding on the floor when a negligibly small push is given to it. Calculate the horizontal velocity of the axis of the cylindrical part of the carpet when its radius decreases to (R/2).

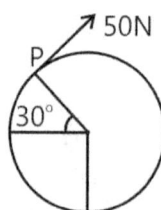

Sol: As the carpet unrolls, the radius and mass of cylindrical part decreases and center of mass descends. Thus loss in the gravitational potential energy is equal to gain in rotational plus translational kinetic energy.

If ρ is the density of material of the carpet, initial mass of the carpet (cylinder) M will be $\pi R^2 L\rho$. When its radius becomes half, the mass of cylindrical part will be

$$M_F = \pi(R/2)^2 L\rho = M/4$$

3.46

So initial PE of the carpet is MgR while final

$(M/4) g(R/2) = MgR/8$

So loss in potential energy when due to unrolling radius changes from R to R/2

$MgR (1-(1/8)) = (7/8)MgR$ ... (i)

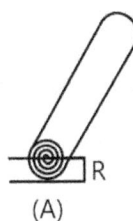 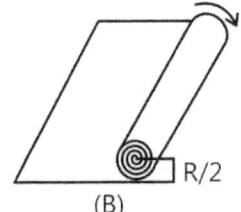

(A)   (B)

This loss in potential energy is equal to increase in rotation KE which is

$K = K_T + K_R = \frac{1}{2} Mv^2 + \frac{1}{2} I\omega^2$

If v is the velocity when half the carpet has unrolled, then as

$v = \frac{R}{2}\omega, M \to \frac{M}{4}$ and $I = \frac{1}{2}\left[\frac{M}{4}\right]\left[\frac{R}{2}\right]^2$

$K = \frac{1}{2}\left[\frac{M}{4}\right]v^2 + \frac{1}{2}\left[\frac{MR^2}{32}\right]\left[\frac{2v}{R}\right]^2$

i. e., $K = \frac{1}{8}Mv^2 + \frac{1}{16}Mv^2 = \frac{3}{16}Mv^2$ ... (i)

So from equation (i) and (ii)

$\left(\frac{3}{16}\right)Mv^2 = \left(\frac{7}{8}\right)MvR$

i.e., $v = \sqrt{(14gR)/3}$

Example 2: A uniform solid cylinder of radius R = 15 cm rolls over a horizontal plane passing into an inclined plane forming an angle α = 30° with the horizontal. Find the maximum value of the velocity v_0 which still permits the cylinder to roll onto the inclined plane section without a jump. The sliding is assumed to be absent.

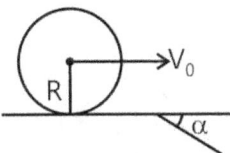

Sol: As the cylinder rolls into the inclined plane section, its center of mass descends, thus loss in gravitational potential energy will be equal to increase in kinetic

energy. At the edge the COM moves in circular arc during the time interval when the vertical radius through the point of contact turns by angle α to become normal to the inclined plane. During this interval normal reaction from edge should always be greater than zero.

Initial energy $E_1 = \frac{1}{2}mv_0^2 + \frac{1}{2}I_{c.m}\omega^2 + mgR$

For rolling $\frac{v_0}{R} = \omega$

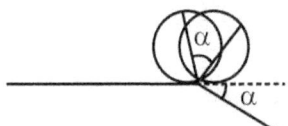

$\Rightarrow E_1 = \frac{1}{2}mv_0^2 + \frac{1}{2}\cdot\frac{1}{2}mR^2\frac{v_0^2}{R^2} + mgR$

$= \frac{3}{4}mv_0^2 + mgR$

$E_2 = \frac{1}{2}mv^2 + \frac{1}{2}I_{c.m}\omega^{'2} + mgR\cos\alpha$

$= \frac{3}{4}mv^2 + mgR\cos\alpha$

From COE

$\frac{3}{4}mv^2 + mgR\cos\alpha = \frac{3}{4}mv_0^2 + mgR$

$mv^2 = mv_0^2 + \frac{4}{3}mgR(1-\cos\alpha)$ .... (i)

F.B.D. of the cylinder when it is at the edge.

Center of mass of the cylinder describes circular motion about P.

Hence $mg\cos\alpha - N = mv^2/R$

$\Rightarrow N = mg\cos\alpha - mv^2/R$

$= mg\cos\alpha - \frac{mv_0^2}{R} - \frac{4}{3}mg + \frac{4}{3}mg\cos\alpha$

For no jumping, $N \geq 0$

$\Rightarrow \frac{7}{3}mg\cos\alpha - \frac{4}{3}mg - \frac{mv_0^2}{R} \geq 0$

$\Rightarrow v_0 \leq \sqrt{\frac{7gR}{3}\cos\alpha - \frac{4}{3}g}$

Example 3: Two thin circular disc of the mass 2 kg each and radius 10 cm each are joined by a rigid massless rod of length 20 cm. The axis of the rod is perpendicular to the plane of the disc through their centers as shown in the figure. The object is kept at the center as shown in the figure. The object is kept on a truck in such a way that the axis of the object is horizontal and perpendicular to the direction of motion of the truck. Its friction with the floor of the truck is large enough to prevent slipping. If the truck has an acceleration of 9 m/s² calculate.

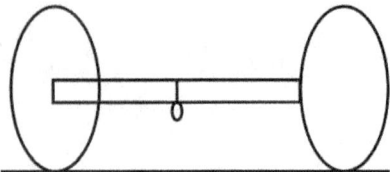

(a) The force of friction on each disc.

(b) The magnitude and direction of the frictional torque acting on each disc about the center of mass 'O' of the object. Take x-axis along the direction of the motion of the truck, and z-axis along vertically upwards direction. Express the torque in the vector form in terms of unit vectors $\hat{i}$, $\hat{j}$ and $\hat{k}$ in the x, y and z directions.

(c) Find the minimum value of the co-efficient of friction μ between the object and the floor of the truck which makes rolling of the object possible.

Sol: This problem is best solved in the reference frame of truck. Each disc will experience pseudo force as well as frictional force. Get two equations, one by applying Newton's second law at the center of mass of the object, and the other relating the torques of forces about the center of mass and the moment of inertia about axis passing through the center of mass.

F.B.D. of the object with respect to truck.

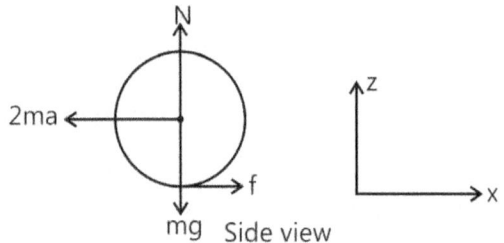

mg   Side view

In the reference frame of truck it experiences a pseudo force $F = -2ma\,\hat{i}$

where a = acceleration of the truck.

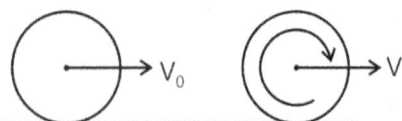

Pseudo force does not provide torque about the center of the disc. Because of this force object has tendency to slide along – Ve x –axis, hence frictional force will act along + Ve x – axis. For translational motion.

$2ma - f = ma'$ .... (i)

Here a' = acceleration of the center of mass of the object.

For rotational motion

$fR = I\alpha$

$= 2.\dfrac{mR^2}{2}.\dfrac{a'}{R}$ for no slipping $\alpha = a/R$

$\Rightarrow a' = \dfrac{f}{m}$ ..... (ii)

From (i) and (ii) we get

$F = \dfrac{2}{3}ma\,\hat{i} \Rightarrow$ Force of friction on each disc is

$\dfrac{f}{2} = \dfrac{ma}{3}\hat{i} = 6\hat{i}\,N$

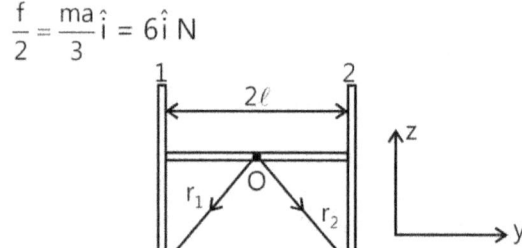

$\vec{f_1} = \dfrac{ma}{3}\hat{i}$

$\vec{r_1} = \ell\hat{j} - R\hat{k}$

$\vec{\tau}_{f_1} = \vec{r_1} \times \vec{f_1} = -(\ell\hat{j} + R\hat{k}) \times \dfrac{ma}{3}\hat{i}$

$= -\dfrac{maR}{3}\hat{j} + \dfrac{ma\ell}{3}\hat{k}$

$= 6 \times 0.1\,\hat{i} + 6 \times 0.1\,\hat{k}$

$= -0.6\,\hat{j} + 0.6\,\hat{k}$

$\vec{\tau}(f_2) = -0.6\,\hat{j} - 0.6\,\hat{k}$

(c) Maximum value of frictional force is $2\mu\,mg$

$\Rightarrow \dfrac{2}{3}ma \le 2\mu mg \qquad \Rightarrow \mu > \dfrac{a}{3g}.$

Example 4: A uniform disc of mass m and radius r is projected horizontal with velocity v_0 on a rough horizontal floor so that it starts of with a purely sliding motion at t = 0. At t = t_0 second it acquires a purely rolling motion.

(a) Calculate the velocity of the center of mass of the disc at t = t_0.

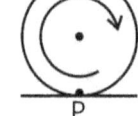

(b) Assuming coefficient of friction to be μ calculate t_0.

(c) The work done by the frictional force as a function of time.

(d) Total work done by the friction over a time t much longer then t_0.

Sol: This problem can be solved either by applying Newton's laws of motion or by law of conservation of angular momentum about the point of contact of the disc with the floor. The force of friction will act opposite to the direction of motion, and the work done by friction will be equal to loss in kinetic energy. The friction will stop doing any work once pure rolling starts.

F.B.D. of the disc.

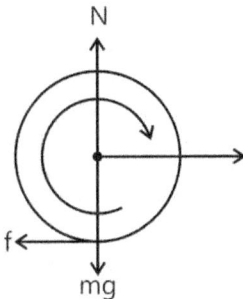

When the disc is projected it starts sliding and hence there is a relative motion between the points of contact. Therefore frictional force acts on the disc in the direction opposite to the motion.

(a) Now for translational motion

$$a_{c.m} = \frac{f}{m}$$

$f = \mu N$ (as it slides) $= \mu mg$

$\Rightarrow a_{c.m} = -\mu g$, negative sign indicates that

$a_{c.m}$ is opposite to $v_{c.m}$

$$\Rightarrow v_{c.m(t)} = v_0 - \mu g t_0$$

$$\Rightarrow t_0 = \frac{(v_0 - v)}{\mu g} \qquad \text{...(i)}$$

where $v_{c.m(t_0)} = v$

For rotational motion about center

$$\tau_f + \tau_{mg} = I_{c.m}\alpha \Rightarrow \mu mgr = \frac{mr^2}{2}\alpha$$

$$\Rightarrow \alpha = \frac{2\mu g}{r} \qquad \text{...(ii)}$$

Therefore $\omega_{(t_0)} = 0 + \frac{2\mu g}{r} t_0$

Using $\omega_t = \omega_0 + \alpha t$

$$\Rightarrow \omega = \frac{2(v_0 - v)}{r} \qquad \text{... (iii)}$$

(Using (i)) $V_{c.m} = \omega r$

$\Rightarrow v = 2(v_0 - v)$ (using (iii))

$$\Rightarrow v = \frac{2}{3}v_0$$

Alternative method: Since frictional force passes through the point of contact, hence about this point no external torque is acting.

Therefore angular momentum of the disc about point of contact does not change.

Initial angular momentum about p is given by

$L_1 = 0 + mv_0 r$ (Using $L_P = L_{c.m} + \bar{r} \times P_{c.m}$)

When it starts pure rolling its angular momentum about P is given by

$L_2 = L_{c.m} + \omega + mvr$

For rolling $v = \omega r$

$$\Rightarrow L_2 = \frac{mr^2}{2}\frac{v}{r} + mvr = \frac{3}{2}mvr$$

From COAM

$$L_1 = L_2 \Rightarrow v = \frac{2}{3}v_0$$

(b) Putting the value of v in equation (i)

We get $t_0 = \dfrac{v_0}{3\mu g}$

(c) Work done by the frictional force is equal to change in K.E.

$\Rightarrow W_{friction}$

$$= \frac{1}{2}m(v_0 - \mu g t)^2 + \frac{1}{2}\left(\frac{mr^2}{2}\right)\left(\frac{2\mu g t}{r}\right)^2 - \frac{1}{2}mv_0^2$$

$$= m\left(\frac{3}{2}\mu^2 g^2 t^2 - v_0\mu g t\right), \text{ For } t \le t_0$$

(d) For time $t \ge t_0$ work done by the friction is zero.

For longer time total work done is same as that in part

$(c) \Rightarrow W = m\left(\frac{3}{2}\mu^2 g^2 t^2\left(\frac{v_0}{3\mu g}\right)^2 - v_0\mu g t\frac{v_0}{3\mu g}\right)$

$$= -\frac{mv_0^2}{6}$$

Example 5: In the shown figure a mass m slides down a frictionless surface from height h and collides with a uniform vertical rod of length L and mass M. After collision the mass m sticks to the rod. The rod is free to rotate in a vertical plane about fixed axis through O. find the maximum angular deflection of the rod from its initial position.

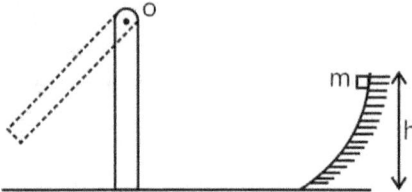

Sol: During the collision between rod and the mass, the linear momentum is not conserved because of the reaction force acting on the rod due to the hinge at the fixed point O. The torque of reaction force at the hinge will be zero about the hinge itself, i.e. about point O. So we can conserve the angular momentum of the "rod and mass system" before and after collision. As the rod rotates, the gain in gravitational potential energy is equal to the loss in the kinetic energy.

Just before collision, velocity of the mass m is along the horizontal and is equal to $v_0 = \sqrt{2gh}$. In the process of collision only angular momentum of the system will be conserved about the point O.

If L_1 and L_2 are the angular momentum of the system just before and just after the collision then $L_1 = mv_0L$

And $L_2 = I\omega = \left(\dfrac{ML^2}{3} + mL^2\right)\omega$

From Conservation of Angular Momentum

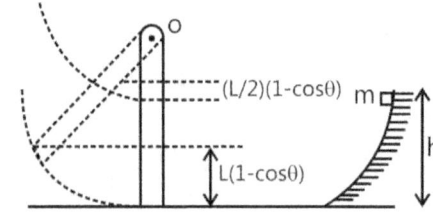

$$\left(\dfrac{M}{3} + m\right)L^2\omega = mv_0L$$

$$\Rightarrow \omega = \dfrac{mv_0}{\left(\dfrac{M}{3} + m\right)L}$$

Let the rod deflect through an angle θ.

Initial energy of rod and mass system $= \dfrac{1}{2}I\omega^2$

Where $I = \left(\dfrac{ML^2}{3} + mL^2\right)$

Gain in potential energy of the system

∵ From conservation of energy

$$\dfrac{1}{2}I\omega^2 = \left(m + \dfrac{M}{2}\right)gL(1 - \cos\theta)$$

$$\Rightarrow \dfrac{1}{2}\left(\dfrac{ML^2}{3} + mL^2\right) \times \dfrac{m^2v_0^2}{\left(\dfrac{M}{3} + m\right)^2 L^2}$$

$$= \left(m + \dfrac{M}{2}\right)gL(1 - \cos\theta)$$

$$\dfrac{1}{2}\dfrac{m^2v_0^2}{\left(\dfrac{M}{3} + m\right)} = \left(m + \dfrac{M}{2}\right)gL(1 - \cos\theta)$$

$$\cos\theta = 1 - \dfrac{1}{2}\dfrac{m^2v_0^2}{\left[\dfrac{M}{3} + m\right]\left[\dfrac{M}{2} + m\right]gL}.$$

Example 6: A billiard ball, initially at rest, is given a sharp impulse by a cue. The cue is held horizontally a distance h above the center line as shown in figure. The ball leaves the cue with a speed v_0 and because of its forward rotation (backward slipping) eventually acquires a final speed $\dfrac{9}{7}v_0$ show that.

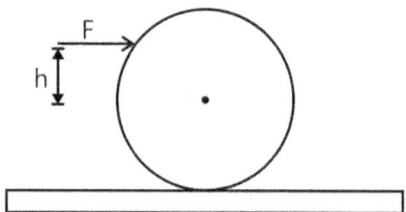

$h = \dfrac{4}{5}R$ where R is the radius of the ball.

Sol: Initial linear and angular velocity of ball is found by calculating the linear and angular impulse delivered by the cue. The angular momentum of the ball about the point of contact with ground surface, during its combined translational and rotational motion remains conserved.

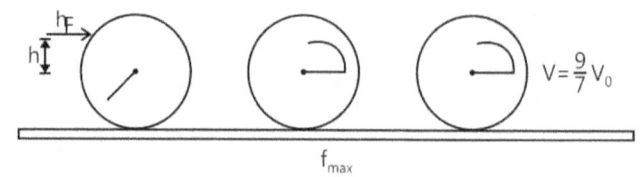

3.50

Let ω_0 be the angular speed of the ball just after it leaves the cue. The maximum friction acts in forward direction till the slipping continues. Let v be linear speed and ω the speed when slipping is ceased.

$$\therefore v = R\omega \text{ or } \omega = \frac{v}{R}$$

Given, $v = \frac{9}{7} v_0$ ..... (i)

$$\therefore \omega = \frac{9}{7}\frac{v_0}{R}$$ ..... (ii)

Applying Linear impulse = change in linear momentum

$F \, dt = V_0$ .... (iii)

Applying Angular impulse = change in angular momentum

or $\quad Fh \, dt = \frac{2}{5} mR^2 \omega_0$ .... (iv)

Angular momentum about bottommost point will remain conserved.

i.e., $\quad L_i = L_f$

or $\quad I\omega_0 + mRv_0 = I\omega + mRv$

$$\therefore \frac{2}{5}mR^2\omega_0 + mRv_0$$

$$= \frac{2}{5}mR^2\left(\frac{9}{7}\frac{v_0}{R}\right) + \frac{9}{7}mRv_0$$ ...... (v)

Solving Eqs. (iii), (iv) and (v), we get $\quad h = \frac{4}{5}R$

Example 7: Determine the maximum horizontal force F that may be applied to the plank of mass m for which the solid does not slip as it begins to roll on the plank. The sphere has a mass M and radius R (see figure). The coefficient of static and kinetic friction between the sphere and the plank are μ_s and μ_k respectively.

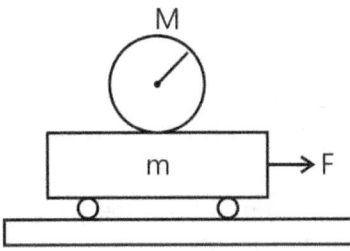

Sol: As the plank moves forward, the sphere, due to its inertia, has a tendency to slip backwards relative to the plank. So the force of friction acts on the sphere in the forward direction. For maximum force F, the friction will

be limiting. Write the equations of Newton's second law and torque about center of mass for the sphere and the equation of Newton's second law for the plank.

The free body diagram of the sphere and the plank are as shown below:

Writing equation of motion:

For sphere: Linear acceleration $a_1 = \frac{\mu_s Mg}{M} = \mu_s g$ ....(i)

Angular acceleration: $\alpha = \frac{(\mu_s Mg)R}{\frac{2}{5}MR^2} = \frac{5}{2}\frac{\mu_s g}{R}$ ....(ii)

For plank: Linear acceleration

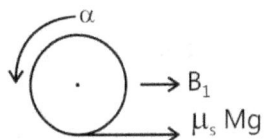

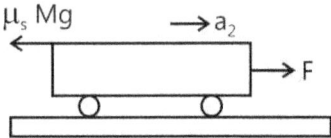

$\alpha_2 = \dfrac{F - \mu_s Mg}{m}$

For no slipping $\alpha_2 = a_1 + R\alpha$

Solving the above four equations, we get

$$F = \mu_s g\left(M + \frac{7}{2}m\right)$$

Thus, maximum value of F can be $\mu_s g\left(M + \frac{7}{2}m\right)$

Example 8: A uniform disc of radius r_0 lies on a smooth horizontal plane. A similar disc spinning with the angular velocity ω_0 is carefully lowered onto the first disc. How soon do both discs spin with the same angular velocity if the friction coefficient between them is equal to μ ?

Sol: The initial angular momentum about its center of the disc being lowered will be equal to the combined angular momentum of both the discs about their centers once they start rotating together. Each disc can be thought of made-up of elementary rings of infinitesimal thickness. The torque about the center of disk due to friction force on each ring will be different from that on the other rings in the disc as the radii of rings are different, varying from 0 to r_0. So use the method of integration to find the torque on the entire disc.

From the law of conservation of angular momentum.
$$I\omega_0 = 2I\omega$$

Here, I = moment of inertia of each disc relative to common rotation axis

$$\therefore \ \omega = \frac{\omega_0}{2} = \text{steady state angular velocity}$$

The angular velocity of each disc varies due to the torque τ of the frictional forces. To calculate τ, let us take an elementary ring with inner and outer radii r and r + dr. The torque of the friction acting on the given is equal to.

$$d\tau = \mu r \left(\frac{mg}{\pi r_0^2}\right) 2\pi r \, dr = \left(\frac{2\mu mg}{r_0^2}\right) r^2 \, dr$$

where m is the mass each disc. Integrating this respect to r between 0 and r_0, we get

$$\tau = \frac{2}{3}\mu mg r_0$$

The angular velocity of the lower disc increases by $d\omega$ over the time interval

$$= \left(\frac{3r_0}{4\mu g}\right) d\omega$$

Integrating this equation with respect to ω between 0 and $\frac{\omega_0}{2}$, we find the desired time $t = \dfrac{3r_0\omega_0}{8\mu g}$

Example 9: A solid sphere of radius r is gently placed on a rough horizontal ground with an initial angular speed ω_0 and no linear velocity. If the coefficient of friction is μ, find the time when the slipping stops. In addition, state the linear velocity v and angular velocity ω at the end of slipping.

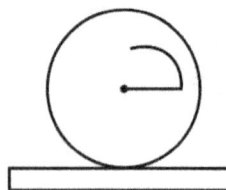

Sol: Due to backward slipping the force of friction will act forwards and torque due to friction will be anti-clockwise. This problem can be solved either by Newton's second law and torque about center of mass method or by applying the law of conservation of angular momentum about the point of contact of the sphere with the ground.

Let m be the mass of the sphere.

Since, it is a case of backward slipping, force of friction is in forward direction. Limiting friction will act in this case.

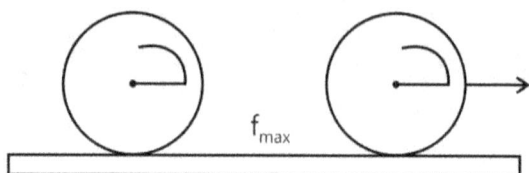

Linear acceleration $\qquad a = \dfrac{f}{m} = \dfrac{\mu mg}{m} = \mu g$

Angular retardation $\qquad \alpha = \dfrac{\tau}{I} = \dfrac{f.r}{\frac{2}{5}mr^2} = \dfrac{5}{2}\dfrac{\mu g}{r}$

Slipping ceases when $v = r\omega$

Or $\quad (at) = r(\omega_0 - \alpha t)$

Or $\quad \mu g t = r\left(\omega_0 - \dfrac{5}{2}\dfrac{\mu g}{r}\right)$

$\dfrac{7}{2}\mu g t = r\omega_0 ; \ t = \dfrac{2}{7}\dfrac{r\omega_0}{\mu g}$

$v = at = \mu g t = \dfrac{2}{7}r\omega_0$

And $\omega = \dfrac{v}{r} = \dfrac{2}{7}r\omega_0$

Alternative solution: Net torque on the sphere about the bottommost point is zero. Therefore, angular momentum of the sphere will remain conserved about the bottommost point.

$$L_t = L_f$$

$$\therefore \ I\omega_0 = I\omega + mrv$$

Or $\quad \dfrac{2}{5}mr^2\omega_0 = \dfrac{2}{5}mr^2\omega + mr(\omega r)$

$$\therefore \qquad \omega = \dfrac{2}{7}\omega_0$$

And $v = r\omega = \dfrac{2}{7}r\omega_0$

Example 10: A thin massless thread is wound on reel of mass 3 kg and moment of inertia 0.6 kg-m². The hub radius is R = 10 cm and peripheral radius is 2R = 20 cm. The reel is placed on a rough table and the friction is enough to prevent slipping. Find the acceleration of the center of reel and of hanging mass of 1 kg (see figure).

Sol: Apply Newton's second law at the center of mass of reel in horizontal direction. Find relation between net torque about center of mass of reel and moment of inertia about axis passing through the center of mass. Apply Newton's second law for hanging mass in vertical direction.

$$\alpha = \frac{\tau}{I} = \frac{f(2R) - T.R}{I} = \frac{0.2f - 0.1T}{0.6}$$

$$= \frac{f}{3} - \frac{T}{6} \qquad \text{...(ii)}$$

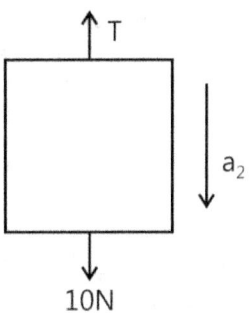

Free body diagram of mass is

Equation of motion is,

$$10 - T = a_2 \qquad \text{...(iii)}$$

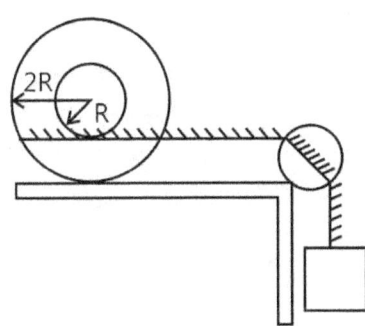

Let, a_1 = acceleration of center of mass of reel

a_2 = acceleration of 1 kg block

α = angular acceleration of reel (clockwise)

T = tension in the string

and f = force of friction

Free body diagram of reel is as shown below; (only horizontal forces are shown.).

Equations of motion are: $T - f = 3a_1$ ... (i)

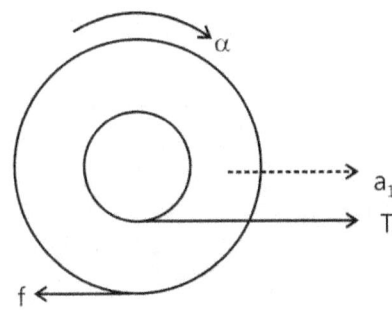

For no slipping condition,

$$a_1 = 2R\alpha \text{ or } a_1 = 0.2\alpha \qquad \text{...(iv)}$$

And $a_2 = a_1 - R\alpha$ or $a_2 = a_1 - 0.1\alpha$ .....(v)

Solving the above five equations, we get

$$a_1 = 0.27 \text{ m/s}^2; a_2 = 0.135 \text{ m/s}^2$$

JEE Main/Boards

Exercise 1

Q.1 What are the units and dimensions of moment of inertia? Is it a vector?

Q.2 What is rotational analogue of force?

Q.3 What is rotational analogue of mass of a body?

Q.4 What are two theorems of moment of inertia?

Q.5 What is moment of inertia of a solid sphere about its diameter?

Q.6 What is moment of inertia of a hollow sphere about an axis passing through its center.

Q.7 What are the factors on which moment of inertia of a body depends?

Q.8 Is radius of gyration of a body a constant quantity?

Q.9 There are two spheres of same mass and same radius, one is solid and other is hollow. Which of them has a larger moment of inertia about its diameter?

Q.10 Two circular discs A and B of the same mass and same thickness are made of two different metals whose densities are d_A and d_B ($d_A > d_B$). Their moments of inertia about the axes passing through their centers of gravity and perpendicular to their planes are I_A and I_B. Which is greater, I_A or I_B?

Q.11 The moments of inertia of two rotating bodies A and B are I_A and I_B ($I_A > I_B$) and their angular moments are equal. Which one has a greater kinetic energy?

Q.12 Explain the physical significance of moment of inertia and radius of gyration.

Q.13 Obtain expression of K.E. for rolling motion.

Q.14 State the laws of rotational motion.

Q.15 Establish a relation between torque and moment of inertia of a rigid body.

Q.16 State and explain the principle of conservation of angular momentum. Give at least two examples.

Q.17 Derive an expression for moment of inertia of a thin circular ring about an axis passing through its center and perpendicular to the plane of the ring.

Q.18 The moment of inertia of a circular ring about an axis passing through the center and perpendicular to its plane is 200 g cm². If radius of ring is 5 cm, calculate the mass of the ring.

Q.19 Calculate moment of inertia of a circular disc about a transverse axis through the center of the disc. Given, diameter of disc is 40 cm, thickness = 7 cm and density of material of disc = 9 g cm^{-3}

Q.20 A uniform circular disc and a uniform circular ring each has mass 10kg and diameter 1m. Calculate their moment of inertia about a transverse axis through their center.

Q.21 Calculate moment of inertia of earth about its diameter, taking it to be a sphere of radius 6400 km and mass 6 ×10²⁴ kg.

Q.22 Calculate moment of inertia of a uniform circular disc of mass 700 g and diameter 20 cm about

(i) An axis through the center of disc and perpendicular to its plane,.

(ii) A diameter of disc

(iii) A tangent in the plane of the disc,

(iv) A tangent perpendicular to the plane of the disc.

Q.23 Three particles, each of mass m, are situated at the vertices of an equilateral triangle ABC of side L. Find the moment of inertia of the system about the line AX perpendicular to AB in the plane of ABC, in the given figure.

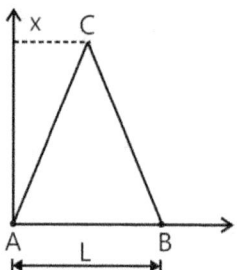

Q.24 Calculate K.E. of rotation of a circular disc of mass 1 kg and radius 0.2 m rotating about an axis passing through its center and perpendicular to its plane. The disc is making 30/π rpm.

Q.25 A circular disc of mass M and radius r is set into pure rolling on a table. If ω be its angular velocity, show that it's total K.E. is given by (3/4) Mv², where v is its linear velocity. M.I. of circular disc = (1/2) mass × (radius)².

Q.26 The sun rotates around itself once in 27 days. If it were to expand to twice its present diameter, what would be its new period of revolution?

Q.27 A 40 kg flywheel in the form of a uniform circular disc 1 meter in radius is making 120 r.p..m. Calculate its angular momentum about transverse axis passing through center of fly wheel.

Q.28 A body is seated in a revolving chair revolving at an angular speed of 120 r.p.m. By some arrangement, the body decreases the moment of inertia of the system from 6 kg m² to 2 kg m². What will be the new angular speed?

Exercise 2

Single Correct Choice Type

Q.1 Thee bodies have equal masses m. Body A is solid cylinder of radius R, body B is square lamina of side R, and body C is a solid sphere of radius R. Which body has the smallest moment of inertia about an axis passing through their center of mass and perpendicular to the plane (in case of lamina)

(A) A   (B) B   (C) C   (D) A and C both

Q.2 For the same total mass which of the following will have the largest moment of inertia about an axis passing through its center of mass and perpendicular to the plane of the body.

(A) A disc of radius a

(B) A ring of radius a

(C) A square lamina of side 2a

(D) Four rods forming a square of side 2a

Q.3 A thin uniform rod of mass M and length L has its moment of inertia I_1 about its perpendicular bisector. The rod is bend in the form of semicircular arc. Now its moment of inertia perpendicular to its plane is I_2. The ratio of $I_1 : I_2$ will be

(A) < 1   (B) >1   (C) =1   (D) Can't be said

Q.4 Moment of inertia of a thin semicircular disc (mass = M & radius = R) about an axis through point O and perpendicular to plane of disc, is given by:

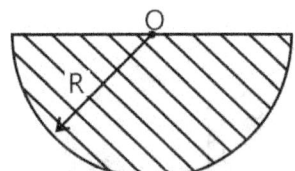

(A) $\frac{1}{4}MR^2$ (B) $\frac{1}{2}MR^2$ (C) $\frac{1}{8}MR^2$ (D) MR^2

Q.5 A rigid body can be hinged about any point on the x-axis. When it is hinged such that the hinge is at x, the moment of inertia is given by I = 2x² - 12x + 27

The x-coordinate of center of mass is

(A) x = 2   (B) x = 0   (C) x = 1   (D) x = 3

Q6. A weightless rod is acted upon by upward parallel forces of 2N and 4N at ends A and B respectively. The total length of the rod AB=3m. To keep the rod in equilibrium, a force of 6N should act in the following manner:

(A) Downwards at any point between A and B.

(B) Downwards at mid-point of AB

(C) Downwards at a point C such that AC=1m

(D) Downwards at a point D such that BD= 1m.

Q.7 A heavy rod of length L and weight W is suspended horizontally by two vertical ropes as shown. The first rope is attached to the left end of rod while the second rope is attached a distance L/4 from right end. The tension in the second rope is:

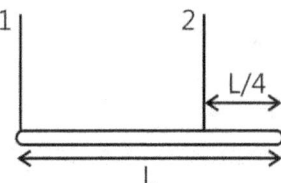

(A) (1/2) W   (B) (1/4) W

(C) (1/3) W   (D) (2/3) W   (E) W

Q.8 A right triangular plate ABC of mass m is free to rotate in the vertical plane about a fixed horizontal axis though A. It is supported by a string such that the side AB is horizontal. The reaction at the support A in equilibrium is:

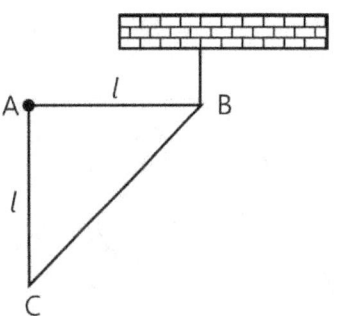

(A) $\frac{mg}{3}$ (B) $\frac{2mg}{3}$ (C) $\frac{mg}{2}$ (D) mg

Q.9 A rod is hinged at its center and rotated by applying a constant torque from rest. The power developed by the external torque as a function of time is:

(A)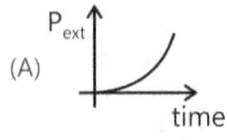

(B)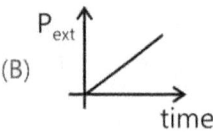

(C)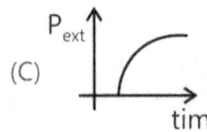

(D)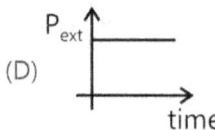

Q.10 Two uniform spheres of mass M have radii R and 2R. Each sphere is rotating about a fixed axis through its diameter. The rotational kinetic energies of the spheres are identical. What is the ratio of the angular moments of these sphere? That is, $\dfrac{L^2 2R}{L_R} =$

(A) 4   (B) $2\sqrt{2}$ (C) 2   (D) $\sqrt{2}$ (E) 1

Q.11 A spinning ice skater can increase his rate of rotation by bringing his arms and free leg closer to his body. How does this procedure affect the skater's momentum and kinetic energy?

(A) Angular momentum remains the same while kinetic energy increases.

(B) Angular momentum remains the same while kinetic energy decreases

(C) Both angular momentum and kinetic energy remains the same.

(D) Both angular momentum and kinetic energy increase.

Q.12 A child with mass m is standing at the edge of a disc with moment of inertia I, radius R, and initial angular velocity ω. See figure given below. The child jumps off the edge of the disc with tangential velocity v with respect to the ground. The new angular velocity of the disc is

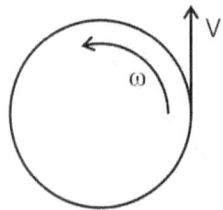

(A) $\sqrt{\dfrac{I\omega^2 - mv^2}{I}}$

B) $\sqrt{\dfrac{(I + mR^2)\omega^2 - mv^2}{I}}$

(C) $\dfrac{I\omega - mvR}{I}$

(D) $\dfrac{(I + mR^2)\omega - mvR}{I}$

Q.13 A uniform rod of length l and mass M is rotating about a fixed vertical axis on a smooth horizontal table. It elastically strikes a particle placed at a distance $l/3$ from its axis and stops. Mass of the particle is

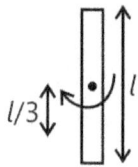

(A) 3M   (B) $\dfrac{3M}{4}$ (C) $\dfrac{3M}{2}$ (D) $\dfrac{4M}{3}$

Q.14 a disc of radius R is rolling purely on a flat horizontal surface, with a constant angular velocity. The angle between the velocity and acceleration vectors of point P is

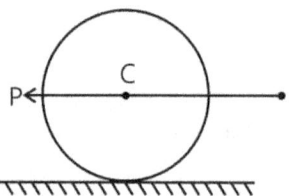

(A) Zero   (B) 45°

(C) 135°   (D) $\tan^{-1}(1/2)$

Q.15 A particle starts from the point (0m, 8m) and moves with uniform velocity of 3m/s. After 5 second, the angular velocity of the particle about the origin will be:

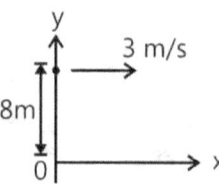

(A) $\dfrac{8}{289}$ rad/s   (B) $\dfrac{3}{8}$ rad/s

(C) $\dfrac{24}{289}$ rad/s   (D) $\dfrac{8}{17}$ rad/s

Q.16 Two points of a rigid body are moving as shown. The angular velocity of the body is:

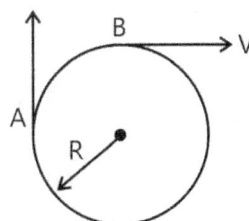

(A) $\dfrac{v}{2R}$ (B) $\dfrac{v}{R}$ (C) $\dfrac{2v}{R}$ (D) $\dfrac{2v}{3R}$

Q.17 A yo-yo is released from hand with the string wrapped around your finger. If you hold your hand still, the acceleration of the yo-yo is

(A) Downward, much greater than g

(B) Downward much greater than g

(C) Upward, much less than g

(D) Upward, much greater than g

(E) Downward, at g

Q.18 Inner and outer radii of N a spool are r and R respectively. A thread is wound over its inner surface and placed over a rough horizontal surface. Thread is pulled by a force F as shown in figure. Then in case of pure rolling.

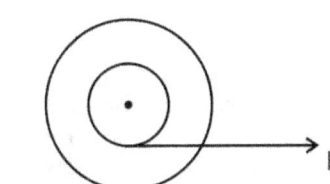

(A) Thread unwinds, spool rotates anticlockwise and friction acts leftwards

(B) Thread unwinds, spool rotates clockwise and friction acts leftwards

(C) Thread wind, spool moves to the right and friction acts rightwards.

(D) Thread winds, spool moves to the right and friction does not come into existence.

Q.19 A sphere is placed rotating with its center initially at rest in a corner as shown in figure (a) & figure (b). Coefficient of friction between all surfaces and the sphere is $\dfrac{1}{3}$. Find the ratio of the frictional force $\dfrac{f_a}{f_b}$ by ground in situations (a) & (b).

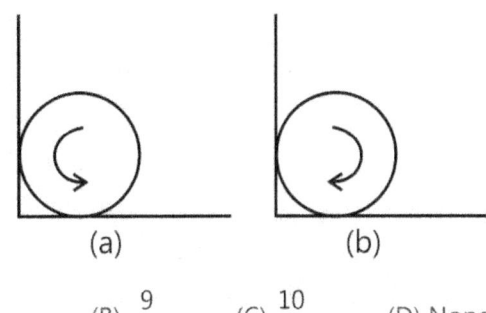

(A) 1 (B) $\dfrac{9}{10}$ (C) $\dfrac{10}{9}$ (D) None

Q.20 A body kept on a smooth horizontal surface is pulled by a constant horizontal force applied at the top point of the body. If the body rolls purely on the surface, its shape can be:

(A) Thin pipe (B) Uniform cylinder

(C) Uniform sphere (D) Thin spherical shell

Q.21 A uniform rod AB of mass m and length l is at rest on a smooth horizontal surface. An impulse j is applied to the end B, perpendicular to the rod in the horizontal direction. Speed of point P at A distance $\dfrac{l}{6}$ from the center towards a of the rod after time t = $\dfrac{\pi ml}{12J}$ is

(A) $2\dfrac{J}{m}$ (B) $\dfrac{J}{\sqrt{2}m}$

(C) $\dfrac{J}{m}$ (D) $\sqrt{2}\dfrac{J}{m}$

Q.22 The moment of inertia of a solid cylinder about its axis is given by (1/2) MR². If this cylinder rolls without slipping, the ratio of its rotational kinetic energy to its translational kinetic energy is

(A) 1: 1 (B) 2: 2 (C) 1: 2 (D) 1: 3

Q.23 A force F is applied to a dumbbell for a time interval t, first as in (i) and then as in (ii). In which case does the dumbbell acquire the greater center-of-mass speed?

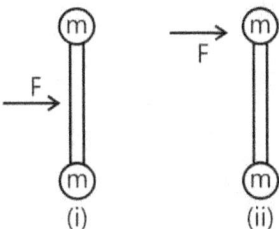

(A) (i)

(B) (ii)

(C) There is no difference

(D) The answer depends on the rotation inertia of the dumbbell

Q.24 A hoop and a solid cylinder have the same mass and radius. They both roll, without slipping on a horizontal surface. If their kinetic energies are equal

(A) The hoop has a greater translational speed then the cylinder

(B) The cylinder has a greater translational speed then the hoop

(C) The hoop and the cylinder have the same translational speed

(D) The hoop has a greater rotational speed then the cylinder.

Q.25 A ball rolls down an inclined plane, as shown in figure. The ball is first released from rest from P and then later from Q. Which of the following statement is /are correct?

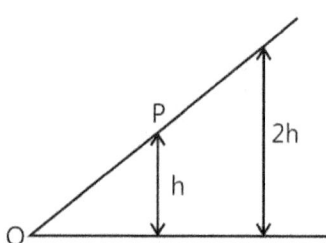

(i) The ball takes twice as much time to roll from Q to O as it does to roll from P to O.

(ii) The acceleration of the ball at Q is twice as large as the acceleration at P.

(iii) The ball has twice as much K.E.at O when rolling from Q as it does when rolling from P.

(A) i, ii only   (B) ii, iii only

(C) i only   (D) iii only

Q.26 If a person is sitting on a rotating stool with his hands outstretched, suddenly lowers his hands, then his

(A) Kinetic energy will decrease

(B) Moment of inertia will decrease

(C) Angular momentum will increase

(D) Angular velocity will remain constant

Q.27 Choose the correct statement(s)

(A) The momentum of the ring is conserved

(B) The angular momentum of the ring is conserved about its center of mass

(C) The angular momentum of the ring is conserved about a point on the horizontal surface.

(D) The mechanical energy of the ring is conserved.

Previous Years' Questions

Q.1 Let I be moment of inertia of a uniform square plate about an axis AB that passes through its center and is parallel to two of its sides. CD is a line in the plane of the plate that passes through the center of the plate and makes an angle è with AB. The moment of inertia of the plate about the axis CD is then equal to. *(1998)*

(A) I   (B) $I \sin^2\theta$

(C) $I \cos^2\theta$ (D) $I \cos^2(\theta/2)$

Q.2 A smooth sphere is moving on a frictionless horizontal plane with angular velocity ω and center of mass velocity v. It collides elastically and head on with an identical sphere B at rest, Neglect friction everywhere. After the collision their angular speeds are ω_A and ω_B respectively. Then, *(1999)*

(A) $\omega_A < \omega_B$ (B) $\omega_A = \omega_B$

(C) $\omega_A = \omega$ (D) $\omega_B = \omega$

Q.3 A particle of mass m is projected with a velocity v making an angle of 45° with the horizontal. The magnitude of the angular momentum of the projectile about the point of projection when the particle is at its maximum height h is *(1990)*

(A) Zero   (B) $mv^3/(4\sqrt{2}g)$

(C) $mv^3/(\sqrt{2}g)$ (D) $m\sqrt{2gh^3}$

Q.4 Consider a body, shown in figure, consisting of two identical balls, each of mass M connected by a light rigid rod. If an impulse J = Mv is imparted to the body at one of its end, what would be its angular velocity? *(2003)*

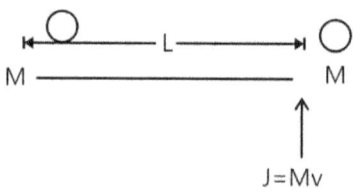

(A) v/L   (B) 2v/L   (C) v/3L   (D) v/4L

Q.5 A tube of length L is filled completely with an incompressible liquid of mass M and closed at both the ends. The tube is then rotated in a horizontal plane about one of its ends with a uniform angular velocity ω. The force exerted by the liquid at the other end is *(1992)*

(A) $\dfrac{M\omega^2 L}{2}$ (B) $M\omega^2 L$ (C) $\dfrac{M\omega^2 L}{4}$ (D) $\dfrac{M\omega^2 L^2}{2}$

Q.6 A cylinder rolls up an inclined plane, reaches some height and then rolls down (without slipping throughout these motions.) The directions of the frictional force acting on the cylinder are *(2002)*

(A) Up the incline while ascending and down the incline while descending.

(B) Up the incline while ascending as well as descending

(C) Down the incline while ascending and up the inline while descending.

(D) Down the incline while ascending as well as descending.

Q.7 Two point masses of 0.3 kg and 0.7 kg fixed at the ends of a rod of length 1.4 m and of negligible mass. The rod is set rotating about an axis perpendicular to its length with a uniform angular speed. The point on the rod through which rotation of the rod is minimum, is located at a distance of *(1995)*

(A) 0.42 m from mass of 0.3kg

(B) 0.70 m from mass of 0.7 kg

(C) 0.98 m form mass of 0.3 kg

(D) 0.98 m form mass of 0.7 kg

Q.8 A disc of mass M and radius R is rolling with angular speed ω on a horizontal plane as shown. The magnitude of angular momentum of the about the origin O is *(1999)*

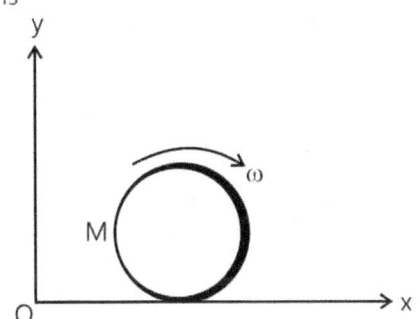

(A) $\left(\dfrac{1}{2}\right)MR^2\omega$ (B) $MR^2\omega$

(C) $\left(\dfrac{3}{2}\right)MR^2\omega$ (D) $2MR^2\omega$

Q.9 A cubical block of side L rests on a rough horizontal surface with coefficient of friction μ. A horizontal force F is applied sufficient high, so that the block does not slide before toppling, the minimum force required to topple the block is *(2000)*

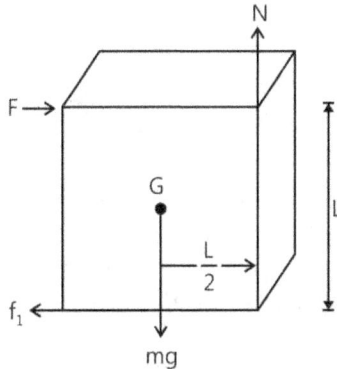

(A) Infinitesimal   (B) mg/4

(C) mg/2   (D) mg $(1-\mu)$

Q.10 An equilateral triangle ABC formed from a uniform wire has two small identical beads initially located at A. The triangle is set rotating about the vertical axis AO. Then the beads are released from rest simultaneously and allowed to slide down, one along AB and other along AC as shown. Neglecting frictional effects, the quantities that are conserved as beads slide down are *(2000)*

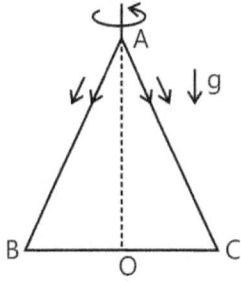

(A) Angular velocity and total energy (kinetic and potential)

(B) Total angular momentum and total energy

(C) Angular velocity and moment of inertia about the axis of rotation

(D) Total angular momentum and moment of inertia about the axis of rotation.

Q.11 One quarter section is cut from a O uniform circular disc of radius R. This section has a mass M. It is made to rotate about a line perpendicular to its plane and passing through the center of the original disc. Its moment of inertia about the axis of rotation is *(2001)*

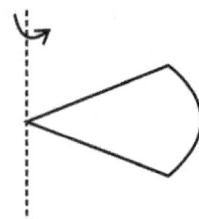

(A) $\frac{1}{2}MR^2$ (B) $\frac{1}{4}MR^2$ (C) $\frac{1}{8}MR^2$ (D) $\sqrt{2}MR^2$

Q.12 A circular platform is free to rotate in a horizontal plane about a vertical axis passing through its center. A tortoise is sitting at the edge of the platform. Now the platform is given an angular velocity ω_0

When the tortoise moves along a chord of the platform with a constant velocity (with respect to the platform). The angular velocity of the platform $\omega(t)$ will vary with time t as *(2002)*

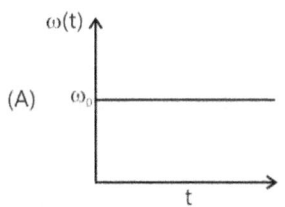
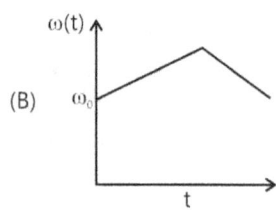
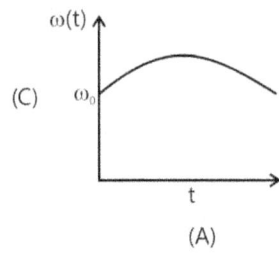
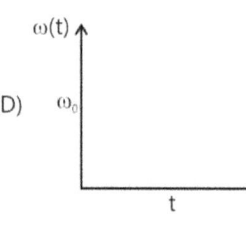

(A)

Q.13 A thin circular ring of mass M and radius r is rotating about its axis with a constant angular velocity ω. Two objects, each of mass m_2 are attached gently to the opposite ends of a diameter of the ring. The wheel now rotates with an angular velocity *(2006)*

(A) $\omega M/(M + m)$

(B) $\omega(M - 2m)/(M + 2m)$

(C) $\omega M/(M + 2m)$

(D) $\omega(M + 2m)/M$

Q.14 A cubical block of side a moving with velocity v on a horizontal smooth plane as shown. It hits at point O. The angular speed of the block after it hits O is *(1999)*

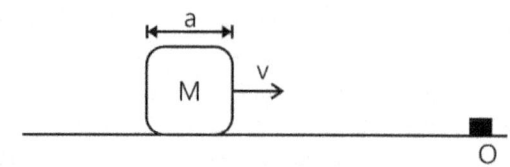

(A) 3v / 4a   (B) 3v / 2a

(C) $\sqrt{3}/\sqrt{2}a$ (D) Zero

Q.15 A thin wire of length L and uniform linear mass density ρ is bent into a circular loop with center at O as shown. The moment of inertia of the loop about the axis XX' is *(2000)*

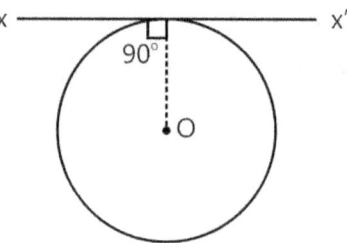

(A) $\frac{\rho L^3}{8\pi^2}$ (B) $\frac{\rho L^3}{16\pi^2}$ (C) $\frac{5\rho L^3}{16\pi^2}$ (D) $\frac{3\rho L^3}{8\pi^2}$

Q.16 A diatomic molecule is made of two masses m_1 and m_2 which are separated by a distance r. If we calculate its rotational energy by applying Bohr's rule of angular momentum quantization, its energy will be given by (n is an integer) *(2012)*

(A) $\frac{(m_1 + m_2)^2 n^2 \hbar^2}{2m_1^2 m_2^2 r^2}$ (B) $\frac{n^2 \hbar^2}{2(m_1 + m_2)r^2}$

(C) $\frac{2n^2 \hbar^2}{(m_1 + m_2)r^2}$ (D) $\frac{(m_1 + m_2)n^2 \hbar^2}{2m_1 m_2 r^2}$

Q.17 A hoop of radius r and mass m rotating with an angular velocity ω_0 is placed on a rough horizontal surface. The initial velocity of the centre of the hoop is zero. What will be the velocity of the centre of the hoop when it ceases to slip? *(2013)*

(A) $\frac{r\omega_0}{3}$ (B) $\frac{r\omega_0}{2}$ (C) $r\omega_0$ (D) $\frac{r\omega_0}{4}$

Q.18 A bob of mass m attached to an inextensible string of length ℓ is suspended from a vertical support. The bob rotates in a horizontal circle with an angular speed ω rad/s about the vertical. About the point of suspension: *(2014)*

(A) Angular momentum changes in direction but not in magnitude.

(B) Angular momentum changes both in direction and magnitude.

(C) Angular momentum is conserved.

(D) Angular momentum changes in magnitude but not in direction.

Q.19 The current voltage relation of diode is given by $I = \left(e^{1000\,V/T} - 1\right)$ mA, where the applied voltage V is in volts and the temperature T is in degree Kelvin. If a student makes an error measuring ± 0.01 V while measuring the current of 5 mA at 300 K, what will be the error in the value of current in mA? *(2014)*

(A) 0.5 mA   (B) 0.05 mA

(C) 0.2 mA   (D) 0.02 mA

Q.20 From a solid sphere of mass M and radius R a cube of maximum possible volume is cut. Moment of inertia of cube about an axis passing through its center and perpendicular to one of its faces is: *(2015)*

(A) $\dfrac{MR^2}{16\sqrt{2}\,\pi}$ (B) $\dfrac{4MR^2}{9\sqrt{3}\,\pi}$ (C) $\dfrac{4MR^2}{3\sqrt{3}\,\pi}$ (D) $\dfrac{MR^2}{32\sqrt{2}\,\pi}$

Q.21 A particle of mass m is moving along the side of a square of side 'a', with a uniform speed v in the x-y plane as shown in the figure:

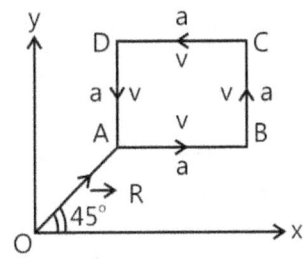

Which of the following statements is false for the angular momentum $\vec{L}$ about the origin? *(2016)*

(A) $\vec{L} = mv\left[\dfrac{R}{\sqrt{2}} - a\right]\hat{k}$ when the particle is moving from C to D.

(B) $\vec{L} = mv\left[\dfrac{R}{\sqrt{2}} + a\right]\hat{k}$ when the particle is moving from B to C.

(C) $\vec{L} = \dfrac{mv}{\sqrt{2}}R\hat{k}$ when the particle is moving from D to A.

(D) $\vec{L} = -\dfrac{mv}{\sqrt{2}}R\hat{k}$ when the particle is moving from A to B.

Q.22 A roller is made by joining together two cones at their vertices O. It is kept on two rails AB and CD which are placed asymmetrically (see figure), with its axis perpendicular to CD and its centre O at the centre of line joining AB and CD (see figure). It is given a light push so that it starts rolling with its centre O moving parallel to CD in the direction shown. As it moves, the roller will tend to: *(2016)*

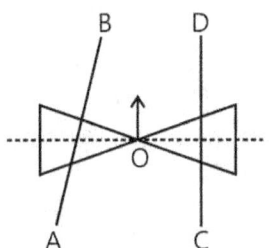

(A) Turn right

(B) Go straight

(C) Turn left and right alternately

(D) Turn left

Exercise 1

Q.1 A thin uniform rod of mass M and length L is hinged at its upper end, and released from rest in a horizontal position. Find the tension at a point located at a distance L/3 from the hinge point, when the rod becomes vertical.

Q.2 A rigid body in shape of a triangle has V_A = 5 m/s downwords, V_B = 10 m/s downwords. Find velocity of point C.

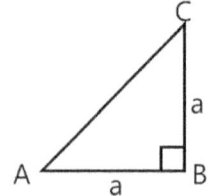

Q.3 A rigid horizontal smooth rod AB of mass 0.75 kg and length 40 cm can rotate freely about a fixed vertical axis through its mid-point O. Two rings each of mass 1 kg are initially at rest at a distance of 10 cm from O on either side of the rod. The rod is set in rotation with an angular velocity of 30 rad per second. Find the velocity of each ring along the length of the rod in m/s when they reach the ends of the rod.

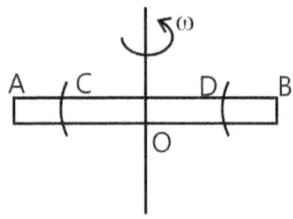

Q.4 A straight rod AB of mass M and length L is placed on a frictionless horizontal surface. A horizontal force having constant magnitude F and a fixed direction starts acting at the end A. The rod is initially perpendicular to the force. Find the initial acceleration of end B.

Q.5 A wheel is made to roll without slipping, towards right, by pulling a string wrapped around a coaxial spool as shown in figure. With what velocity the string should be pulled so that the center of wheel moves with a velocity of 3 m/s?

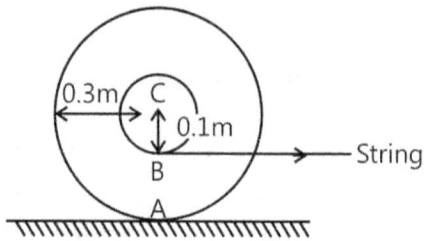

Q.6 A uniform wood door has mass m, height h, and width w. It is hanging from two hinges attached to one side; the hinges are located h/3 and 2h/3 from the bottom of the door.

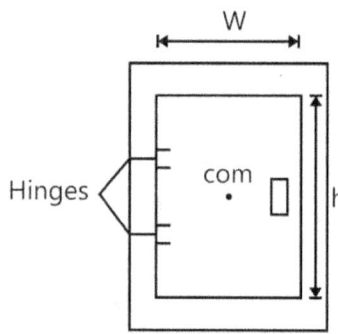

Suppose that m = 20.0 kg, h = 2.20m, and W = 1.00 m and the bottom smooth hinge is not screwed into the door frame, find the forces acting on the door.

Q.7 A thin rod AB of length a has variable mass per unit length $\rho_0\left(1+\dfrac{x}{a}\right)$ where x is the distance measured from a and ρ_0 is a constant

(a) Find the mass M of the rod.

(b) Find the position of center of mass of the rod.

(c) Find moment of inertia of the rod about an axis passing through A and perpendicular to AB. Rod is freely pivoted at A and is hanging in equilibrium when it is struck by a horizontal impulse of magnitude P at the point B.

(d) Find the angular velocity with which the rod begins to rotate.

(e) Find minimum value of impulse P if B passes through a point vertically above A.

Q.8 Two separate cylinders of masses m (=1 kg) and 4m and radii R (=10cm) and 2R are rotating in clockwise direction with ω_1 = 100rad/sec and ω_2 = 200 rad/sec. Now they are held in contact with each other as in figure Determine their angular velocity after the slipping between the cylinders stops.

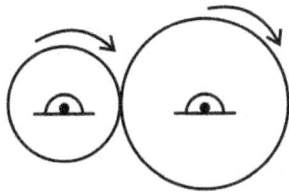

Q.9 A spool of inner radius R and outer radius 3R has a moment of inertia = MR^2 about an axis passing through its geometric center, where M is the mass of the spool. A thread wound on the inner surface of the spool is pulled horizontally with a constant force = Mg. Find the acceleration of the point on the thread which is being pulled assuming that the spool rolls purely on the floor.

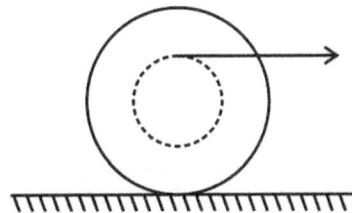

Q.10 A sphere of mass m and radius r is pushed onto a fixed horizontal surface such that it rolls without slipping from the beginning. Determine the minimum speed v of its mass center at the bottom so that it rolls completely around the loop of radius (R + r) without leaving the track in between.

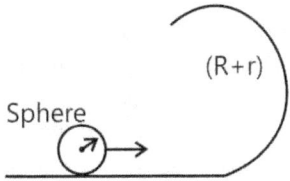

Q.11 Two uniform cylinders each of mass m = 10 kg and radius r = 150 mm, are connected by a rough belt as shown. If the system is released from rest, determine

(a) The tension in the portion of the belt connecting the two cylinder.

(b) The velocity of the center of cylinder a after it has moved through 1.2 m.

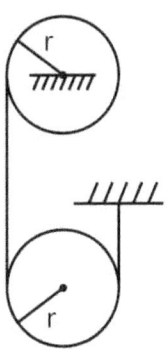

Q.12 A thin wire of length L and uniform linear mass density ρ is bent into a circular loop with center at O as shown in the figure. The moment of inertia of the loop about the axis XX` is

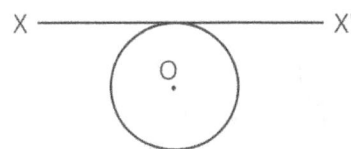

Q.13 A block X of mass 0.5 kg is held by a long massless string on a frictionless inclined plane of inclination 30° to the horizontal. The string is wound on a uniform solid cylindrical drum Y of mass 2 kg and of radius 0.2 m as shown in the figure. The drum is given an initial angular velocity such that the block X stands moving up the plane. (g = 9.8 m/s²)

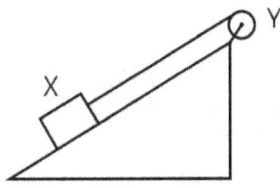

(i) Find the tension in the string during the motion.

(ii) At a certain instant of time the magnitude of the angular velocity of Y is 10rad/sec. Calculate the distance travelled by X form that instant of time unit it comes to rest.

Q.14 A uniform rod AB of length L and mass M is lying on a smooth table. A small particle of mass m strikes the rod with a velocity v_0 at point at distance from the center O. The particle comes to rest after collision. Find the value of x, so that of the rod remains stationary just after collision.

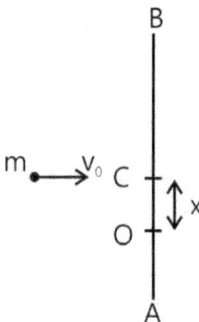

Q.15 A uniform plate of mass m is suspended in each of the ways shown. For each case determine immediately after the connection at B has been released:

Pin support

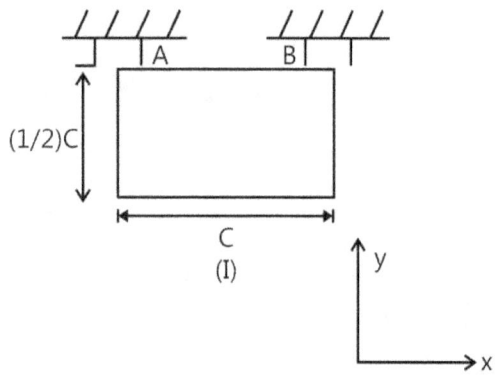

(I)

(a) The angular acceleration of the plate.

(b) The acceleration of its mass center.

Q.16 A carpet of mass 'M' made of inextensible material is rolled along its length in the form of cylinder of radius 'R' and is kept on a rough floor. The carpet starts unrolling without standing on the floor when a negligibly small push is given to it. The horizontal velocity of the axis of the cylindrical parts of the carpet when its radius decreases to R/2 will be:

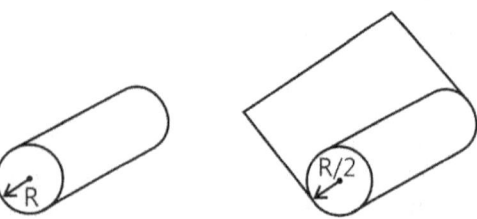

Q.17 A uniform disk of mass m and radius R is projected horizontally with velocity v_0 on a rough horizontal floor so that it starts off with a purely sliding motion at t = 0. After t_0 seconds it acquires a purely rolling motion as shown in figure.

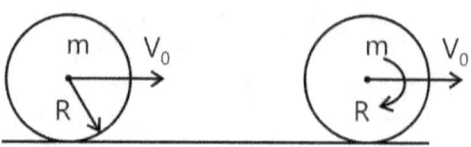

(i) Calculate the velocity of the center of mass of the disk at t_0

(ii) Assuming the coefficient of friction to be μ calculate t_0. Also calculate the work done by the frictional force as a function of time and the total work done by it over a time t much longer then t_0.

Q.18 A circular disc of mass 300 gm and radius 20 cm can rotate freely about a vertical axis passing through its center of mass o. A small insect of mass 100 gm is initially at a point A on the disc (which is initially stationary). The insect starts walking from rest along the rim of the disc with such a time varying relative velocity that the disc rotates in the opposite direction with a constant angular acceleration = 2π rad/s². After some time T, the insect is back at the point A. By what angle has the disc rotated till now, as seen by a stationary earth observer? Also find the time T.

Q.19 A uniform disc of mass m and radius R rotates about a fixed vertical axis passing through its center with angular velocity ω. A particle of same mass m and having velocity $2\omega R$ towards center of the disc collides with the disc moving horizontally and stick to its rim. Find

(a) The angular velocity of the disc

(b) The impulse on the particle due to disc.

(c) The impulse on the disc due to hinge.

Q.20 The door of an automobile is open and perpendicular to the body. The automobile starts with an acceleration of 2 ft/sec², and the width of the door is 30 inches. Treat the door as a uniform rectangle, and neglect friction to find the speed of its outside edge as seen by the driver when the door closes.

Q.21 A 20 kg cabinet is mounted on small casters that allow it to move freely (μ = 0) on the floor. If a 100 N force is applied as shown, determine.

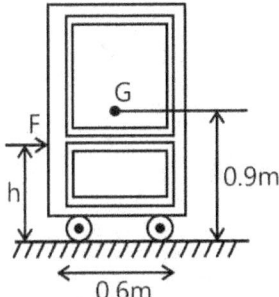

(a) The acceleration of the cabinet,

(b) The range of values of h for which the cabinet will not tip.

Q.22 Two thin circular disks of mass 2 kg and radius 10 cm each are joined by a rigid massless rod of length 20 cm. The axis of the rod is along the perpendicular to the planes of the disk through their center. The object is kept on a truck in such a way that the axis of the object is horizontal and perpendicular to the direction of motion of the truck. Its friction with the floor of the truck is large enough so that the object can roll on the truck without slipping. Take x-axis as the vertically upwards direction. If the truck has an acceleration of 9m/s² calculate.

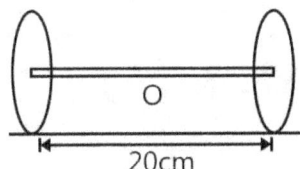

(a) The force of friction on each disk.

(b) The magnitude and the direction of the frictional torque acting on each disk about the center of mass O of the object. Express the torque in the vector form of unit vectors in the x-y and z direction.

Q.23 Three particles A, B, C of mass m each are joined to each other by mass less rigid rods to form an equilateral triangle of side a. Another particle of mass m hits B with a velocity v_0 directed along BC as shown. The colliding particle stops immediately after impact.

(i) Calculate the time required by the triangle ABC to complete half-revolution in its subsequent motion. (ii) What is the net displacement of point B during this interval?

Exercise 2

Single Correct Choice Type

Q.1 Let I_1 and I_2 be the moment of inertia of a uniform square plate about axes APC and OPO' respectively as shown in the figure. P is center of square. The ratio $\dfrac{I_1}{I_2}$ of moment of inertia is

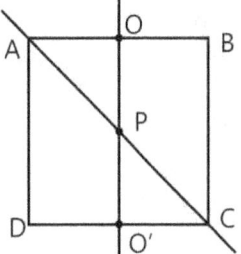

(A) $\dfrac{1}{\sqrt{2}}$ (B) 2   (C) $\dfrac{1}{2}$ (D) 1

Q.2 Moment of inertia of a rectangular plate about an axis passing through P and perpendicular to the plate is I. Then moment of PQR about an axis perpendicular to the plane.

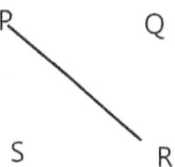

(A) About P = I/2   (B) About R = I/2

(C) About P > I/2   (D) About R > I/2

Q.3 Find the moment of inertia of a plate cut in shape of a right angled triangle of mass M, AC=BC=a about an axis perpendicular to plane, side the plane of the plate and passing through the mid-point of side AB.

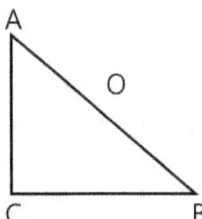

(A) $\dfrac{Ma^2}{12}$ (B) $\dfrac{Ma^2}{6}$ (C) $\dfrac{Ma^2}{3}$ (D) $\dfrac{2Ma^2}{3}$

Q.4 Let I be the moment of inertia of a uniform square plate about an axis AB that passes through its center

3.65

and is parallel to two of its sides. CD is a line in the plane of the plate that passes through the center of the plate and makes an angle θ with AB. The moment of inertia of the plate about the axis CD is then equal to

(A) I

(B) I sin² θ

(C) I cos² θ

(D) I cos² (θ/2)

Q.5 A heavy seesaw (i.e., not mass less) is out of balance. A light girl sits on the end that is tilted downward, and a heavy body sits on the other side so that the seesaw now balances. If they both move forward so that they are one-half of their original distance from the pivot point (the fulcrum) what will happen to the seesaw?

(A) The side the body is sitting on will tilt downward

(B) The side the girl is sitting on will once again tilt downward

(C) Nothing; the seesaw will still be balanced

(D) It is impossible to say without knowing the masses and the distances.

Q.6 A pulley is hinged at the center and a mass less thread is wrapped around it. The thread is pulled with a constant force F starting from rest,. As the time increases,

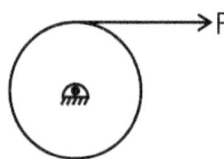

(A) Its angular velocity increases, but force on hinge remains constant

(B) Its angular velocity remains same, but force on hinge increases

(C) Its angular velocity increases and force of hinge increases

(D) Its angular velocity remains same and force on hinge is constant

Q.7 A uniform flag pole of length L and mass M is pivoted on the ground with a frictionless hinge. The flag pole makes an angle θ with the horizontal. The moment of inertia of the flag pole about one end is (1/3) ML². If it starts falling from the position shown in the accompanying figure, the linear acceleration of the free end of the flag pole – labeled P – would be:

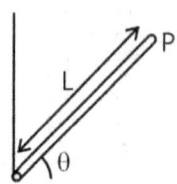

(A) (2/3) gcos θ

(B) (2/3)g

(C) g

(D) $\left(\dfrac{3}{2}\right)$gcos θ

(E) (3/2)g

Q.8 A mass m is moving at speed v perpendicular to a rod of length d and mass M = 6m which pivots around a frictionless axle running through its center. It strikes and sticks to the end of the rod. The moment of inertia of the rod about its center is Md²/12. Then the angular speed of the system right after the collision is.

(A) 2v / d

(B) 2v / (3d)

(C) v / d

(D) 3v / (2d)

Q.9 A sphere of mass M and radius R is attached by a light rod of length I to a point P. The sphere rolls without slipping on a circular track as shown. It is released from the horizontal position. The angular momentum of the system about P when the rod becomes vertical is:

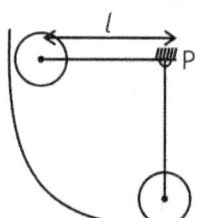

(A) $M\sqrt{\dfrac{10}{7}gl}[I+R]$

(B) $M\sqrt{\dfrac{10}{7}gl}[I+\dfrac{2}{5}R]$

(C) $M\sqrt{\dfrac{10}{7}gl}[I+\dfrac{7}{5}R]$

(D) None of the above

Q.10 A ladder of length L is slipping with its ends against a vertical wall and a horizontal floor. At a certain moment, the speed of the end in contact with the horizontal floor is v and the ladder makes an angle $\alpha = 30°$ with the horizontal. Then the speed of ladder's center must be

(A) 2v $\sqrt{3}$

(B) v/2

(C) v

(D) None

Q.11 In the previous question, if $dv/dt = 0$, then the angular acceleration of the ladder when $\alpha = 45°$ is

(A) $2v^2/L^2$

(B) $v^2/2L^2$

(C) $\sqrt{2}[v^2\ L^2]$

(D) None

Q.12 A uniform circular disc placed on a rough horizontal surface has initially a velocity v_0 and an angular velocity ω_0 as shown in the figure. The disc comes to rest after moving some distance in the direction of motion. Then $\dfrac{v_0}{r\omega_0}$ is

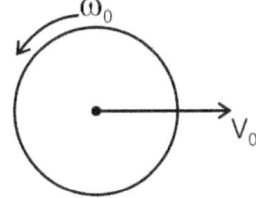

(A) 1/2   (B) 1   (C) 3/2   (D) 2

Q.13 An ice skater of mass m moves with speed 2v to the right, while another of the same mass m moves with speed v toward the left, as shown in figure I. Their paths are separated by a distance b. At t = 0, when they are both at x = 0, they grasp a pole of length

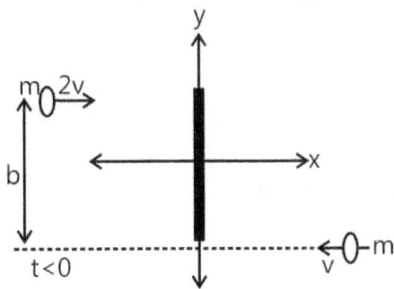

b and negligible mass. For r > 0 consider the system as a rigid body of two masses m separated by distance b, as shown in figure II. Which of the following is the correct formula for motion after t = 0 of the skater initially at y = b/2?

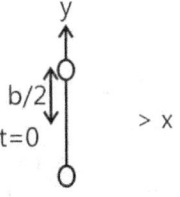

(A) $x = 2vt$, $y = b/2$

(B) $x = vt + 0.5\,b \sin(3vt/b)$, $y = 0.5b \cos(3vt/b)$

(C) $x = 0.5c = vt + 0.5b \sin(3vt/b)$, $y = 0.5b \cos(3vt/b)$

(D) $x = 0.5vt + 0.5b \sin(3vt/b)$, $y = 0.5b \cos(3vt/b)$

Q.14 A yo-yo is resting on a perfectly rough horizontal table. Forces F_1, F_2 and F_3 are applied separately as shown. The correct statement is

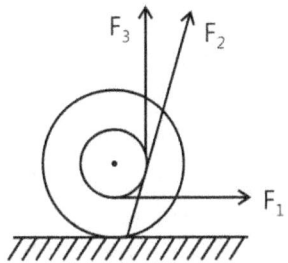

(A) When F_3 is applied the center of mass will move to the right.

(B) When F_2 is applied the center of mass will move to the right.

(C) When F_1 is applied the center of mass will move to the right.

(D) When F_2 is applied the center of mass will move to the right.

Multiple Correct Choice Type

Q.15 A rod of weight w is supported by two parallel knife edges and B and is in equilibrium in a horizontal position. The knives are at a distance d from each other. The center of mass of the rod is at a distance x from A.

(A) The normal reaction at a is $\dfrac{wx}{d}$

(B) The normal reaction at a is $\dfrac{w(d-x)}{d}$

(C) The normal reaction at B is $\dfrac{wx}{d}$

(D) The normal reaction at B is $\dfrac{w(d-x)}{d}$

Q.16 A block with a square base measuring and height h, is placed on an inclined plane. The coefficient of friction is μ. The angle of inclination (α) of the plane is gradually increased. The block will

(A) Topple before sliding if $\mu > \dfrac{a}{h}$

(B) Topple before sliding if $\mu < \dfrac{a}{h}$

(C) Slide before toppling if $\mu > \dfrac{a}{h}$

(D) Slide before toppling if $\mu < \dfrac{a}{h}$

Q.17 A particle falls freely near the surface of the earth. Consider a fixed point O (not vertically below the particle) on the ground.

(A) Angular momentum of the particle about O is increasing.

(B) Torque of the gravitational force on the particle about O is decreasing.

(C) The moment of inertia of the particle about O is decreasing.

(D) The angular velocity of the particle about O is increasing.

Q.18 The torque τ on a body about a given point is found to be equal to a × L where a constant vector is and L is the angular momentum of the body about that point. From this it follows that

(A) dL/dt is perpendicular to L at all instants of time

(B) The components of l in the direction of a does not change with time.

(C) The magnitude of l does not change with time.

(D) L does not change with time.

Q.19 In the given figure. a ball strikes a uniform rod of same mass elastically and rod is hinged at point A. Then which of the statement (S) is /are correct?

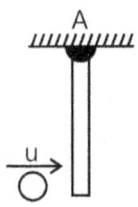

(A) Linear momentum of system (ball + rod) is conserved.

(B) Angular momentum of system (ball + rod) about the hinged point A is conserved.

(C) Kinetic energy of system (ball + rod) before the collision is equal to kinetic energy of system just after the collision

(D) Linear momentum of ball is conserved.

Q.20 A hollow sphere of radius R and mass m is fully filled with non-viscous liquid of mass m. It is rolled down a horizontal plane such that its center of mass moves with a velocity v. If it purely rolls

(A) Kinetic energy of the sphere is $\dfrac{5}{6}mv^2$

(B) Kinetic energy of the sphere is $\dfrac{4}{3}mv^2$

(C) Angular momentum of the sphere about a fixed point on ground is $\dfrac{8}{3}mvR$

(D) Angular momentum of the sphere about a fixed point on ground is $\dfrac{14}{5}mvR$

Q.21 In the figure shown, the plank is being pulled to the right with a constant speed v. If the cylinder does not slip then:

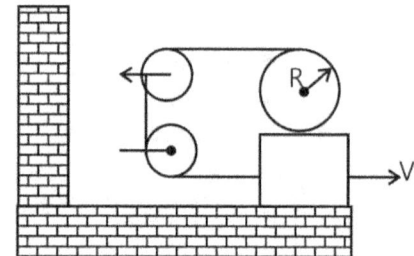

(A) The speed of the center of mass of the cylinder is 2v.

(B) The speed of the center of mass of the cylinder is zero.

(C) The angular velocity of the cylinder is v/R.

(D) The angular velocity of the cylinder is zero.

Q.22 A disc of circumference s is at rest at a point A on a horizontal surface when a constant horizontal force begins to act on its center. Between A and B there is sufficient friction to prevent slipping and the surface is smooth to the right of B. AB = s. The disc moves from A to B in time T. To the right of B,

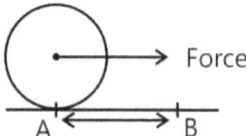

(A) The angular acceleration of disc will disappear, linear acceleration will remain unchanged

(B) Linear acceleration of the disc will increase

(C) The disc will make one rotation in time T/2.

(D) The disc will cover a distance greater then s in further time T.

Q.23 A rigid object is rotating in a counterclockwise sense around a fixed axis. If the rigid object rotates though more than$180°$ but less than $360°$, which of the following pairs of quantities can represent an initial angular position and a final angular position of the rigid object.

(A) 3 rad, 6 rad   (B) –1 rad, 1 rad

(C) 1 rad, 5 rad   (D –1rad, 2.5 rad

Q.24 ABCD is a square plate with center O. The moments of inertia of the plate about the perpendicular axis through O is I and about the axes 1, 2, 3 & 4 are I_1, I_2, I_3, & I_4 respectively. If follows that:

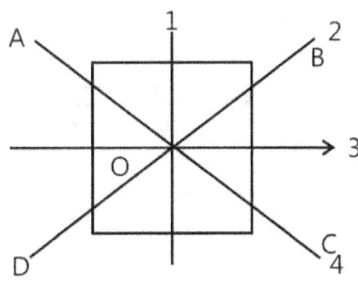

(A) $I_2 = I_3$ (B) $I = I_1 + I_4$

(C) $I = I_2 + I_4$ (D) $I_1 = I_3$

Q.25 A body is in equilibrium under the influence of a number of forces. Each force has a different line of action. The minimum number of forces required is

(A) 2, if their lines of action pass through the center of mass of the body.

(B) 3, if their lines of action are not parallel.

(C) 3, if their lines of action are parallel.

(D) 4, if their lines of action are parallel and all the forces have the same magnitude.

Q.26 A block of mass m moves on a horizontal rough surface with initial velocity v. The height of the center of mass of the block is h from the surface. Consider a point a on the surface in line with the center of mass.

(A) Angular momentum about a is mvh initially

(B) The velocity of the block decreases as time passes.

(C) Torque of the forces acting on block is zero about a.

(D) Angular momentum is not conserved about A.

Q.27 A man spinning in free space changes the shape of his body, eg. By spreading his arms or curling up. By doing this, he can change his

(A) Moment of inertia

(B) Angular momentum

(C) Angular velocity

(D) Rotational kinetic energy

Q.28 A ring rolls without slipping on the ground. Its center C moves with a constant speed u. P is any point on the ring. The speed of P with respect to the ground is v.

(A) $0 \leq v \leq 2u$

(B) v = u, if CP is horizontal

(C) v = u is CP makes an angle of $30°$ with the horizontal and P is below the horizontal level of c.

(D) $v = \sqrt{2}\,u$, if CP is horizontal

Q.29 A small ball of mass m suspended from the ceiling at a point O by a thread of length ℓ moves along a horizontal circle with a constant angular velocity ω.

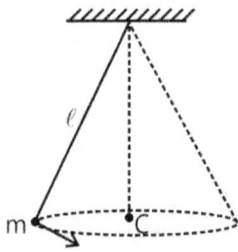

(A) Angular momentum is constant about O

(B) Angular momentum is constant about C

(C) Vertical component of angular momentum about O is constant

(D) Magnitude of angular momentum about O is constant.

Q. 30 If a cylinder is rolling down the incline with sliding.

(A) After some time it may start pure rolling

(B) After sometime it will start pure rolling

(C) It may be possible that it will never start pure rolling

(D) None of these.

Q.31 Which of the following statements are correct.

(A) Friction acting on a cylinder without sliding on an

inclined surface is always upward along the incline irrespective of any external force acting on it

(B) Friction acting on a cylinder without sliding on an inclined surface may be upward may be downwards depending on the external force acting on it

(C) Friction acting on a cylinder rolling without sliding may be zero depending on the external force acting on it

(D) Nothing can be said exactly about it as it depends on the frictional coefficient on inclined plane.

Q. 32 A plank with a uniform sphere placed on it rests on a smooth horizontal plane. Plank is pulled to right by a constant force F. If sphere does not slip over the plank. Which of the following is correct?

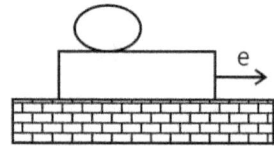

(A) Acceleration of the center of sphere is less than that of the plank

(B) Work done by friction acting on the sphere is equal to its total kinetic energy

(C) Total kinetic energy of the system is equal to work done by the force F

(D) None of the above.

Q. 33 a uniform disc is rolling on a horizontal surface. At a certain instant B is the point of contact and A is at height 2R from ground, where R is radius of disc.

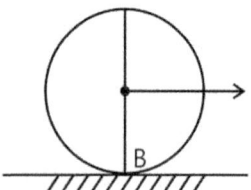

(A) The magnitude of the angular momentum of the disc about B is thrice that about A

(B) The angular momentum of the disc about A is anticlockwise

(C) The angular momentum of the disc about B is clockwise

(D) The angular momentum of the disc about A is equal to that about B.

Q. 34 A wheel of radius r is rolling on a straight line, the velocity of its center being v. At a certain instant the point of contact of the wheel with the grounds is M and N is the highest point on the wheel (diametrically opposite to M). The incorrect statement is:

(A) The velocity of any point P of the wheel is proportional to MP

(B) Points of the wheel moving with velocity greater than v form a larger area of the wheel than points moving with velocity less than v

(C) The point of contact M is instantaneously at rest

(D) The velocities of any two parts of the wheel which are equidistant from center are equal.

Q.35 A ring of mass M and radius R sliding with a velocity v_0 suddenly enters into a rough surface where the coefficient of friction is μ, as shown in figure.

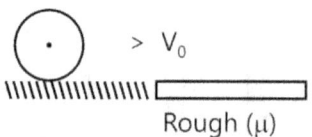

Rough (μ)

Choose the correct statement(s)

(A) As the ring enters on the rough surface, the limiting frictional force acts on it

(B) The direction of friction is opposite to the direction of motion.

(C) The frictional force accelerates the ring in the clockwise sense about its center of mass

(D) As the ring enters on the rough surface it starts rolling.

Q.36 Choose the correct statement (s)

(A) The ring starts its rolling motion when the center of mass is stationary

(B) The ring starts rolling motion when the point of contact becomes stationary

(C) The time after which the ring starts rolling is $\dfrac{v_0}{2\mu g}$

(D) The rolling velocity is $\dfrac{v_0}{2}$

Q.37 Choose the correct alternative (s)

(A) The linear distance moved by the center of mass before the ring starts rolling is $\dfrac{3v_0^2}{8\mu g}$

(B) The net work done by friction force is $-\dfrac{3}{8}\,mv_0^2$

(C) The loss is kinetic energy of the ring is $\dfrac{mv_0^2}{4}$

(D) The gain in rotational kinetic energy is $+\dfrac{mv_0^2}{8}$

Q.38 A tightrope walker in a circus holds a long flexible pole to help stay balanced on the rope. Holding the pole horizontally and perpendicular to the rope helps the performer.

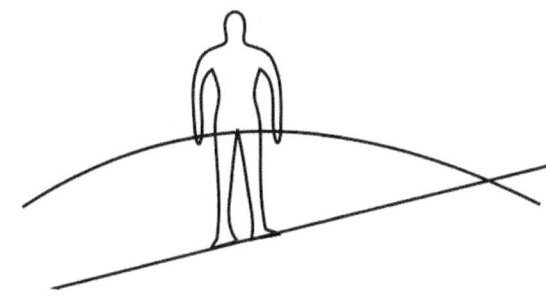

(A) By lowering the overall center-of- gravity

(B) By increasing the rotation inertia

(C) In the ability to adjust the center- of -gravity to be over the rope.

(D) In achieving the center of gravity to be under the rope.

Assertion Reasoning Type

(A) Statement-I is true, statement-II is true and statement-II is correct explanation for statement-I.

(B) Statement-I is true, statement-II is true and statement-II is NOT the correct explanation for statement-I.

(C) Statement-I is true, statement-II is false.

(D) Statement-I is false, statement-II is true.

Q.39 Consider the following statements

Statement-I: a cyclist always bends inwards while negotiating a curve

Statement-II: By bending he lowers his center of gravity of these statements.

Q.40 Statement-I A disc A moves on a smooth horizontal plane and rebounds elastically from a smooth vertical wall (Top view is shown in Fig 7.166), in this case about any point on line XY the angular momentum of the disc remains conserved.

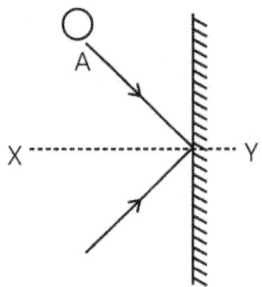

Statement-II: About any point in the plane, the torque of gravity force and normal contact force by ground balance each other

Q.41 Statement-I: The angular velocity of all the points on the laminar rigid body lying in the plane of a body as seen from any other point on it is the same.

Statement-II: The distance between any 2 points on the rigid body remains constant.

Q.42 Consider the following statements:-

Statement-I: The moment of inertia of a rigid body reduces to its minimum value as compared to any other parallel axis when the axis of rotation passes through its center of mass.

Statement-II: The weight of a rigid body always acts through its center of mass in uniform gravitational field.

Q.43 Statement-I: The moment of inertia of any rigid body is minimum about axis which passes through its center of mass as compared to any other parallel axis.

Statement-II: The entire mass of a body can be assumed to be concentrated at its center of mass for applying Newton's force Law.

Q.44 A uniform thin rod of length L is hinged about one of its ends and is free to rotate about the hinge without friction, Neglect the effect of gravity. A force F is applied at a distance x from the hinge on the rod such that force is always perpendicular to rod. As the value of x is increased from zero to L,

Statement-I: The component of reaction force by hinge on the rod perpendicular to length of rod increases.

Statement-II: The angular acceleration of rod increases.

Q.45 Statement-I: For a round shape body of radius R rolling on a fixed ground, the magnitude of velocity of its center is given by ωR, where ω is its angular speed.

Statement-II: When distribution of mass is symmetrical then center of round shape body is its center of mass.

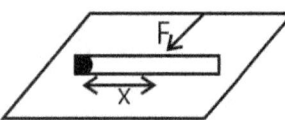

Q.46 Statement-I: a body cannot roll on a smooth horizontal Surface.

Statement-II: when a body rolls purely, the point of contact should be at rest with respect to surface.

Comprehension Type

Paragraph 1:

The figure shows an isosceles triangular plate of mass M and base L. The angle at the apex is 90°. The apex lies at the origin and base is parallel to X – axis

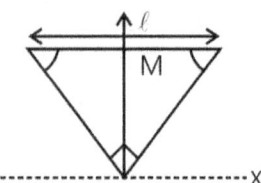

Q.47 The moment of inertia of the plate about the z – axis is

(A) $\dfrac{ML^2}{12}$ (B) $\dfrac{ML^2}{24}$ (C) $\dfrac{ML^2}{6}$ (D) None of these

Q.48 The moment of inertia of the plate about the x axis is

(A) $\dfrac{ML^2}{8}$ (B) $\dfrac{ML^2}{32}$ (C) $\dfrac{ML^2}{24}$ (D) $\dfrac{ML^2}{6}$

Q.49 The moment of inertia of the plate about its base parallel to the x – axis is

(A) $\dfrac{ML^2}{18}$ (B) $\dfrac{ML^2}{36}$ (C) $\dfrac{ML^2}{24}$ (D) None of these

Q.50 The moment of inertia of the plate about the y – axis is

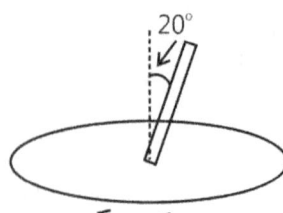

(A) $\dfrac{ML^2}{6}$ (B) $\dfrac{ML^2}{8}$ (C) $\dfrac{ML^2}{24}$ (D) None of these

Paragraph 2:

A uniform rod is fixed to a rotating turntable so that its lower end is on the axis of the turntable and it makes an angle of N20°to the vertical. (The rod is thus rotating with uniform angular velocity about a vertical axis passing through one end.) If the turntable is rotating clockwise as seen from above.

Q.51 What is the direction of the rod's angular momentum vector (calculated about its lower end)?

(A) Vertically downwards

(B) Down at 20°to the horizontal

(C) Up at 20° to the horizontal

(D) Vertically upwards

Q.52 Is there torque acting on it, and if so in what direction?

(A) Yes, vertically

(B) Yes, horizontal

(C) Yes at 20°to the horizontal

(D) No

Paragraph 3:

In the following problems, indicate the correct direction of friction force acting on the cylinder, which is pulled on a rough surface by a constant force F.

Q.53 A cylinder of mass M and radius R is pulled horizontal by a force F. The frictional force can be given by which of the following diagrams

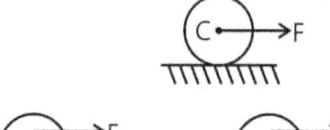

(A)  (B)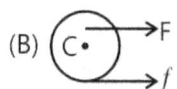

(C) 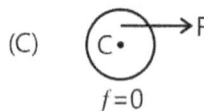 (D) Cannot be interpret

Q.54 A cylinder is pulled horizontally by a force F acting at a point below the center of mass of the cylinder, as shown in figure. The frictional force can be given by which of the following diagrams?

(A) 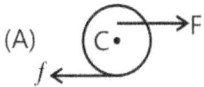 (B)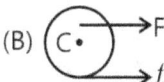

(C) 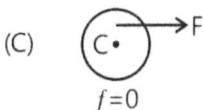 (D) Cannot be interpret

Q.55 A cylinder is pulled horizontally by a force F acting at a point above the center of mass of the cylinder, as shown in figure. The frictional force can be given by which of the following diagrams

(A) 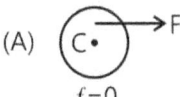 (B)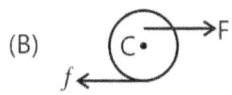

(C) 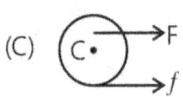 (D) Cannot be interpret

Q.56 A cylinder is placed on a rough plank which in turn is placed on a smooth surface. The plank is pulled with a constant force F. The frictional force can be given by which of the following diagrams.

(A) 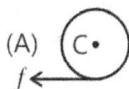 (B)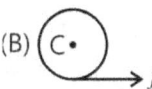

(C) 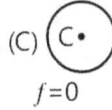 (D) Cannot be interpreted.

Previous Years' Questions

Q.1 A thin uniform angular disc (See figure) of mass M has outer radius 4R and inner radius 3R. The work required to take a unit mass from point P on its axis to infinity is *(2010)*

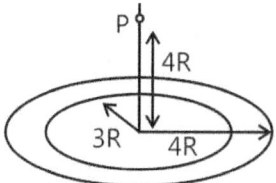

(A) $\dfrac{2GM}{7R}(4\sqrt{2}-5)$ (B) $-\dfrac{2GM}{7R}(4\sqrt{2}-5)$

(C) $\dfrac{GM}{4R}$ (D) $\dfrac{2GM}{5R}(\sqrt{2}-1)$

Q.2 A solid sphere of radius R has moment of inertia I about its geometrical axis. If it's moment of inertia about the tangential axis (which is perpendicular to plane of the disc), is also equal to I, then the value of r is equal to *(2006)*

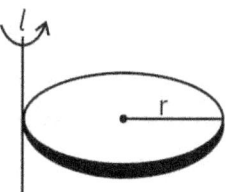

(A) $\dfrac{2}{\sqrt{15}}R$ (B) $\dfrac{2}{\sqrt{5}}R$ (C) $\dfrac{3}{\sqrt{15}}R$ (D) $\dfrac{\sqrt{3}}{\sqrt{15}}R$

Q.3 A block of base 10 cm × 10 cm and height 15 cm is kept on an inclined plane. The coefficient of friction between them is $\sqrt{3}$. The inclination θ of this inclined plane from the horizontal plane is gradually increased from 0°. Then, *(2009)*

(A) At θ = 30°, the block will start sliding down the plane

(B) The block will remains at rest on the plane up to certain θ and then it will topple

(C) At θ = 60°, the block will start sliding down the plane and continue to do so at higher angles

(D) At θ = 60°, the block will start sliding down the plane and on further increasing θ, it will topple at certain θ.

Q.4 From a circular disc of radius R and mass 9M, a small disc of radius R/3 is removed from the disc. The moment of inertia of the remaining disc about axis perpendicular to the plane of the disc and passing through O is *(2010)*

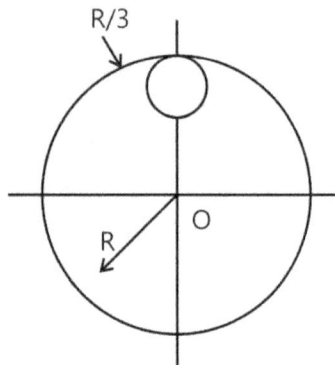

(A) 4 MR²   (B) $\frac{40}{9}$MR²   (C) 10MR²   (D) $\frac{37}{9}$MR²

Q.5 Let I be moment of inertia of a uniform square plate about an axis AB that passes through its center and is parallel to two of its sides. CD is a line in the plane and makes an angle θ with AB. The moment of inertia of the plate about the axis CD is then equal to *(1998)*

(A) I

(B) $I \sin^2 \theta$

(C) $I \cos^2 \theta$

(D) $I \cos^2 (\theta/2)$

Q.6 A solid sphere is in pure rolling motion on an inclined surface having inclination θ *(2006)*

(A) Frictional force acting on sphere is $f = \mu$

(B) f Is dissipative force

(C) Friction will increase its angular velocity and decrease its linear velocity

(D) If θ decreases, friction will decrease

Q.7 A ball moves over a fixed track as shown in figure. From A to B the ball rolls without slipping. If surface BC is frictionless and K_A, K_B and K_C are kinetic energies of the ball at A, B and C respectively, then *(2006)*

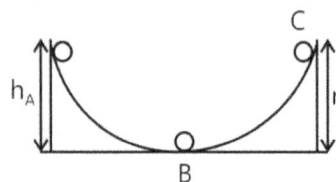

(A) $h_A > h_C$; $K_B > K_C$

(B) $h_A > h_C$; $K_C > K_A$

(C) $h_A = h_C$; $K_B = K_C$

(D) $h_A < h_C$; $K_B > K_C$

Q.8 A child is standing with folded hands at the center of platform rotating about its central axis. The kinetic energy of the system is K. The child now stretches his arms so that the moment of inertia of the system doubles. The kinetic energy of system now is *(2004)*

(A) 2K   (B) $\frac{K}{2}$ (C) $\frac{K}{4}$ (D) 4 K

Q.9 Consider a body, shown in figure, consisting of two identical balls, each of mass M connected by a light rigid rod. If an impulse J = Mv is imparted to the body at one of its end, what would be its angular velocity? *(2003)*

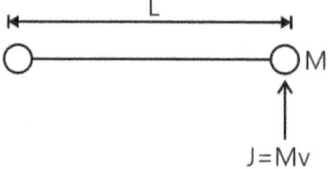

(A) v / L   (B) 2v / L   (C) v / 3L   (D) v / 4L

Q.10 A disc is rolling (without slipping) on a horizontal surface. C is its center and Q and P are two points equidistant from C. Let v_P, v_Q and v_C be the magnitude of velocity of points P, Q, and C respectively, then *(2004)*

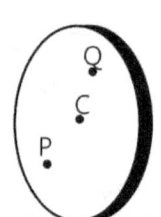

(A) $v_Q > v_C > v_P$

(B) $v_Q < v_C < v_P$

(C) $v_Q = v_P, v_C = \frac{1}{2}v_P$

(D) $v_Q < v_C > v_P$

Paragraph 1: Two discs A and B are mounted coaxially on a vertical axle. The discs have moments of inertia I and 2I respectively about the common axis. Disc A is imparted an initial angular velocity 2ω using the entire potential energy of a spring compressed by a distance x_1. Disc B is imparted an angular velocity ω by a spring having the same spring constant and compressed by a distance x_2. Both the discs rotate in the clockwise direction. *(2007)*

Q.11 The ratio $\frac{x_1}{x_2}$ is

(A) 2   (B) $\frac{1}{2}$ (C) $\sqrt{2}$ (D) $\frac{1}{\sqrt{2}}$

Q.12 When disc B is brought in contact with disc A, they acquire a common angular velocity in time t. The average frictional torque on one disc by the other during this period is.

(A) $\dfrac{2I\omega}{3t}$ (B) $\dfrac{9I\omega}{2t}$ (C) $\dfrac{9I\omega}{4t}$ (D) $\dfrac{3I\omega}{2t}$

Q.13 The loss of kinetic energy during the above process is

(A) $\dfrac{I\omega^2}{2}$ (B) $\dfrac{I\omega^2}{3}$ (C) $\dfrac{I\omega^2}{4}$ (D) $\dfrac{I\omega^2}{6}$

Q.14 A small object of uniform density rolls up a curved surface with an initial velocity v. If reaches up to a maximum height of $3v^2/4g$ with respect to the initial position. The object is

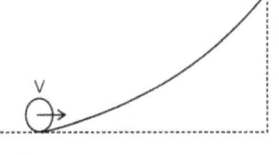

(A) Ring   (B) Solid sphere

(C) Hollow sphere   (D) Disc

Paragraph 2: A uniform thin cylindrical disk of mass M and radius R is attached to two identical mass less springs of spring constant k which are fixed to the wall as shown in the figure. The springs are attached to the disk diametrically on either side at a distance d from its center. The axle is mass less and both the springs and the axle are in a horizontal plane. The un-stretched length of each spring is L. The disk is initially at its equilibrium position with its center of mass (CM) at a distance L from the wall. The disk rolls without slipping with velocity $\vec{V}_0 = v_0\,\hat{i}$. The coefficient of friction is μ.   **(2008)**

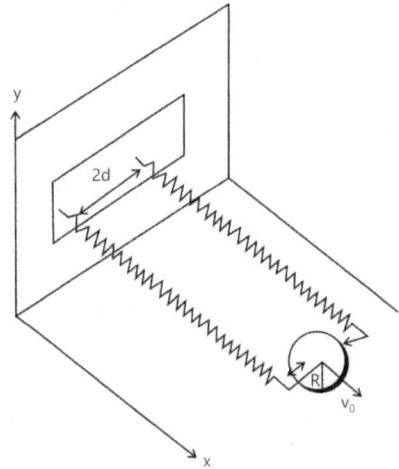

Q.15 The net external force acting on the disk when its center of mass is at displacement x with respect to its equilibrium position is

(A) –kx   (B) –2kx   (C) $-\dfrac{2kx}{3}$ (D) $-\dfrac{4kx}{3}$

Q.16 The center of mass of the disk undergoes simple harmonic motion with angular frequency ω equal to

(A) $\sqrt{\dfrac{k}{M}}$ (B) $\sqrt{\dfrac{2k}{M}}$ (C) $\sqrt{\dfrac{2k}{3M}}$ (D) $\sqrt{\dfrac{4k}{3M}}$

Q.17 The maximum value of v_0 for which the disk will roll without slipping is

(A) $\mu g\sqrt{\dfrac{M}{k}}$ (B) $\mu g\sqrt{\dfrac{M}{2k}}$ (C) $\mu g\sqrt{\dfrac{3M}{k}}$ (D) $\mu g\sqrt{\dfrac{5M}{2k}}$

Q.18 A thin uniform rod, pivoted at O, is rotating in the horizontal plane with constant angular speed ω, as shown in the figure. At time t = 0, a small insect starts from O and moves with constant speed v, with respect to the rod towards the other end. It reaches the end of the rod at t = T and stops. The angular speed of the system remains ω throughout. The magnitude of the torque $\left(|\vec{\tau}|\right)$ about O, as a function of time is best represented by which plot?   **(2012)**

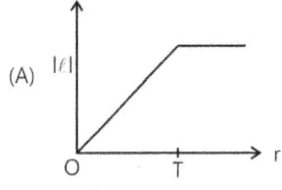 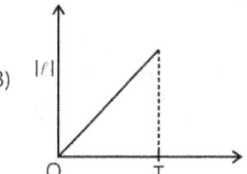

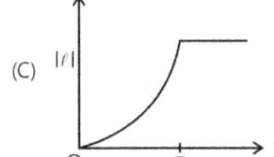 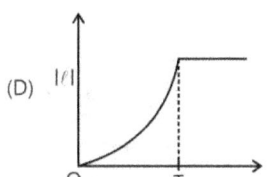

Q.19 A small mass m is attached to a massless string whose other end is fixed at P as shown in the figure. The mass is undergoing circular motion in the x-y plane with centre at O and constant angular speed ω. If the angular momentum of the system, calculated about O and P are denoted by $\vec{L}_O$ and $\vec{L}_P$ respectively, then   **(2012)**

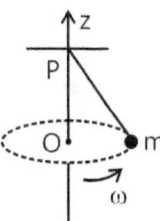

(A) $\vec{L}_O$ and $\vec{L}_P$ do not vary with time.

(B) $\vec{L}_O$ varies with time while $\vec{L}_p$ remains constant.

(C) $\vec{L}_O$ remains constant while $\vec{L}_p$ varies with time.

(D) $\vec{L}_O$ and $\vec{L}_p$ both vary with time.

Q.20 A lamina is made by removing a small disc of diameter 2R from a bigger disc of uniform mass density and radius 2R, as shown in the figure. The moment of inertia of this lamina about axes passing though O and P is I_0 and I_p respectively. Both these axes are perpendicular to the plane of the lamina. The ratio I_p / I_0 to the the nearest integer is **(2012)**

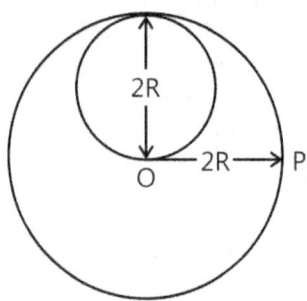

Q.21 Two identical discs of same radius R are rotating abouttheir axes in opposite directions with the same constant angular speed ω. The discs are in the same horizontal plane. At time $t = 0$, the points P and

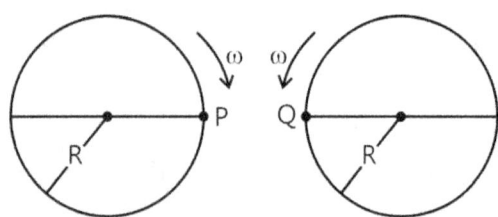

Q are facing each other as shown in the figure. The relative speed between the two points P and Q is v_r In one time period (T) of rotation of the discs, v_r as a function of time is best represented by - **(2012)**

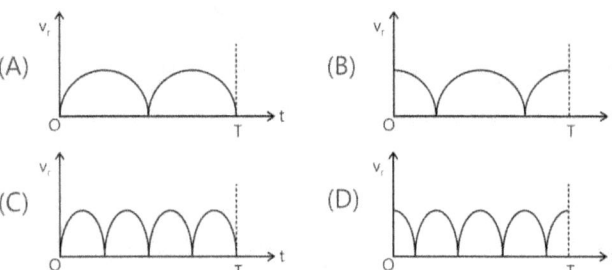

Q.22 Consider a disc rotating in the horizontal plane with a constant angular speed ω about its centre O. The disc has a shaded region on one side of the diameter and an unshaded region on the other side as shown in the figure. When the disc is in the orientation as shown, two pebbles P and Q are simultaneously projected at an angle towards R. The velocity of projection is in the y-z plane and is same for both pebbles with respect to the disc. Assume that (i) they land back on the disc before the disc has completed $\frac{1}{8}$ rotation (ii) their range is less than half the disc radius and (iii) ω remains constant throughout. Then **(2012)**

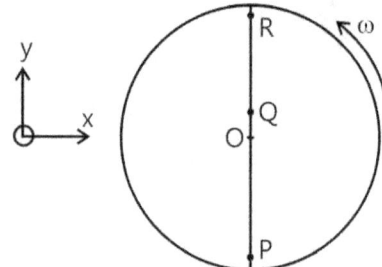

(A) P lands in the shaded region and Q in the unshaded region.

(B) P lands in the unshaded region and Q in the shaded region.

(C) Both P and Q land in the unshaded region.

(D) Both P and Q land in the shaded region.

Paragraph for Questions 23 and 24

The general motion of a rigid body can be considered to be a combination of (i) a motion of its centre of mass about an axis, and (ii) its motion about an instantaneous axis passing through the centre of mass. These axes need not be stationary. Consider, for example, a thin uniform disc welded (rigidly fixed) horizontally at its rim to a massless stick, as shown in the figure. When the disc-stick system is rotated about the origin on a horizontal frictionless plane with angular speed ω, the motion at any instant can be taken as a combination of

(i) a rotation of the centre of mass of the disc about the z-axis, and (ii) a rotation of the disc through an instantaneous vertical axis passing through its centre of mass (as is seen from the changed orientation of points P and Q). Both these motions have the same angular speed ω in this chase.

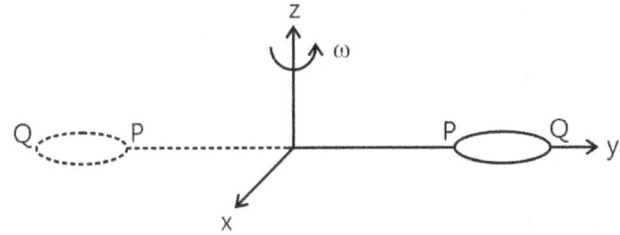

Now consider two similar systems as shown in the figure: Case (a) the disc with its face vertical and parallel to x-y plane; Case (b) the disc with its face making an angle of 45° with x-y plane and its horizontal diameter parallel to x-axis. In both the cases, the disc is welded at point P, and the systems are rotated with constant angular speed ω about the z-axis.

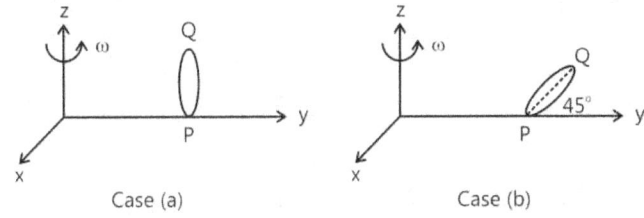

| Case (a) | Case (b) |

Q.23 Which of the following statements about the instantaneous axis (passing through the centre of mass) is correct? *(2012)*

(A) It is vertical for both the cases (a) and (b) (B) It is vertical for case (a); and is at 45° to the x-z plane and lies in the plane of the disc for case (b)

(C) It is horizontal for case (a); and is at 45° to the x-z plane and is normal to the plane of the disc for case (b)

(D) It is vertical for case (a); and is at 45° to the x-z plane and is normal to the plane of the disc for case (b)

Q.24 Which of the following statements regarding the angular speed about the instantaneous axis (passing through the centre of mass) is correct? *(2012)*

(A) It is $\sqrt{2}\,\omega$ for both the cases

(B) It is ω for case (a); and $\dfrac{\omega}{\sqrt{2}}$ for case (b)

(C) It is ω for case (a); and $\sqrt{2}\,\omega$ for case (b)

(D) It is ω for both the cases

Q.25 The figure shows a system consisting of (i) a ring of outer radius 3R rolling clockwise without slipping on a horizontal surface with angular speed ω and (ii) an inner disc of radius 2R rotating anti-clockwise with angular speed $\omega/2$. The ring and disc are separated by frictionless ball bearings. The system is in the x-z plane. The point P on the inner disc is at a distance R from the origin, where OP makes an angle of 30°

with the horizontal. Then with respect to the horizontal surface, *(2012)*

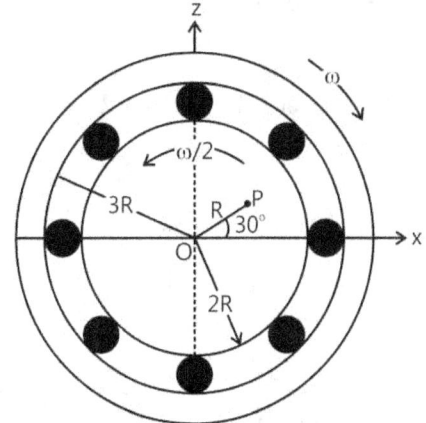

(A) The point O has a linear velocity $3R\omega\hat{i}$

(B) The point P has a linear velocity $\dfrac{11}{4}R\omega\hat{i} + \dfrac{\sqrt{3}}{4}R\omega\hat{k}$

(C) The point P has a linear velocity $\dfrac{13}{4}R\omega\hat{i} - \dfrac{\sqrt{3}}{4}R\omega\hat{k}$

(D) The point P has a linear velocity $\left(3 - \dfrac{\sqrt{3}}{4}\right)R\omega\hat{i} + \dfrac{1}{4}R\omega\hat{k}$

Q.26 Two solid cylinders P and Q of same mass and same radius start rolling down a flixed inclined plane from the same height at the same time. Cylinder P has most of its mass concentrated near its surface, while Q has most of its mass concentrated near the axis. Which statement(s) is (are) correct? *(2012)*

(A) Both cylinders P and Q reach the ground at the same time.

(B) Cylinder P has larger linear acceleration than cylinder Q.

(C) Both cylinders reach the ground with same translational kinetic energy

(D) Cylinder Q reaches the ground with larger angular speed

Q.27 A uniform circular disc of mass 50 kg and radius 0.4 m is rotating with an angular velocity of 10 rad s⁻¹ about its own axis, which is vertical. Two uniform circular rings, each of mass 6.25 kg and radius 0.2 m, are gently placed symmetrically on the disc in such a manner that they are touching each other along the axis of the disc and are horizontal. Assume that the friction is large enough such that the rings are at rest relative to the disc and the system rotates about the original axis. The new angular velocity (in rad s⁻¹) of the system is *(2013)*

3.77

Q.28 In the figure, a ladder of mass m is shown leaning against a wall. It is in static equilibrium making an angle θ with the horizontal floor. The coefficient of friction between the wall and the ladder is μ_1 and that between the floor and the ladder is μ_2. The normal reaction of the wall on the ladder is N_1 and that of the floor is N_2. If the ladder is about to slip, then **(2014)**

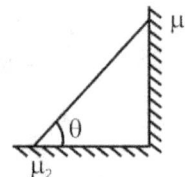

(A) $\mu_1 = 0 \ \mu_2 \neq 0$ and $N_2 \tan\theta = \dfrac{mg}{2}$

(B) $\mu_1 \neq 0 \ \mu_2 = 0$ and $N_1 \tan\theta = \dfrac{mg}{2}$

(C) $\mu_1 \neq 0 \ \mu_2 \neq 0$ and $N_2 = \dfrac{mg}{1 + \mu_1 \mu_2}$

(D) $\mu_1 = 0 \ \mu_2 \neq 0$ and $N_1 \tan\theta = \dfrac{mg}{2}$

Q.29 A uniform circular disc of mass 1.5 kg and radius 0.5 m is initially at rest on a horizontal frictionless surface. Three forces of equal magnitude F = 0.5 N are applied simultaneously along the three sides of an equilateral triangle XYZ with its vertices on the perimeter of the disc (see figure). One second after applying the forces, the angular speed of the disc in rad s⁻¹ is **(2014)**

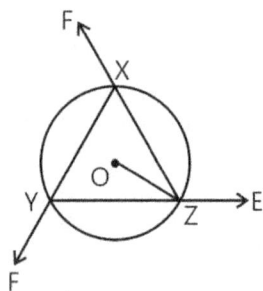

Q.30 A horizontal circular platform of radius 0.5 m and mass 0.45 kg is free to rotate about its axis. Two massless spring toy -guns, each carrying a steel ball of mass 0.05 kg are attached to the platform at a distance 0.25 m from the centre on its either sides along its diameter (see figure).Each gun simultaneously fires the balls horizontally and perpendicular to the diameter in opposite directions. After leaving the platform, the balls have horizontal speed of 9ms⁻¹ with with respect to the ground. The rotational speed of the platform in rad s⁻¹ after the balls leave the platform is **(2014)**

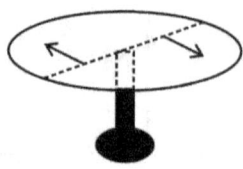

Q.31 Two identical uniform discs roll without slipping on two different surfaces AB and CD (see figure) starting at A and C with linear speeds v_1 and v_2, respectively ,and always remain in contact with the surfaces. If they reach B and D with the same linear speed and $v_1 = 3\,m/s$, then v_2 in m/s is (g = 10 m/s²) **(2015)**

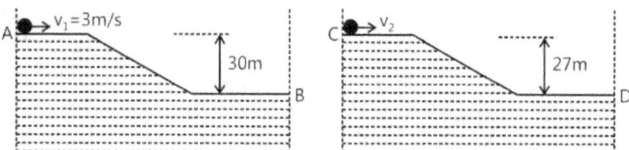

Q.32 A ring of mass M and radius R is rotating with angular speed ω about a fixed vertical axis passing through its centre O with two point masses each of mass $\dfrac{M}{8}$ at rest at O. These masses can move radially outwards along two massless rods fixed on the ring as shown in the figure. At some instant the angular speed of the system is $\dfrac{8}{9}\omega$ and one of the masses is at a distance of $\dfrac{3}{5}R$ from O. At this instant the distance of the other mass from O is **(2015)**

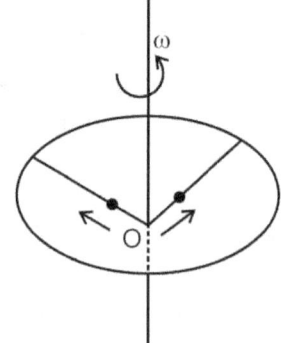

(A) $\dfrac{2}{3}R$ (B) $\dfrac{1}{3}R$ (C) $\dfrac{3}{5}R$ (D) $\dfrac{4}{5}R$

Q.33 The densities of two solid spheres A and B of the same radii R vary with radial distance r as $\rho_A(r) = k\left(\dfrac{r}{R}\right)$ and $\rho_B(r) = k\left(\dfrac{r}{R}\right)^5$, respectively, where k is a constant. The moments of inertia of the individual spheres about axes passing through their centres are I_A and I_B, respectively. If $\dfrac{I_B}{I_A} = \dfrac{n}{10}$, the value of n is **(2015)**

Q.34 A uniform wooden stick of mass 1.6 kg and length ℓ rests in an inclined manner on a smooth, vertical wall of height $h\,(<\ell)$ such that a small portion of the stick extends beyond the wall. The reaction force of the wall on the stick is perpendicular to the stick. The stick makes an angle of 30° with the wall and the bottom of the stick is on a rough floor. The reaction of the wall on the stick is equal in magnitude to the reaction of the floor on the stick. The ratio h/ℓ and the frictional force fat the bottom of the stick are $\left(g = 10 \text{ ms}^{-2}\right)$ **(2016)**

(A) $\dfrac{h}{\ell} = \dfrac{\sqrt{3}}{16}, f = \dfrac{16\sqrt{3}}{3}$ N

(B) $\dfrac{h}{\ell} = \dfrac{3}{16}, f = \dfrac{16\sqrt{3}}{3}$ N

(C) $2\dfrac{h}{\ell} = \dfrac{3\sqrt{3}}{16}, f = \dfrac{8\sqrt{3}}{3}$ N

(D) $\dfrac{h}{\ell} = \dfrac{3\sqrt{3}}{16}, f = \dfrac{16\sqrt{3}}{3}$ N

Q.35 The position vector $\vec{r}$ of a particle of mass m is given by the following equation $\vec{r}(t) = \alpha t^3 \hat{i} + \beta t^2 \hat{j}$, where $\alpha = 10/3 \text{ ms}^{-3}$, $\beta = 5 \text{ ms}^{-2}$ and $m = 0.1$ kg. At t = 1 s, which of the following statement(s) is(are) true about the particle? **(2016)**

(A) The velocity $\vec{v}$ is given $\vec{v} = \left(10\,\hat{i} + 10\hat{j}\right) \text{ ms}^{-1}$

(B) The angular momentum $\vec{L}$ with respect to the origin is given by $\vec{L} = -(5/3)\hat{k}$ Nms

(C) The force $\vec{F}$ is given by $\vec{F} = \left(\hat{i} + 2\hat{j}\right)$ N

(D) The torque $\vec{\tau} = -(20/3)\hat{k}$ N m

Q.36 Two thin circular discs of mass m and 4m, having radii of a and 2a, respectively, are rigidly fixed by a massless, rigid rod of length $\ell = \sqrt{24}$ a through their centers. This assembly is laid on a firm and flat surface, and set rolling without slipping on the surface so that the angular speed about the axis of the rod is ω. The angular momentum of the entire assembly about the point 'O' is $\vec{L}$ (see the figure). Which of the following statement(s) is(are) true? **(2016)**

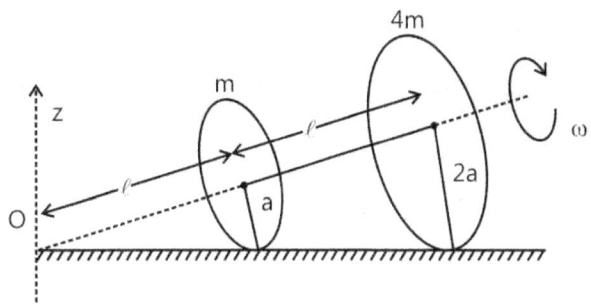

(A) The magnitude of angular momentum of the assembly about its center of mass is $17\,ma^2\,\omega/2$

(B) The magnitude of the z-component of $\vec{L}$ is $55\,ma^2\,\omega$

(C) The magnitude of angular momentum of center of mass of the assembly about the point O is $81\,ma^2\,\omega$

(D) The center of mass of the assembly rotates about the z-axis with an angular speed of $\omega/5$

Important Questions

JEE Main/Boards

Exercise 1

Q.19   Q.22   Q.23

Q.27   Q.28

Exercise 2

Q.12   Q.25

Previous Years' Questions

Q.4   Q.7   Q.9

Q.12

JEE Advanced/Boards

Exercise 1

Q.5   Q.7   Q.8

Q.10   Q.11   Q.21

Q.24

Exercise 2

Q.7   Q.9   Q.21

Q.22   Q.28   Q.41

Q.54

Previous Years' Questions

Q.1   Q.3

Answer Key

JEE Main/Boards

Exercise 1

Q.1 kg m^2, [M^{1}L^{2}T^0], No

Q.2 Torque

Q.3 Inertia

Q.4 Theorem of parallel axes and theorem of perpendicular axes

Q.5 $I = \dfrac{2}{5}MR^2$, M = mass & R = radius

Q.6 $I = \dfrac{2}{3}MR^2$, M = mass & R = radius

Q.8 No

Q.9 Hollow sphere

Q.10 $I_B > I_A$

Q.11 $K_B > K_A$

Q.18 8 g

Q.19 1.584×10^7 g cm^2

Q.20 1.25 kg m^2;

Q.21 9.83×10^{37} kg m^2

Q.22 3.5×10^4 g cm^2 ; 1.75×10^4 g cm^2 ; 8.75×10^4 g cm^2 ; 10.5×10^4 g cm^2

Q.23 $\dfrac{5}{4}$mL^2

Q.24 0.01J

Q.26 108 days

Q.27 80π kg m^2 s^{-1}

Q.28 360 r.p.m

Exercise 2

Single Correct Choice Type

Q.1 B	**Q.2** B	**Q.3** A	**Q.4** B	**Q.5** D	**Q.6** D
Q.7 D	**Q.8** B	**Q.9** B	**Q.10** C	**Q.11** A	**Q.12** D
Q.13 B	**Q.14** B	**Q.15** C	**Q.16** B	**Q.17** B	**Q.18** B
Q.19 B	**Q.20** A	**Q.21** D	**Q.22** C	**Q.23** C	**Q.24** B
Q.25 D	**Q.26** B	**Q.27** C			

Previous Years' Questions

Q.1. A	**Q.2** C	**Q.3** B	**Q.4** A	**Q.5** A	**Q.6** B
Q.7 C	**Q.8** C	**Q.9** C	**Q.10** B	**Q.11** A	**Q.12** C
Q.13 C	**Q.14** A	**Q.15** D	**Q.16** D	**Q.17** B	**Q.18** A
Q.19 C	**Q.20** B	**Q.21** A, C	**Q.22** D		

JEE Advanced/Boards

Exercise 1

Q.1 $2Mg$

Q.2 $5\sqrt{5}$ m/s

Q.3 3

Q.4 $2F/M$

Q.5 2m/s

Q.6 $\vec{F_A} = (-133.64\,\hat{i} + 196\,\hat{j})$N and $\vec{F_B} = 133.64\,\hat{i}$

Q.7 (a) $\dfrac{3a\rho_0}{2}$; (b) $\dfrac{5a}{9}$; (c) $\dfrac{7\rho_0 a^3}{12}$; (d) $\dfrac{12}{7\rho_0 a^2}$;

(e) $\sqrt{\dfrac{7}{4}\rho_0^2 g a^3}$

Q.8 300 rad/sec, 150 rad/sec

Q.9 16 m/s^2

Q.10 $v = \sqrt{\dfrac{27}{7}gR}$

Q.11 (a) $\dfrac{200}{7}$ N; (b) $4\sqrt{\dfrac{3}{7}}$ m/s

Q.12 $\dfrac{3L^2\rho}{8\pi^2}$

Q.13 1.65 N, 1.224 m

Q.14 $L/6$

Q.15 (i) (a) $\dfrac{1.2g}{\cdot c}$ (clockwise) (b) $-0.3g(\hat{i} + 2\,\hat{j})$

(ii) (a) $\dfrac{2.4g}{c}$ (clockwise) (b) 0.5g

Q.16 $V = \dfrac{\sqrt{14gR}}{3}$

Q.17 (i) $v = \dfrac{2v_0}{3}$ (ii) $t = \dfrac{v_0}{3\mu g}$

$w = \dfrac{1}{2}[3\mu^2 mg^2 t^2 - 2\mu\, mg\, t\, v_0](t < t_0)$,

$w = -\dfrac{1}{6}mv_0^2 (t > t_0)$

Q.18 $t = 2\sqrt{5}$ sec, $q = 4\pi/5$ rad

Q.19 (a) $\omega/3$, (b) $\dfrac{\sqrt{37}}{3}$ m ω R, (c) $\dfrac{\sqrt{37}}{3}$ m ω R

Q.20 $\sqrt{15}$ ft/sec

Q.21 (a) 5 m/s^2, (b) 0.3 < h < 1.5 m

JEE Main/Boards

Exercise 1

Sol 1: Moment of inertia of any body is given by $I = \int r^2 dm$ where 'dm' is mass of a small element under consideration and 'r' is the distance between the axis of rotation and the element. Then,

S.I. units of moment of inertia will be m^2 kg or $kg.m^2$

DIMENSIONS –

$[M^0LT^0]^2[M^1L^0T^0] = [M^1L^2T^0]$

It is not a vector quantity since direction is nowhere considered.

Sol 2: Torque is the rotational analogue of force.

Sol 3: Moment of inertia is the rotational analogue of mass of a body.

Sol 4: The theorem of parallel axes

$I = I_{cm} + m.d^2$

where I_{cm} = Moment of Inertia about an axis passing through center of mass and parallel to the considered axis

m = mass of the body

d = distance between the axis (about which the value of I is required) and the axis passing through center of mass and parallel to the considered axis.

The theorem of perpendicular axes

$I_z = I_x + I_y$

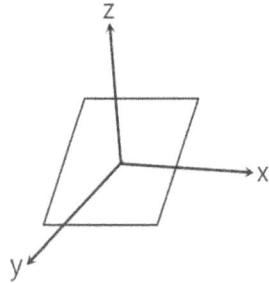

This theorem is applicable for planar bodies only and I_x and I_y should be about two perpendicular axes lying in plane.

Sol 5: Moment of inertia of a solid sphere about its diameter is

$I = \dfrac{2}{5}MR^2$

where M = mass of the sphere

R = Radius of the sphere

Derivation:

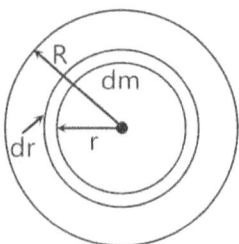

Considering, the solid sphere as a large group of hollow spheres whose radii range from 0 to R with a thickness of dr. Integrating moment of inertia of these dements gives the required value.

$dm = \rho. 4\pi r^2.dr$ (dm = ρ. dv) and $I = \int dI$

$I = \int_0^R \dfrac{2}{3}.r^2 \ \rho. 4\pi r^2 dr$ ($\because I = \int r^2 dm$) and

$\left(I_{hollow\ sphere} = \dfrac{2}{3}MR^2\right)$

$\Rightarrow dI = \dfrac{2}{3}(dm)r^2 = \dfrac{2}{3}. 4\pi\rho\left[\dfrac{r^5}{5}\right]_0^R$

$= 4\pi\rho. \dfrac{R^4}{5} = \dfrac{2}{5}MR^2 \left[\because \rho = \dfrac{M}{\dfrac{4\pi}{3}R^3}\right]$

Sol 6: The moment of inertia of a hollow sphere about an axis passing through its center is $I = \dfrac{2}{3}MR^2$

where M = mass of sphere

R = Radius of sphere

Derivation:

Considering the hollow sphere, as a large group of rings with thickness dr and radii ranging from 0 to R. Integrating the moment of inertia of these rings we will get the required moment of inertia.

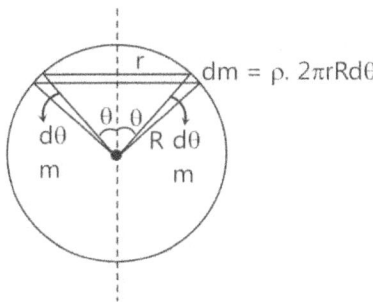

$dm = \rho . 2\pi r R d\theta$

$I = \int dI = \int r^2 dm = \int r . \rho \, 2\pi r R d\theta$

$[\because I_{ring} = MR^2]$

$\Rightarrow dI = dmr^2 \quad \Rightarrow I = \int 2\pi\rho R (R^3 \sin^3\theta)d\theta$

$I = 2\pi\rho R^4 \int\limits_0^\pi \sin^3\theta d\theta$

$\Rightarrow I = \dfrac{MR^2}{2} \int\limits_0^\pi \dfrac{(3\sin\theta - \sin 3\theta)}{4} . d\theta$

$\Rightarrow I = \dfrac{MR^2}{2} \left[-\dfrac{3\cos\theta}{4} + \dfrac{\cos 3\theta}{12} \right]_0^\pi$

$\Rightarrow I = \dfrac{MR^2}{2} \left[\dfrac{3}{4} - \dfrac{1}{12} - \left[-\dfrac{3}{4} + \dfrac{1}{12} \right] \right] \Rightarrow I = \dfrac{2}{3}MR^2$

Sol 7: Factors on which moment of inertia depend

→ Mass of the body

→ Mass distribution of the body

→ Size of the body

→ Axis about which moment of inertia is required

Sol 8: No, it is not a constant

It depends on the axis about which moment of inertia is calculated

Since, $K = \sqrt{\dfrac{I_{axis}}{M}}$

I_{axis} = moment of inertia about a given axis

M = mass of the body (constant)

Sol 9: Given

A solid sphere and hollow sphere have same mass and same radius

$I_{solid\,sphere} = \dfrac{2}{5}MR^2$

$I_{hollow\,sphere} = \dfrac{2}{3}MR^2$

$\therefore I_{hollow\,sphere} > I_{solid\,sphere}$

Sol 10: Given,

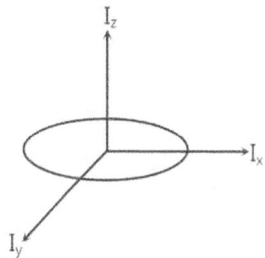

Circular discs A and B of same mass and same thickness but different densities d_A and d_B ($d_A > d_B$)

$M_A = M_B$

$\Rightarrow \pi R_A^2 . d_A t_A = \pi R_B^2 . d_B . t_B$

$\Rightarrow \dfrac{R_A}{R_B} = \left(\dfrac{d_B}{d_A} \right)^{1/2}$

$(I_A)_x = \dfrac{M_A R_A^2}{4} = (I_A)_y$

$\Rightarrow (I_A)_z = (I_A)_x + (I_A)_y = \dfrac{M_A R_A^2}{2}$

[By perpendicular axes theorem]

also

$(I_B)_z = \dfrac{M_B R_B^2}{2}$

$\dfrac{I_A}{I_B} = \left(\dfrac{R_A}{R_B} \right)^2 = \dfrac{d_B}{d_A} \Rightarrow I_A < I_B$ since $d_B < d_A$

Sol 11: Given.

Moment of inertia of two rotating bodies A and B as

I_A and I_B ($I_A > I_B$)

and

Angular moment (L_A and L_B) are equal

$\Rightarrow L_A = L_B$

then kinetic energies

$(K.E.)_A$ and $(K.E.)_B$ will be

$\dfrac{1}{2}L_A^2 / I_A$ and $\dfrac{1}{2}L_B^2 / I_B$ $\left[\because K.E. = \dfrac{1}{2}I\omega^2 = \dfrac{1}{2}L\omega = \dfrac{1}{2}\dfrac{L^2}{I} \right]$

$\Rightarrow (K.E.)_A < (K.E.)_B$ since $I_B < I_A$

Sol 12: Moment of inertia is the measure of tendency

of a body to resist rotational motion if it was at rest or resist being stopped it was rotating.

Radius of Gyration is the radius of the circle is which a zero sized particle of mass M (which is equal to the mass of a body considered) such that the moment of inertia of both the particle and the body about any axis are equal

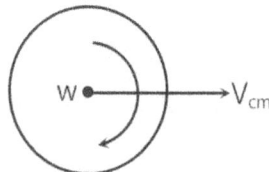

$$\Rightarrow K = \sqrt{\dfrac{I}{M}}$$

Sol 13: Kinetic energy of rolling motion

Consider a body rolling with angular velocity ω and linear center of mass velocity v_{cm}

Velocity of a particle at any point is given by

$(V_{cm} + r\omega \cos\theta)\, \hat{i} + (r\omega \sin\theta)\, \hat{j}$

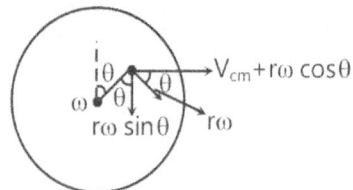

$\text{K.E.} = \int d(\text{K.E.})$

$= \int \dfrac{1}{2} dm \left((V_{cm} + r\omega\cos\theta)^2 + (r\omega\sin\theta)^2 \right)$

$= \dfrac{1}{2} \int_0^M r^2\omega^2 dm + \dfrac{1}{2} \int_0^M V_{cm}^2\, dm +$

$\dfrac{1}{2} \int_0^M 2V_{cm}\, r\omega\cos\theta\, dm$

$= \dfrac{1}{2} I_{cm}\, \omega^2 + \dfrac{1}{2} V_{cm}^2 + V_{cm}\, \omega \int_0^m r\cos\theta\, dm$

$= \dfrac{1}{2} I_{cm}\, \omega^2 + \dfrac{1}{2} MV_{cm}^2 + V_{cm}\, \omega \int_0^R \int_0^{2\pi} r\cos\theta.d\theta.dr$

$= \dfrac{1}{2} I_{cm}\omega^2 + \dfrac{1}{2} MV_{cm}^2 \left[\because \int_0^{2\pi} \cos\theta\, d\theta = 0 \right]$

$\therefore \text{K.E.} = \dfrac{1}{2} I_{cm}\, \omega^2 + \dfrac{1}{2} MV_{cm}^2$

Sol 14: Laws of rotational motion

First law: Every body has tendency to be in rest or is state of rotation unless acted upon by a torque

Second law: The torque applied on a body is moment of inertia times the angular acceleration of the body $J - I\alpha$

Third law: Every action (torque) has an equal and opposite reaction (torque) on the body which gave the action.

Sol 15: Newton's second law of linear motion is

F = ma

where F = force acting on the body

m = mass of body

a = acceleration of body

Now, consider 'dm' part of as body acted upon by a force F and the body is rotating with an angular acceleration of 'α'

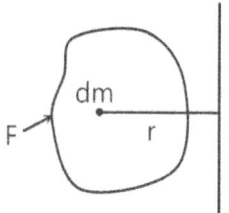

$dT = r \times F$

$= r.\, dm\, (r\alpha)$

$= r^2.\, dm.\, \alpha$

$\int dT = \int r^2 dm.\alpha$

$\Rightarrow T_{axis} = I_{axis}\, \alpha_{axis}$

Sol 16: Principle of conservation of angular momentum.

In absence of a torque, the angular momentum of a body is always conserved (constant)

$T = \dfrac{dL}{dt} = 0$

$\Rightarrow L = \text{constant} \Rightarrow I\omega = \text{constant}$

or $I_1\omega_1 = I_2\omega_2$

Example: By stretching hands a ballet decreases the angular speed of the body by increasing the moment of inertia

Example: A person sitting is a chair which can rotate holds a rotating wheel in hands and when he flips the wheel, the person along with chair rotates conserving angular momentum.

Sol 17: Consider a circular ring of radius R and mass M

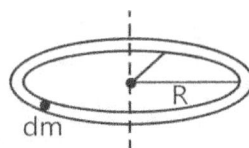

$$I = \int dI = \int R^2 . dm = MR^2$$

Sol 18: Given,

$I = 200 \ g \ cm^2$

$r = 5 \ cm$

We know that for a thin ring the moment of inertia about an axis passing through center is Mr^2

$$\Rightarrow M = \frac{I}{r^2} = 8 \ grams$$

Sol 19: Given,

diameter of disc = 40 cm

thickness of disc = 7 cm

density of disc = 9 gm cm^{-3}

mass of the disc = $\rho.V = \rho.\dfrac{\pi D^2}{4}.t$

= $9 \times \pi \times 400 \times 7 = 79168.13 \ grams$

moment of inertia of disc about a transverse axis through the center of the disc is

$$I = \frac{MR^2}{2}$$

$$\Rightarrow I = \frac{79168.13 \times 20 \times 20}{2} \Rightarrow I = 1.584 \times 10^7 \ g \ cm^2$$

Sol 20: Given,

A uniform circular disc and A Uniform circular ring of same mass of 10 kg and diameter of 1 m

Moment of inertia of disc about a transverse axis through center of the disc is $I = \dfrac{1}{2}MR^2$

$$\Rightarrow I = \frac{1}{2} \times 10 \times \left(\frac{1}{2}\right)^2 = 1.25 \ kg \ m^2$$

Sol 21: Given,

Radius of earth assuming it as a sphere as 6400 km mass of earth is 6×10^{24} kg

Moment of inertia of the earth is

$$I = \frac{2}{5}MR^2 \ (\because I = \frac{2}{5}MR^2 \ for \ a \ sphere \ (solid))$$

$$\Rightarrow I = \frac{2}{5} \times 6 \times 10^{24} \times (6400 \times 10^3)^2$$

$$\Rightarrow I = 9.8304 \times 10^{37} \ kgm^2$$

Sol 22: Given,

A Uniform circular disc of mass 700 gms and diameter 20 cm

(i) Moment of inertia about the transverse axis through center of disc is

$$I = \frac{MR^2}{2}$$

$$\Rightarrow I = \frac{700 \times 10 \times 10}{2} = 3.5 \times 10^4 \ gcm^2$$

(ii) Moment of inertia about the diameter of disc is

$$I = \frac{MR^2}{4}$$

$[\because I_x = I_y$ and $I_x + I_y = I_z$

(perpendicular axis theorem)]

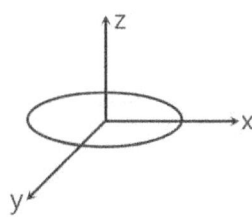

$$\Rightarrow I = \frac{700 \times (10 \times 10)}{4} = 1.75 \times 10^4 gcm^2$$

(iii)

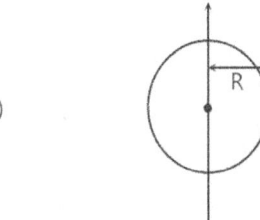

Moment of inertia about a tangent is plane is

$I = I_d + MR^2$ (parllel axis theorem)

$$\Rightarrow I = \frac{5MR^2}{4}$$

$$\Rightarrow I = \frac{5}{4} \times (700) \times (10 \times 10) = 8.75 \times 10^4 \ gcm^2$$

(iv)

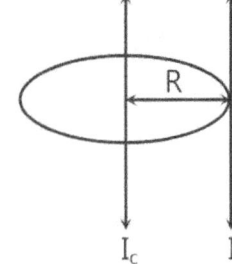

Moment of inertia about a tangent perpendicular to the plane is

$I = I_c + MR^2$

$\Rightarrow I = \dfrac{3}{2} MR^2$

$\Rightarrow I = \dfrac{3}{2} \times 700 \times 10 \times 10 = 10.5 \times 10^4 \text{ gcm}^2$

Sol 23:

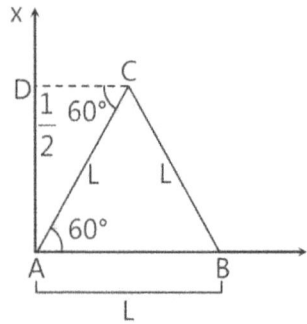

Given, particles of masses m are placed at A, B and C and side of triangle ABC is L

Then,

$CD = AC \cos 60° = \dfrac{AC}{2} = \dfrac{AB}{2} = \dfrac{L}{2}$

moment of inertia of system is

$I = I_A + I_B + I_C$

$\Rightarrow I = m(0)^2 + m(L)^2 + m\left(\dfrac{L}{2}\right)^2$

$\Rightarrow I = \dfrac{5mL^2}{4}$

Sol 24: Given,

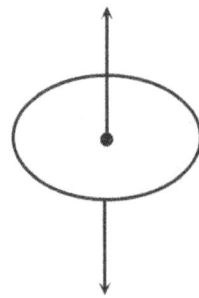

A circular disc of mass 1 kg and radius 0.2 m rotating about transverse axis passing through its center.

Moment of inertia about the given axis as

$I = \dfrac{MR^2}{2} \Rightarrow I = \dfrac{1}{2} \times (0.2)^2 = 0.02 \text{ kg–m}^2$

Given, disc makes $\dfrac{30}{\pi}$ rotations per minute then,

Angular velocity $= 2\pi \times \dfrac{30}{11} \times \dfrac{1}{60}$ rad/s

$= 1$ rad/s

kinetic energy $= \dfrac{1}{2} I\omega^2$

$= \dfrac{1}{2} \times 0.02 \times (1)^2 = 0.01$ joules

Sol 25: Given,

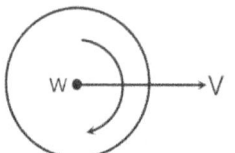

A circular dice of mass M and radius r is set rolling on table the kinetic energy of the disc is given by

$K.E = \dfrac{1}{2} I\omega^2 + \dfrac{1}{2} mv^2$ (refer Q. 14)

$\Rightarrow K.E. = \dfrac{1}{2} \dfrac{mr^2}{2} \omega^2 + \dfrac{1}{2} mv^2$

$\Rightarrow K.T. = \dfrac{mv^2}{4} + \dfrac{mv^2}{2}$

[For pure rolling $v = r\omega$]

$\Rightarrow K.E. = \dfrac{3}{4} mv^2$

Sol 26: Given,

Sun rotates around itself once in 27 days angular velocity of sum $= \dfrac{2\pi}{27}$ rad/days

If it expands to truce its present diameter, then moment of inertia becomes 4

$\therefore I = \dfrac{2}{5} MR^2$ (for sphere)

$\dfrac{I_2}{I_1} = \dfrac{R_2^2}{R_1^2}$

$\Rightarrow I_2 = 4I_1$

$I_1\omega_1 = I_2\omega_2$

(By principle of conservation of angular momentum)

$$\Rightarrow \omega_2 = \frac{\omega_1}{4} = \frac{\omega_1}{108} \text{ rad/days}$$

Then, the new period of revolution

$$= \frac{2\pi}{(2\pi/108)} = 108 \text{ days}$$

$$\because \left(T = \frac{\pi}{\omega} \right)$$

Sol 27: Given,

A 40 kg of flywheel is form of a uniform circular disc 1 meter in radius is making 120 r.p.m.

Angular velocity = $120 \times 2\pi \times \dfrac{1}{60}$ rad/s

= 4π rad/s

Moment of inertia about transverse axis passing through center is

$$I = \frac{MR^2}{2}$$

$$\Rightarrow I = \frac{40 \times (1)^2}{2} = 20 \text{ kg–m}^2$$

angular momentum = $I.\omega$

= 80π kgm²/s = 251.33 kgm²/s

Sol 28: Given,

Angular velocity of the system = 120 r.p.m

Moment of inertia of the system initially = 6 kgm²

Moment of inertia of the system finally = 2 kgm²

$I_1\omega_1 = I_2\omega_2$

(By principle of conservation of angular momentum)

$$\Rightarrow \omega_2 = \frac{I_1}{I_2} \times \omega_1 = 360 \text{ r.p.m.}$$

∴ Final angular velocity = 360 r.p.m

Exercise 2

Single Correct Choice Type

Sol 1: (B)

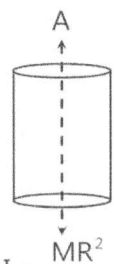 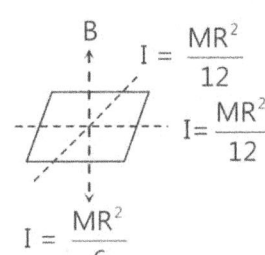

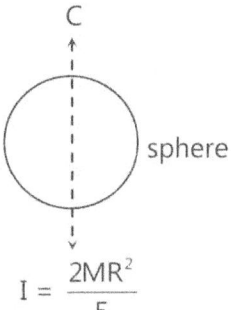

B< C < A

Sol 2: (B)

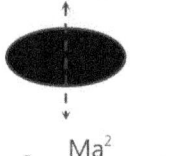

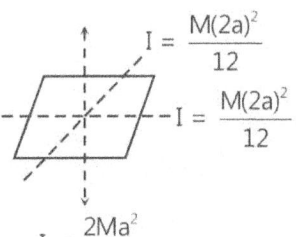

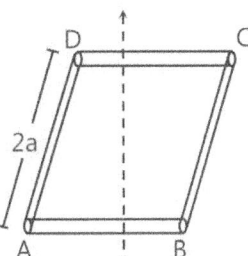

Mass of each rod = $\dfrac{M}{4}$

Moment of inertia of one rod about the given axis is

$$I = \left(\frac{M(2a)^2}{4 \times 12}\right) + \frac{Ma^2}{4} = \frac{4Ma^2}{4 \times 3} = \frac{Ma^2}{3}$$

Total moment of inertia $= 4I = \dfrac{Ma^2}{3}$

Sol 3: (A) Given,

I_1 is the moment of inertia about perpendicular bisector of rod

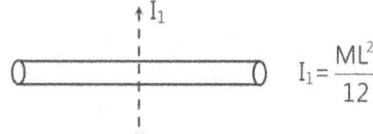

$I_1 = \dfrac{ML^2}{12}$

Now rod is bent into semi-circular arc

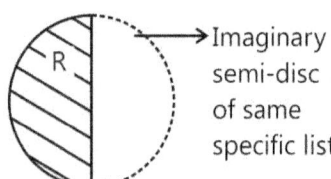

Length of arc $= L = \pi R$

Moment of inertia of the arc $= MR^2$

$$= \frac{ML^2}{\pi^2} = I_2$$

$I_2 > I_1$ (since $\pi^2 < 12$)

Sol 4: (B) So, $2I = \dfrac{1}{2}(M+M)R^2 \Rightarrow 2I = \dfrac{1}{2}(2M)R^2$

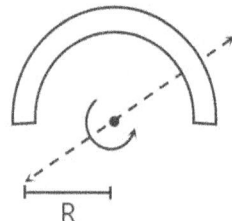

Imaginary semi-disc of same specific list

Mars$=$M
Radius$=$R

$$\Rightarrow \boxed{I = \frac{1}{2}MR^2}$$

Sol 5: (D) Given,

I' as a function of x of a rigid body

$I = 2x^2 - 12x + 27$

The value of moment of inertia is minimum at center of mass point

To calculate min value of I, differentiate w.r.t. x and equate it to 0

$$\frac{dI}{dx} = 0$$

$\Rightarrow 4x - 12 = 0$

$\Rightarrow x = 3$

To check whether it is minimum,

$$\frac{d^2I}{dx^2} > 0$$

$\Rightarrow 4 > 0$

$\therefore$ x = 3 is the x-coordinate of center of mass.

Sol 6: (D)

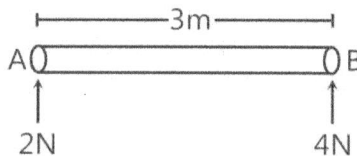

To keep the body in equilibrium a 6 N acts at point x from A

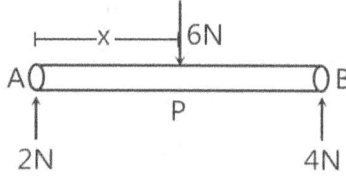

Taking moment of torques about A

$\sum M_A = 0$ (In equilibrium)

$\Rightarrow 6 \times x - 4 \times 3 = 0$

$\Rightarrow x = 2m$

Sol 7: (D)

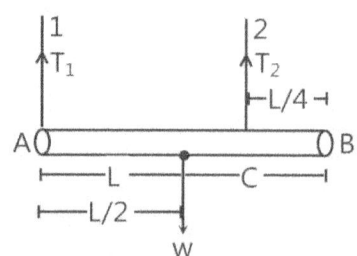

Let T_1 and T_2 be tensions is the strings considering force equilibrium

$T_1 + T_2 = W$

Also Considering moment equilibrium about A

$\sum M_A = 0$

$\Rightarrow T_2 \times \dfrac{3L}{4} - W \times \dfrac{L}{2} = 0$

$\Rightarrow T_2 = \dfrac{2W}{3} \Rightarrow T_1 = \dfrac{W}{3}$

Sol 8: (B)

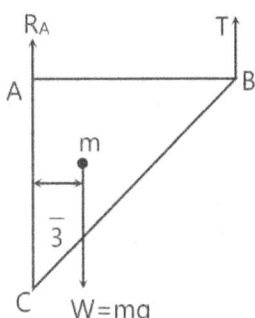

Center of mass of the triangular plate is at a distance of $\dfrac{\ell}{3}$ from AC.

Considering force equilibrium

$R_A + T = W = mg$

Considering torque or moment equilibrium about B

$\Rightarrow R_A(\ell) = W\left(\dfrac{2\ell}{3}\right)$

$\Rightarrow R_A = \dfrac{2mg}{3}$

Sol 9: (B) Given, A constant torque is applied on a rod which hinged the angular acceleration in rod will be

$\alpha = \dfrac{T}{I}$

Angular velocity $\omega = \displaystyle\int_0^t \alpha\, dt = \dfrac{Tt}{I} + \omega_0$

Power developed $= T.\omega = T.\left(\dfrac{Tt}{I} + \omega_0\right)$

If $\omega_0 = 0$

$P = \dfrac{T^2}{I}(t)$

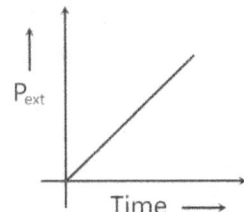

Sol 10: (C) Given,

Two spheres of same mass M and radii R and 2R and also have equal rotational kinetic energies

$\Rightarrow \dfrac{1}{2}I_1\omega_1^2 = \dfrac{1}{2}I_2\omega_2^2 \Rightarrow \dfrac{L_1^2}{I_1} = \dfrac{L_2^2}{I_2}$

[$\because I\omega = L$ (Angular momentum)]

$\Rightarrow \dfrac{L_2}{L_1} = \sqrt{\dfrac{I_2}{I_1}} = \sqrt{\dfrac{R_2^2}{R_1^2}} = \dfrac{R_2}{R_1} = 2$

Sol 11: (A) Angular momentum remains constant since, no torque is acting on the skater.

While kinetic energy increases, since,

K.E. $= \dfrac{1}{2}L.\omega$.

and as L = constant and ω increase K.E. increases

Sol 12: (D) Energy is not conserved in this case because the disc is fixed at its center and a force is acting on it when the child jumps.

But Angular momentum can be conserved since No torque is present in the boy-disc system

$\therefore$ Initial angular momentum = final angular momentum

$(I + mR^2)\omega = I\omega' + MRv$

$\Rightarrow \omega' = \dfrac{(I + mR^2)\omega - mRv}{I}$

Sol 13: (B)

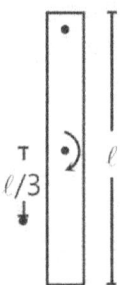

Moment of inertia of the rod about pivot $= \dfrac{M\ell^2}{12}$

Since, the collision is elastic and the rod stops, velocity of the particle is $v = \dfrac{\ell}{3}.\omega$

By principle of Angular Momentum,

$I\omega = m.v.\dfrac{\ell}{3}$

$\Rightarrow \dfrac{M\ell^2}{12}.\omega = m.\dfrac{\ell}{3}.\omega.\dfrac{\ell}{3} \Rightarrow m = \dfrac{3M}{4}$

Sol 14: (B)

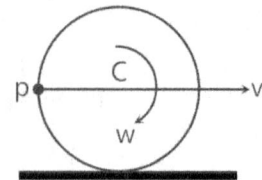

The point P experiences centripetal acceleration towards centre

$$\therefore \ \vec{a} = a\hat{i}$$

the velocity of point P is

$$\vec{v} = v\hat{i} + R\omega\hat{j}$$

$$\Rightarrow \vec{v} = v\hat{i} + v\hat{j}$$

$[\because v = R\omega \text{ pure rolling}]$

$$\Rightarrow \vec{v} = v(\hat{i} + \hat{j})$$

$$\cos\theta = \frac{\vec{a}.\vec{v}}{|\vec{a}||\vec{v}|} = \frac{1}{\sqrt{2}}$$

$$\Rightarrow \theta = 45°$$

Sol 15: (C)

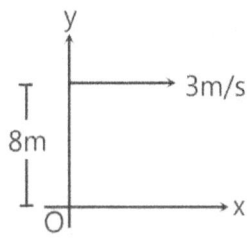

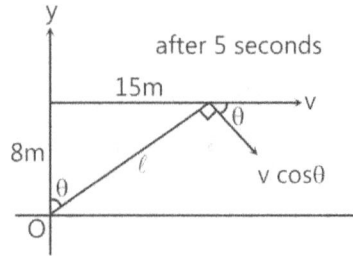

Angular velocity $= \dfrac{v\cos\theta}{\ell} = \dfrac{v8}{\ell^2}$

$$= \frac{24}{8^2 + 15^2} = \frac{24}{289} \text{ rad/s}$$

Sol 16: (B)

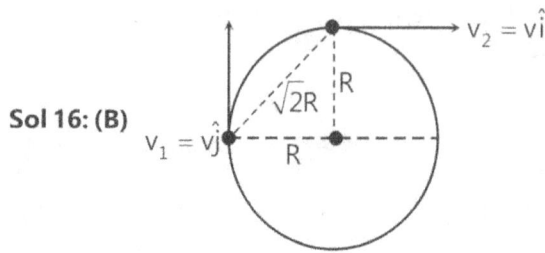

Angular velocity $= \dfrac{\text{relative velocity}}{AB}$

$$= \frac{|v\hat{i} - v\hat{j}|}{\sqrt{2}R} = \frac{v}{R}$$

Sol 17: (B)

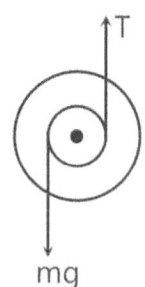

Acceleration of yo-yo $= \dfrac{mg - T}{m} = g - \dfrac{T}{m}$

Sol 18: (B) In case of pure rolling $v = R\omega$

(Bottom–most point has zero velocity)

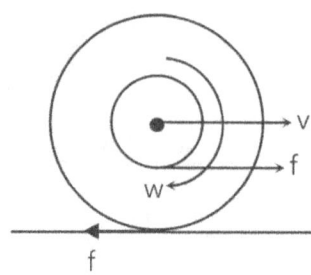

as ω is in clockwise direction, thread winds also friction acts leftwards to increase w.

Sol 19: (B)

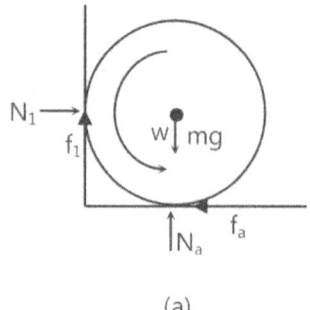

(a)

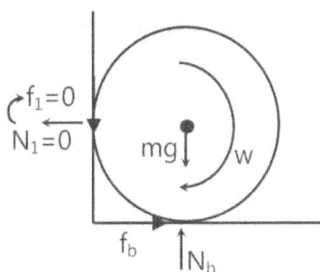

$f_1 + N_a = mg \quad N_b = mg$

$\dfrac{N_1}{3} + N_a = mg \quad f_b = \dfrac{mg}{3}$

$\dfrac{f_a}{3} + Na = mg \quad [\because N_1 = f_a]$

$N_a = \dfrac{9mg}{10} \left[\because f_a = \dfrac{N_a}{3}\right] \Rightarrow f_a = \dfrac{3mg}{10}$

$\dfrac{f_a}{f_b} = \dfrac{9}{10}$

Sol 20: (A)

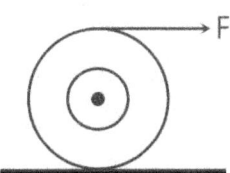

$F = Ma$ (By Newton's second law), also $T = Ta$

$\Rightarrow FR = \dfrac{Ia}{R}$ ($\because \alpha = \dfrac{a}{R}$ for pvre rolling)

$\Rightarrow I = MR^2$

This is satisfied for thin pipe

Sol 21: (D)

By conservation of linear momentum,

$mv_{cm} = J \Rightarrow v_{cm} = \dfrac{\hat{j}}{m}$

By conservation of angular momentum, $I\omega = J.\dfrac{\ell}{2}$

$\Rightarrow \omega = \dfrac{6J}{m\ell}$

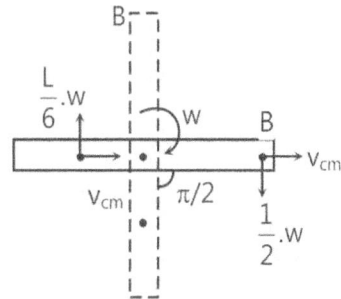

After time $t = \dfrac{\pi m\ell}{12J}$

The angle rotated by rod $\theta = \omega t = \dfrac{\pi}{12} \dfrac{m\ell}{J} . \dfrac{6J}{m\ell} = \dfrac{\pi}{2}$

Velcoity of point P = $\sqrt{\left(\dfrac{\ell\omega}{6}\right)^2 + (v_{cm})^2}$

$= \sqrt{\left(\dfrac{J}{m}\right)^2 + \left(\dfrac{J}{m}\right)^2} = \sqrt{2}\dfrac{J}{m}$

Sol 22: (C) Solid cylinder rolls without slipping

$\Rightarrow v_{cm} = R\omega$

$\dfrac{(K.E.)_{rotational}}{(K.E.)_{translational}} = \dfrac{\dfrac{1}{2}I\omega^2}{\dfrac{1}{2}mv^2}$

$\Rightarrow \dfrac{(K.E.)_r}{(K.E.)_t} = \dfrac{mR^2\omega^2}{2mv^2} = \dfrac{1}{2}$

Sol 23: (C) There will be no diff in velocity of centre of mass

$\Rightarrow F = ma_{cm}$

$\therefore a_{cm}$ = same in both cases

Sol 24: (B)

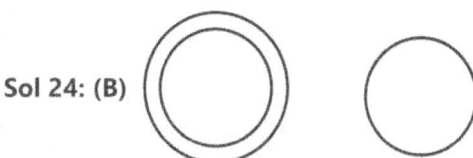

A hoop & solid cylinder of same mass

$\therefore I_{hoop} > I_{cylinder}$ as mass is distributed away from center

$\therefore$ Since gain in potential energy in both case is same

Let P for hoop

$$P = \frac{1}{2}mv^2 + \frac{1}{2}I_{hoop}\left(\frac{V}{R}\right)^2$$

and $P = \frac{1}{2}mv^2 + \frac{1}{2}I_{cy}\left(\frac{V}{R}\right)^2$

$\therefore V_{hoop} < V_{cylinder}$

Sol 25: (D) (i) Upward acceleration = a = same for both case

$\Rightarrow 2h = 0 \times t_Q + \frac{1}{2}a \times t_Q^2 \Rightarrow t_Q = \sqrt{\frac{4h}{g}}$

$h = 0 \times t_p + \frac{1}{2}a \times t_p^2$

$\Rightarrow t_p = \sqrt{\frac{2h}{g}}$

$\therefore t_Q \neq 2t_p$

(ii) The acceleration of both the balls is same = g sin q

(iii) $\Delta KE = \Delta PE$

$\therefore \Delta KE_Q = 2mgH$

$\Delta KE_p = mgH$

$\therefore \Delta KE_Q = 2\Delta KE_p$

Sol 26: (B) Moment of inertia decreases since mass is closer to axis while.

Angular momentum remains constant which implies angular velocity increases and which intern implies increase is kinetic energy.

$(\because L = I\omega$ and $K.E = \frac{1}{2}L\omega)$

Sol 27: (C) Angular momentum is conserved on any point on the ground since the only force present passes through that point making torque zero.

Previous Years' Questions

Sol 1: (A) A'B' $\perp$ AB and C'D' $\perp$ CD

From symmetry $I_{AB} = I_{A'B'}$ and $I_{CD} = I_{C'D'}$

From theorem of perpendicular axes,

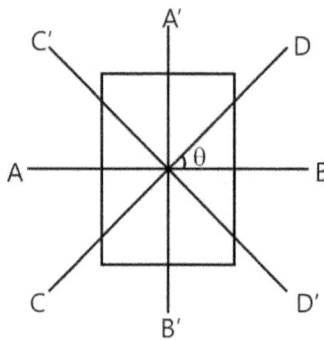

$I_{ZZ} = I_{AB} + I_{A'B'} = I_{CD} + I_{C'D'} = 2I_{AB} = 2I_{CD}$

Alternate:

The relation between I_{AB} and I_{CD} should be true for all value of θ

At $\theta = 0$, $I_{CD} = I_{AB}$

Similarly, at $\grave{e} = \pi/2$, $I_{CD} = I_{AB}$ (by symmetry)

Keeping these things in mind, only option (A) is correct.

Sol 2: (C) Since, it is head on elastic collision between two identical spheres, they will exchange their linear velocities i.e., A comes to rest and B starts moving with linear velocity v, As there is no friction anywhere, torque on both the spheres about their center of mass is zero and their angular velocity remains unchanged. Therefore,

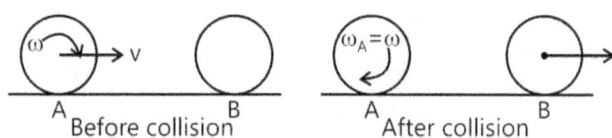

Sol 3: (B) $L = m\frac{v}{\sqrt{2}}r_\perp$

Here, $r_\perp = h = \frac{v^2 \sin^2 45°}{2g} = \frac{v^2}{4g}$

$\therefore L = m\left(\frac{v}{\sqrt{2}}\right)\left(\frac{v^2}{4g}\right) = \frac{mv^3}{4\sqrt{2}g}$

Sol 4: (A) Let ω be the angular velocity of the rod. Applying, angular impulse = change in angular momentum about center of mass of the system

$$J.\frac{L}{2} = I_c\omega$$

$$\therefore (Mv)\left(\frac{L}{2}\right) = (2)\left(\frac{ML^2}{4}\right)\omega$$

$$\therefore \omega = \frac{v}{L}$$

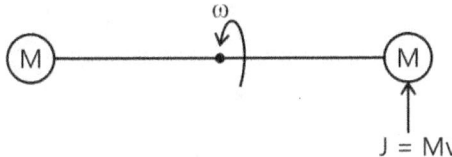

$$J = Mv$$

Sol 5: (A) Mass of the element dx is $m = \dfrac{M}{L}dx$.

This element needs centripetal force for rotation.

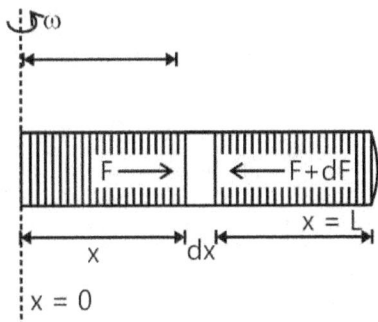

$$\therefore dF = mx\omega^2 = \left(\frac{M}{L}x\omega^2dx\right)$$

$$\therefore F = \int_0^L dF = \frac{m}{L}.\omega^2\int_0^L xdx = \frac{M\omega^2 L}{2}$$

This is the force exerted by the liquid at the other end.

Sol 6: (B) $mg\sin\theta$ component is always down the plane whether it is rolling up or rolling down. Therefore, for no slipping, sense of angular acceleration should also be same in both the cases.

Therefore, force of friction f always act upwards.

Sol 7: (C) Work done $W = \dfrac{1}{2}I\omega^2$

If x is the distance of mass 0.3 kg from the center of mass, we will have

$$I = (0.3)\,x^2 + (0.7)(1.4 - x^2)$$

For work to be minimum, the moment of inertia (I) should be minimum or

$$\frac{dI}{dx} = 0$$

or $2(0.3) - 2\,(0.7)(1.4 - x) = 0$ or $(0.3)x = (0.7)(1.4 - x)$

$$\Rightarrow x = \frac{(0.7)(1.4)}{0.3 + 0.7} = 0.98m$$

Sol 8: (C) From the theorem

$$\vec{L}_0 = \vec{L}_{CM} + M\,(\vec{r}\times\vec{v}) \qquad \text{... (i)}$$

We may write

Angular momentum about O = Angular momentum about CM + Angular momentum of CM about origin

$$\therefore L_0 = I\omega + MRv$$

$$= \frac{1}{2}MR^2\omega + MR(R\omega) = \frac{3}{2}MR^2\omega$$

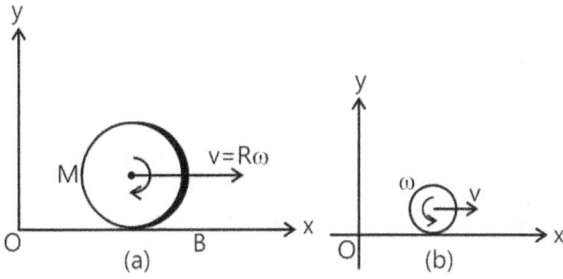

Note that is this case both the terms in

Eq. (i) i.e., $\vec{L}_{CM}$ and $M\,(\vec{r}\times\vec{v})$

Have the same direction. That is why we have used $L_0 = I\omega \sim MRv$ if they are in opposite direction as shown in figure (b).

Sol 9: (C) At the critical condition, normal reaction N will pass through point P. In this condition.

$$\tau_N = 0 = \tau_{fr} \quad \text{(About P)}$$

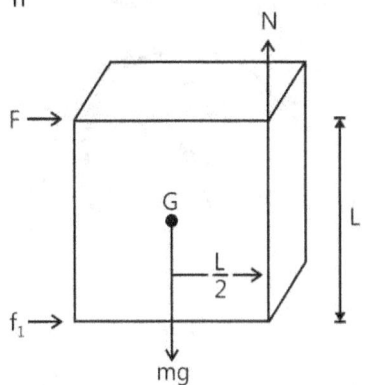

The block will topple when

$$\tau_F > \tau_{mg} \text{ or } FL > (mg)\frac{L}{2}$$

$$\therefore F > \frac{mg}{2}$$

Therefore, the minimum force required to topple the block is

$$F = \frac{mg}{2}$$

Sol 10: (B) Net external torque on the system is zero. Therefore, angular momentum is conserved. Force acting on the system are only conservative. Therefore, total mechanical energy of the system is also conserved.

Sol 11: (A) Mass of the whole disc = 4M

Moment of inertia of the disc about the given axis

$$= \frac{1}{2}(4M)R^2 = 2MR^2$$

$\therefore$ Moment of inertia of quarter section of the disc

$$= \frac{1}{4}(2MR^2) = \frac{1}{2}MR^2$$

Note: These type of questions are often asked in objective. Students generally error in taking mass of the whole disc. They take if M instead of 4 M.

Sol 12: (C) Since, there is no external torque, angular momentum will remains conserved. The moment of inertia will first decrease till the tortoise moves from A to Can then increase as it moves from C and D. Therefore, ω will initially increase and then decrease.

Let R be the radius of platform, m the mass of disc and M is the mass of platform.

Moment of inertia when the tortoise is at a

$$I_1 = mR^2 + \frac{MR^2}{2}$$

And moment of inertia when the tortoise is at B.

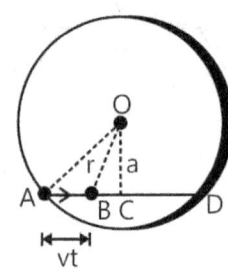

$$I_2 = mr^2 + \frac{MR^2}{2}$$

Here, $r^2 = a^2 + [\sqrt{R^2 - a^2} - vt]^2$

Form conservation of angular momentum

$$\omega_0 I_1 = \omega(t)I_2$$

Substituting the values we can see that variation of ω(t) is non-linear.

Sol 13: (C) $\therefore I_1\omega_1 = I_2\omega_2$

$$\omega_2 = \frac{I_1}{I_2}\omega = \left(\frac{Mr^2}{Mr^2 + 2mr^2}\right)\omega = \left(\frac{M}{M + 2m}\right)\omega$$

Sol 14: (A) $r = \sqrt{2}\dfrac{a}{2}$ or $r^2 = \dfrac{a^2}{2}$

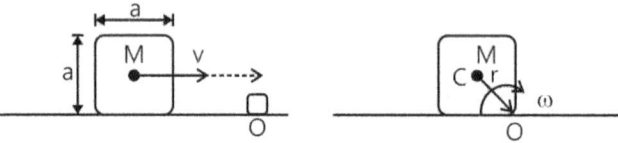

Net torque about O is zero. Therefore, angular momentum (L) about O will be conserved,

Or $L_i = L_f$

$$Mv\left(\frac{a}{2}\right) = I_0\omega = (I_{CM} + Mr^2)\omega$$

$$= \left\{\left(\frac{Ma^2}{6}\right) + M\left(\frac{a^2}{2}\right)\right\}\omega = \frac{2}{3}Ma^2\omega$$

$$\omega = \frac{3v}{4a}$$

Sol 15: (D) Mass of the ring M = ρL

Let R be the radius of the ring, then

$$L = 2\pi R \text{ or } R = \frac{1}{2\pi}$$

Moment of inertia about an axis passing through O and parallel to XX` will be

$$I_0 = \frac{1}{2}MR^2$$

Therefore, moment of inertia about XX` (from parallel axis theorem) will be given by

$$I_{XX'} = \frac{1}{2}MR^2 + MR^2 = \frac{3}{2}MR^2$$

Substituting values of m and R

$$I_{XX'} = \frac{3}{2}(\rho L)\left(\frac{L^2}{4\pi^2}\right) = \frac{3\rho L^3}{8\pi^2}$$

3.94

Sol 16: (D) $r_1 = \dfrac{m_2 r}{m_1 + m_2}$; $r_2 = \dfrac{m_1 r}{m_1 + m_2}$

$$(I_1 + I)\omega = \dfrac{nh}{2\pi} = n\hbar$$

$$\text{K.E.} = \dfrac{1}{2}(I_1 + I_2)\omega^2 = \dfrac{n^2\hbar^2(m_1 + m_2)}{2m_1 m_2 r^2}$$

Sol 17: (B) From conservation of angular momentum about any fix point on the surface

$$mr^2\,\omega_0 = 2mr^2\,\omega$$

$$\therefore \omega = \dfrac{\omega_0}{2}$$

$$\therefore V_{CM} = \dfrac{\omega_0 r}{2}$$

Sol 18: (A)

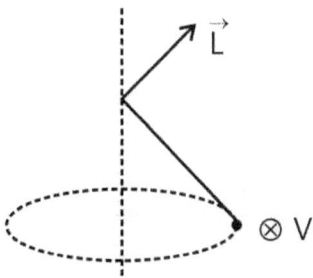

$\vec{L}$ changes in direction not in magnitude

Sol 19: (C) $5 = e^{1000\frac{V}{T}} - 1$

$$\Rightarrow e^{1000\frac{V}{T}} = 6 \qquad \qquad ....(i)$$

Again, $I = e^{1000\frac{V}{T}} - 1$

$$\dfrac{dI}{dV} = e^{\frac{1000V}{T}}\dfrac{1000}{T}$$

$$dI = \dfrac{1000}{T}e^{1000\frac{V}{T}}dV$$

Using (i)

$$\Delta I = \dfrac{1000}{T} \times 6 \times 0.01 = \dfrac{60}{T} = \dfrac{60}{300} = 0.2\,mA$$

Sol 20: (B) For maximum possible volume of cube

$2R = \sqrt{3}\,a$, a is side of the cube.

Moment of inertia about the required axis $= I = \rho a^3 \dfrac{a^2}{6}$,

where $\rho = \dfrac{M}{\dfrac{4}{3}\pi R^3}$

$$I = \dfrac{3M}{4\pi R^3}\dfrac{1}{6}\left(\dfrac{2R}{\sqrt{3}}\right)^5 = \dfrac{3M}{4\pi R^3}\dfrac{1}{6}\dfrac{32R^5}{9\sqrt{3}} = \dfrac{4MR^2}{9\sqrt{3}\pi} = \dfrac{4MR^2}{9\sqrt{3}\pi}$$

Sol 21: (A, C) $\vec{L}_0 = mv\dfrac{R}{\sqrt{2}}(-\hat{k})$ \qquad [D to A]

$$\vec{L}_0 = mv\left[\dfrac{R}{\sqrt{2}} + a\right]\hat{k} \quad \text{[C to D]}$$

Sol 22: (D) From normal reactions of roller, we can conclude it moves towards left.

JEE Advanced/Boards

Exercise 1

Sol 1: Given,

A thin uniform rod of mass M and length L is hinged at its upper end, and is released from rest in a horizontal position.

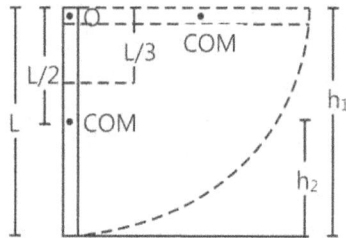

Let angular velocity of the rod about hinge 'O' when it is vertical be 'ω'

Moment of inertia of rod about o is

$$I = I_{com} + M\left(\dfrac{L}{2}\right)^2 \text{ (parallel axis theorem)}$$

$$\Rightarrow I = \dfrac{ML^2}{12} + \dfrac{ML^2}{4} = \dfrac{ML^2}{3}$$

By using principle of conservation of energy

$$\Delta K.E = -\Delta P.E$$

$$\Rightarrow \dfrac{1}{2}I\omega^2 - 0 = -Mg(h_2 - h_1)_{com}$$

$$\Rightarrow \frac{1}{2} \frac{ML^2}{3}.\omega^2 = Mg\frac{L}{2} \Rightarrow \omega = \sqrt{\frac{3g}{L}}$$

The tension in the rod at a point $\frac{L}{3}$ from hinge would be due to weight below that point and centrifugal force of that part.

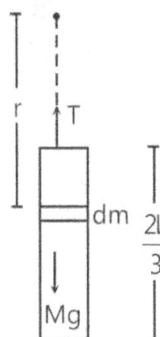

$$m = \rho.A.\frac{2L}{3} = \frac{2M}{3} \left[\because \rho = \frac{M}{A.L} \right]$$

Centrifugal force $= \int_{L/3}^{L} r\omega^2.dm$

$$= \omega^2 \int_{L/3}^{L} r.\rho.Adr = \omega^2 \rho.A \left[\frac{r^2}{2} \right]_{L/3}^{L}$$

$$= \frac{3g}{L} . \frac{M}{L} \left[\frac{1}{2} \right] [L^2] \left[\frac{8}{9} \right] = \frac{4Mg}{3}$$

Tension at the point is $T = mg + F_c$

$$= \frac{2Mg}{3} + \frac{4Mg}{3} = 2Mg$$

Sol 2:

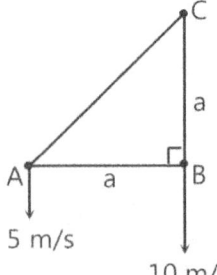

Given,

$V_A = 5m/s$ $V_B = 10$ m/s

In the frame of A

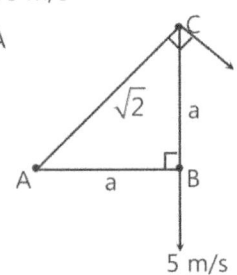

A is stationary

Since, angular velocity of system would be same through

$$\omega = \frac{V_B}{a} = \frac{5}{a} \text{ rad/s}$$

and

$V_C = \sqrt{2} a.\omega = 5\sqrt{2}$ m/s perpendicular to AC in vector form.

$V_C = +5\hat{i} + (-5)\hat{j}$ if co-ordinate system is along AB and BC

Velocity of 'C' in original frame

$$\vec{V_C} = \vec{V_C} + \vec{V_a} = +5\hat{i} -5\hat{j} -5\hat{j} \quad \therefore (V_A = -5\hat{j})$$

$$\Rightarrow \vec{V_C} = 5\hat{i} - 10\hat{j}$$

$$|\vec{V_C}| = \sqrt{5^2 + 10^2} = 5\sqrt{5} \text{ m/s}$$

Sol 3:

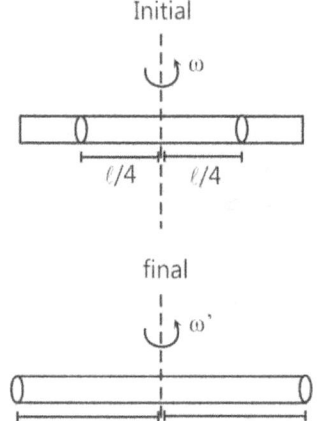

Angular momentum of the system is conserved, since no torque is applied

$L_i = L_f$

$$\Rightarrow I_i \omega = I_f \omega^1$$

$$\Rightarrow \left(\frac{m_r \ell^2}{12} + m \left(\frac{\ell}{4} \right)^2 \times 2 \right) \omega$$

$$= \left(\frac{m_r \ell^2}{12} + m \left(\frac{\ell}{2} \right)^2 \times 2 \right) \omega^1$$

$$\Rightarrow \frac{0.03}{0.09} \times 30 = \omega^1$$

$$\Rightarrow \omega_1 = 10 \text{ rad/s}$$

By energy conservation,

$$\frac{1}{2}I_i\omega^2 = \frac{1}{2}I_f(\omega_1)^2 + \left(\frac{1}{2}mv^2\right) \times 2$$

$$\Rightarrow \frac{L}{2m}(\omega - \omega^1) = V^2 \Rightarrow \frac{0.03 \times 30 \times (20)}{0.2} = V^2$$

$$\Rightarrow v = 3 \text{ m/s}$$

$\therefore$ Velocity of ring along rod = 3m/s

Sol 4: Given,

A straight rod AB of mass M and length L, a horizontal force F starts on A

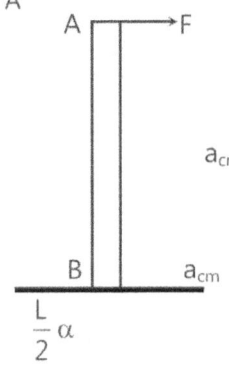

F = Ma_{cm} (by newton's second law)

T = $I\alpha$

$$\Rightarrow F \cdot \frac{L}{2} = \frac{ML^2}{12} \times a$$

$$\Rightarrow \alpha = \frac{6F}{ML}$$

Acceleration of end B= $a_{cm}\hat{i} - \frac{1}{2}x\hat{i}$

$$= -\frac{2F}{M}\hat{i} \quad \left(\because a_{cm} = \frac{F}{M}\right)$$

$\therefore$ Magnitude of acceleration of end B = $\frac{2F}{M}$

Sol 5:

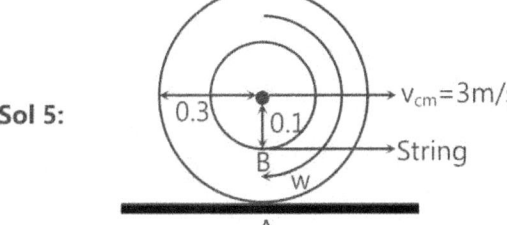

Given, the wheel is rolling without slipping

$r_A\omega = V_{cm} (\because V_A = 0)$

(pure rolling)

The velocity of the string should be

$$V_B = V_{cm} - r_B\omega = V_{cm}\left(1 - \frac{r_B}{r_A}\right) = 2\text{m/s}$$

Sol 6: Given,

A uniform wood door of mass m, height h and width w.

Location of hinges are $\frac{h}{3}$ and $\frac{2h}{3}$ from the bottom of the door.

Let the hinges be named A and B.

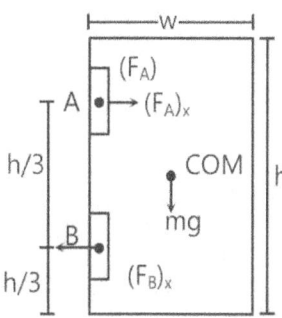

Given, hinge A is screwed while B is not, So, the upward component of force by hinge B is absent.

By equilibrium equations,

$$\sum F_x = 0 \Rightarrow (F_A)_x = (F_B)_x$$

$$\sum F_y = 0 \Rightarrow (F_A)_y = mg$$

$$\sum M_{COM} = 0$$

(Moment about center of mass)

$$\Rightarrow (F_A)_y\left(\frac{w}{2}\right) + (F_A)_x\left(\frac{h}{6}\right) + (F_B)_x\left(\frac{h}{6}\right) = 0$$

$$\Rightarrow mg \cdot \frac{\omega}{2} + (F_B)_x\left(\frac{h}{3}\right) = 0 \Rightarrow (F_B)_x = -\frac{3mg\omega}{2h}$$

and $\vec{F_A} = (F_A)_x\hat{i} + (F_B)\hat{j}, \vec{F_B} = F_B(-\hat{i})$

$$\vec{F_A} = -\frac{3mg\omega}{2h}\hat{i} + mg\hat{j}, \vec{F_B} = \frac{3mg\omega}{2h}\hat{i}$$

Given m = 20 kg, h = 2.2 m, ω = 1m

$$\Rightarrow \vec{F_A} = (-133.64\hat{i} + 196\hat{j})N \text{ and } \vec{F_B} = 133.64\hat{i}$$

Sol 7: Given, A thin rod of length 'a' with variable mass per unit length $\rho = \rho_0\left(1 + \frac{x}{a}\right)$ where x is distance from A.

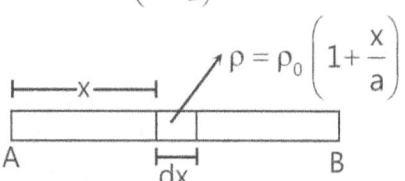

(a) Mass of the elemental part is $dm = \rho.dx$

$$\Rightarrow dm = \rho_0\left(1+\frac{x}{a}\right).dx$$

Mass of the rod

$$m = \int_0^M dm = \int_0^a \rho_0\left(1+\frac{x}{a}\right).dx$$

$$\Rightarrow m = \rho_0\left[x+\frac{x^2}{2a}\right]_0^a \quad m = \rho_0\left(\frac{3a}{2}\right)$$

$$\therefore \text{ Mass of rod} = \frac{3a\rho_0}{2}$$

(b) Center of mass is situated at distance of C from A where

$$C = \frac{\int_0^M x.dm}{\int_0^M dm}$$

Value of $\int_0^M x.dm = \int_0^a x(\rho_0)\left(1+\frac{x}{a}\right)dx$

$$= \rho_0\left[\frac{x^2}{2}+\frac{x^3}{3a}\right]_0^a \Rightarrow \rho_0\left(\frac{5a^2}{6}\right)$$

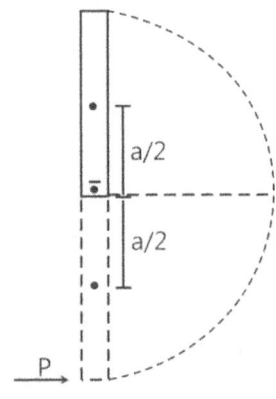

for minimum value of P, the angular velocity rod in the final position should be zero

by applying conservation of energy

K.E. $= -\Delta$P.E.

$$\Rightarrow K.E._f - K.E._i = -mg(h_f - h_i)$$

$$0 - \frac{1}{2}I\omega^2 = -mg\,(a)$$

$$\Rightarrow C = \frac{\frac{5a^2}{6}}{\frac{3a^2}{2}} \Rightarrow C = \frac{5a}{9}$$

(c) Given, to find the moment of inertia about axis perpendicular to rod and passing through A.

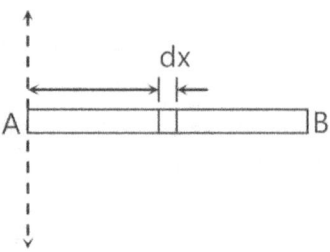

$$I = \int_0^M x^2 dm \quad (\because dI = x^2\,dm)$$

but $dm = \rho_0\left(1+\frac{x}{a}\right).dx$

$$\Rightarrow I = \int_0^a \rho_0\left(x^2+\frac{x^3}{a}\right).dx$$

$$\Rightarrow I = \rho_0\left[\frac{x^3}{3}+\frac{x^4}{4a}\right]_0^a \Rightarrow I = \frac{7\rho_0 a^3}{12}$$

(d) We know that,

Angular momentum $L = I.\omega$

$$P.a = \frac{7\rho_0 a^2}{12}.\omega$$

$$\omega = \frac{12}{7\rho_0 a^2}$$

(e) Given, an impulse of 'P' is applied at point B, then

Angular impulse about the axis will be

$L = P.a$

$$\Rightarrow (I\omega).\omega = 2mga$$

$$\Rightarrow \frac{P.a.12.P}{7\rho_0 a^2} = 2\,mga$$

$$\Rightarrow P^2 = \frac{7}{6}\rho_0 ga^2\left(\frac{3}{2}\rho_0 a\right) = \sqrt{\frac{7}{4}\rho_0^2 ga^3}$$

Sol 8: Given, two cylinders of mass 1 kg and 4 kg with radii 10 cm and 20 cm respectively.

also initial angular velocities as

$\omega_1 = 100$ rad/s and $\omega_2 = 200$ rad/s

final angular velocities will be such that there is no slip at point of contact

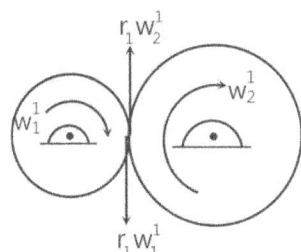

$$\Rightarrow V_{contact} = 0 \Rightarrow r_1\omega_1^1 - r_2\omega_2^1 = 0$$

$$\Rightarrow \omega_2^1 = \frac{r_1\omega_1^1}{r_2}$$

Angular impulse on one cylinder due to other is

$$I_1(\omega_1^1 - \omega_1) = (\Delta P)(R_1)$$

where ΔP = linear impulse while for the other

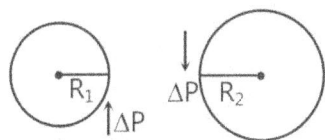

$$I_2(\omega_2^1 - \omega_2) = (\Delta P)(R_2)$$

$$\Rightarrow \frac{I_1}{I_2} \frac{(\omega_1^1 - \omega_1)}{(\omega_2^1 - \omega_2)} = \frac{R_1}{R_2}$$

$$\Rightarrow 8(\omega_2^1 - \omega_2) = \omega_1^1 - \omega_1$$

$$\Rightarrow \omega_1^1 = \frac{\omega_1 - 8\omega_2}{5} - 300 \text{ rad/s}$$

while $\omega_2^1 = -\dfrac{\omega_1}{2} = 150$ rad/s

Sol 9:

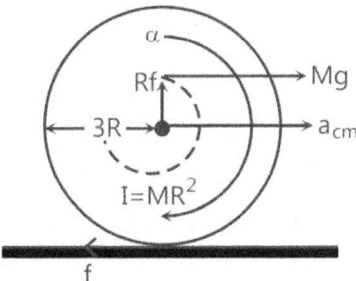

Let,

a_{cm} be acceleration of center of mass α be angular acceleration

by Newton's second law,

$$Mg - f = Ma_{cm} \qquad \text{.... (i)}$$

and by considering torque

$$Mg \times R + f \times 3R = I \cdot \alpha \qquad \text{.... (ii)}$$

Since, the body is in pure rolling

$$(V_{cm} = 3R\omega) \Rightarrow (a_{cm} = 3R\alpha)$$

Solving (i) and (ii) we get,

$$a_{cm} = \frac{6}{5} \cdot g$$

Acceleration of the point where force is applied B

$$\vec{a} = a_{cm}\hat{i} + R(\alpha)\hat{i} = \frac{3}{4} a_{cm}\hat{i} = \frac{8}{5} g\hat{i}$$

$$|\vec{a}| \cong 16 \text{ m/s}^2$$

Sol 10:

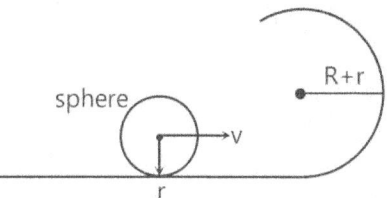

Given, A sphere of mass m and radius r and radius of loop as R + r the velocity of the sphere at the top most point should be such that the centrifugal force balances the weight of sphere

$$\Rightarrow \frac{mv_f^2}{R} = mg$$

($\because$ Center of mass makes circle of radius R)

$$\Rightarrow v_f = \sqrt{Rg}$$

Since, the sphere is in pure rolling at every point of time,

$$v_{cm} = r\omega$$

By principle of conservation of energy

$$K.E._1 + P.E._1 = K.E._f + P.E._f$$

$$\Rightarrow \frac{1}{2}mv^2 + \frac{1}{2}I\omega^2 + mg(r)$$

$$= \frac{1}{2}mv_f^2 + \frac{1}{2}I\omega_f^2 + mg(2R + r)$$

$$\Rightarrow \frac{1}{2}mv^2 + \frac{1}{2}\left(\frac{2}{5}\right)mv^2 = \frac{1}{2}mv_f^2 +$$

$$\frac{1}{2}\left(\frac{2}{5}\right)\left(\frac{2}{5}\right)mv_f^2 + mg(2R)$$

$$\Rightarrow v^2 = v_f^2 + g(2R) \times \frac{10}{7}$$

$$\Rightarrow v = \sqrt{Rg + \frac{20Rg}{7}} = \sqrt{\frac{27Rg}{7}}$$

$$\therefore v = \sqrt{\frac{27Rg}{7}}$$

Sol 11: $r.T_2 = I_2.\alpha_2;$

$\alpha_2 = \dfrac{a_2}{r_2}$

$mg - T_1 - T_2 = ma_1;$

$\alpha_1 = \dfrac{a_1}{r_1}$

(a) $(T_1 - T_2).r = I_1\alpha_1$ also,

$a_2 = 2a_1$ by constraint relations.

$\Rightarrow a_1 = \dfrac{2g}{7}$

$\Rightarrow T_2 = \dfrac{2mg}{7} = \dfrac{200}{7}$ N

(b) And also velocity 'v' of the body after traveling 1.2 m

$v = \sqrt{2 \times a \times s} = 4\sqrt{\dfrac{3}{7}}$ m/s

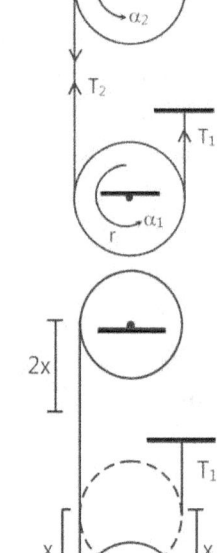

Sol 12: The moment of inertia of a thin hoop about it's diameter is

$\dfrac{1}{2}MR^2$

Here $M = L\rho$

Also we have $2\pi R = L$

$\Rightarrow R = \dfrac{1}{2\pi}$

So we have, $I = \dfrac{1}{2}MR^2 = \dfrac{1}{2}L\rho\left(\dfrac{L}{2\pi}\right)^2 = \dfrac{L^3\rho}{8\pi^2}$

Now using parallel axis theorem we have

$I_{xx'} = I_{cm} + MR^2 = \dfrac{L^3\rho}{8\pi^2} + L\rho\left(\dfrac{L}{2\pi}\right)^2 = \dfrac{3L^2\rho}{8\pi^2}$

Sol 13:

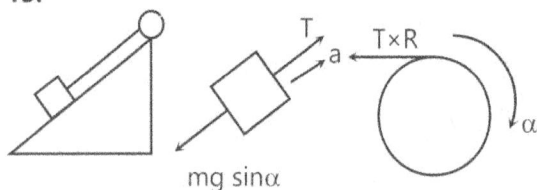

mg sinα

(a) Now $T - mg\sin\alpha = ma$ ... (i)

$T\times R = -I\alpha = -\dfrac{MR^2}{2}\times\alpha = -\dfrac{MR\alpha}{2}$ ... (ii)

$R\alpha = a$ ... (iii)

$\Rightarrow T = -\dfrac{Ma}{2}$

$-\dfrac{Ma}{2} - mg\sin\alpha = ma$

$\Rightarrow -mg\sin\alpha = \left(m + \dfrac{M}{2}\right)a$

$\Rightarrow a = -\dfrac{mg\sin\alpha}{\left(m + \dfrac{M}{2}\right)}$

So, $T = -\dfrac{Ma}{2} = \dfrac{M}{2}\times\dfrac{(+mg\sin\alpha)}{\left(m + \dfrac{M}{2}\right)}$

$= \dfrac{M.m.g\sin\alpha}{(2m + M)} = \dfrac{2\times(1/2)\times(9.8)(1/2)}{(2\times0.5 + 2)} = \dfrac{9.8}{6} = 1.65$ N

(b) $\omega = 10$ rad/s

$\Rightarrow \alpha = a/R = -\dfrac{mg\sin\alpha}{R\left(m + \dfrac{M}{2}\right)} = -\dfrac{mg}{R(2m + M)} = -\dfrac{(0.5)(9.8)}{0.2(1 + 2)}$

$= -\dfrac{5}{6}.9.8$ rad/s²

$= -8.166$ rad/s²

Now, $\omega^2 = \omega_0^2 + 2\alpha\theta$

$\Rightarrow 0 = (10)^2 + 2(-8.166)\times\theta$

$\Rightarrow \theta = \dfrac{100}{2\times8.166} = 6.123$ rad,

So distance = $\theta R = 6.123\times0.2 = 1.224$ m

Sol 14:

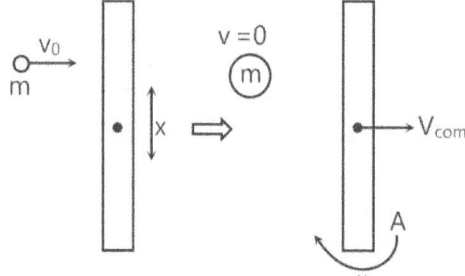

$v_A = V_{com} - \dfrac{\omega L}{2};$ so for $v_A = 0$

$\Rightarrow V_{com} = \dfrac{\omega L}{2}$

Now, $mv_0 = M.v_{com}$ (moment cons.)

$\Rightarrow V_{com} = \dfrac{mv_0}{M}$

and $mv_{0x} = I\omega$

[Angular momentum conservation about O]

So $mv_0x = \dfrac{ML^2}{12} \times \omega$

$\Rightarrow \omega = \dfrac{12mv_0x}{ML^2}$

So $V_{com} = \omega L/2'$

$\Rightarrow \dfrac{mv_0}{M} = \dfrac{12mv_0x}{ML^2} \times \dfrac{L}{2} \Rightarrow x = L/6$

Sol 15:

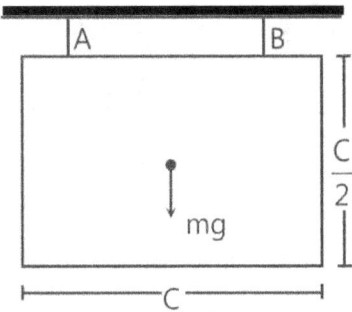

Before connection B is released

$T_A + T_B = mg$ (By force equilibrium)

and

$T_A = T_B$ (for torque equilibrium)

$T_A = T_B = \dfrac{Mg}{2}$

$\Rightarrow$ Just after B is released

T_A is mg but $T_B = 0$

$\Rightarrow$ FBD

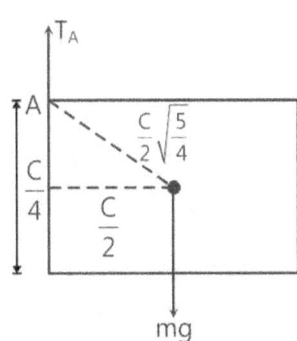

Moment of inertia about

$A = \dfrac{m}{12}\left(C^2 + \left(\dfrac{C}{2}\right)^2\right) + m\left(\dfrac{C}{2}\sqrt{\dfrac{5}{4}}\right)^2$

$\Rightarrow I = \dfrac{mc^2}{12}\left(\dfrac{5}{4}\right) + \dfrac{5}{16}mc^2 \Rightarrow I = \dfrac{5}{12}mc^2$

Torque about A, $T = (mg)\left(\dfrac{c}{2}\right) = I\alpha$

$\Rightarrow \alpha = \dfrac{mg\dfrac{c}{2}}{\dfrac{5mc^2}{12}} = \dfrac{6g}{5c} = \dfrac{1.2g}{c}$

Acceleration of the center is $a = \dfrac{\sqrt{5}c}{4} \times \left(\dfrac{1.2g}{c}\right)$

$= 0.3(\sqrt{5}g)$

or

$\vec{a} = -0.3(\hat{i} + 2\hat{j})$

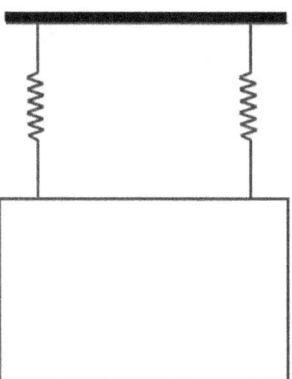

After connection B is released

TA is still $\dfrac{mg}{2}$, while $T_B = 0$

linear acceleration $a = \dfrac{g}{2}\left(\dfrac{mg - \dfrac{mg}{2}}{m}\right)$

angular acceleration $= 0.5\,g$

Sol 16: By energy conservation,

$K.E_i + P.E_i = K.E_f + P.E_f$

$\Rightarrow 0 + mgR$

$= \dfrac{1}{2}\left(\dfrac{m}{4}\right)(v^2)\left(1 + \dfrac{1}{2}\right) + \left(\dfrac{m}{4}\right)(g)\left(\dfrac{R}{2}\right)$

$\Rightarrow v = \sqrt{\dfrac{14gR}{3}}$

$\therefore$ Velocity of the axis of cylinder $= \sqrt{\dfrac{14gR}{3}}$

Sol 17:

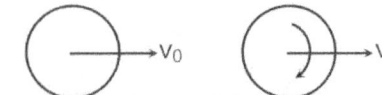

Angular momentum about any point on ground is conserved

$$\Rightarrow mv_0R = I\omega + mvR$$

$$mv_0R = \frac{mR^2\omega}{2} + mvR$$

$$\Rightarrow v = \frac{2}{3}v_0 \ [\because v = R\omega]$$

Work done by frictional for time t_0

$$W = -\frac{1}{2}mv_0^2 + \left(\frac{1}{2}mv^2 + \frac{1}{2}I\omega^2\right)$$

$$W = -\frac{mv_0^2}{2} + \left(\frac{3}{4}m.\frac{4}{9}v_0^2\right)$$

$$W = -\frac{1}{6}mv_0^2$$

Also for $t > t_0$ No frictional force exists

$$\Rightarrow W = -\frac{1}{6}mv_0^2 \text{ for } t \geq t_0$$

Also $ma = -\mu mg \Rightarrow a = -\mu g$

and $v = v_0 - \mu gt$

$$t_0 = \frac{v_0}{3\mu g}$$

$$\alpha = \frac{T}{I} = \frac{f.R}{\dfrac{MR^2}{2}} = \frac{2\mu g}{R}$$

Work done by friction for $t < t_0 = K.E_f - K.E_i$

$$\Rightarrow W = \left(\frac{1}{2}m(v_0 - \mu gt)^2 + \frac{1}{2}I\left(\frac{2\mu gt}{R}\right)^2\right) - \left(\frac{1}{2}mv_0^2\right)$$

$$\Rightarrow W = \left(\frac{1}{2}m\mu^2g^2t^2 + \frac{1}{2}m(2\mu^2g^2t^2) - mv_0\mu gt\right)$$

$$\Rightarrow W = \frac{1}{2}(3m\mu^2g^2t^2 - 2mv_0\mu gt)$$

Sol 18:

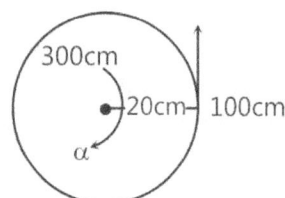

FBD of disc is

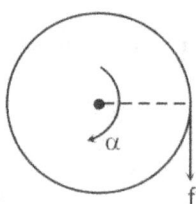

$$f.R = I\alpha \Rightarrow f.R = \frac{MR^2}{2}\alpha \Rightarrow \alpha = \frac{2f}{MR}$$

FBD of ant, in frame of disc

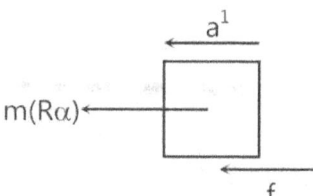

$$\Rightarrow mR\alpha + f = ma^1 \Rightarrow mR\alpha + \frac{mR\alpha}{2} = a^1$$

$$\Rightarrow a^1 = \frac{5R\alpha}{2} \quad \therefore M = 3m$$

Given, after time T, the ant reaches same point

$$\Rightarrow \frac{1}{2}\left(\frac{5R\alpha}{2}\right)T^2 = 2\pi.R \Rightarrow T = \frac{2}{\sqrt{5}} \text{ seconds}$$

Also the angle moved by disc $= \frac{1}{2}\alpha T^2 = \frac{4\pi}{5}$ radians

Sol 19:

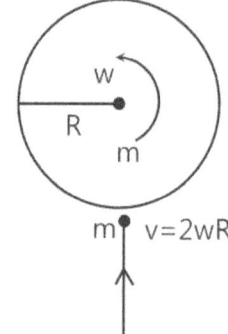

Angular momentum of the system is conserved

$$L_i = L_f$$

$$\Rightarrow \frac{mR^2\omega}{2} = \left(\frac{mR^2}{2} + mR^2\right)\omega^1$$

$$\Rightarrow \omega^1 = \frac{\omega}{3}$$

Impulse in the direction of velocity

$$= m(v_f - v_i) = -mv = -2m\omega R$$

Impulse perpendicular to direction of velocity

$= m(v_f - m_i) = m(R\omega^1)$

$= \dfrac{mR\omega}{3}$

Net Impulse $= -2m\omega R\,\hat{j} + \dfrac{mR\omega}{3}\,\hat{i}$

$|\overrightarrow{\Delta p}| = \dfrac{\sqrt{37}}{3}\,mR\omega$

Impulse on particle due to disc =

Impulse on hinge due to disc

$\Rightarrow |\overrightarrow{\Delta P}|_{\text{disc due to hinge}} = \dfrac{\sqrt{37}}{3}\,mR\omega$

Sol 20:

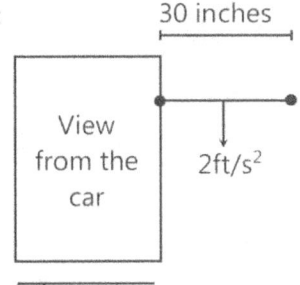

30 inches

View from the car   2ft/s²

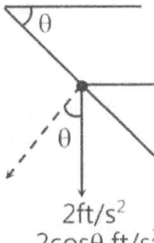

θ

θ

2ft/s²
2cosθ ft/s²

Angular acceleration of door $= \dfrac{2\cos\theta}{\left(\dfrac{w}{2}\right)}$ ft/s²

w is the width of the door

We know that,

$\dfrac{d\omega}{dt} = \alpha \Rightarrow \dfrac{d\omega}{d\theta}\cdot\dfrac{d\theta}{dt} = \alpha$

$\Rightarrow \omega.d\omega = \alpha . d\theta$

$\Rightarrow \dfrac{\omega^2}{2} - 0 = \dfrac{4}{w}\displaystyle\int_0^{\pi/2}\cos\theta.d\theta$

$\Rightarrow \dfrac{\omega^2}{2} = \dfrac{4}{w}$ [1]

$\Rightarrow \omega = 2$

Velocity of the outer edge $= (\omega)(w) = \sqrt{8w}$

Sol 21:

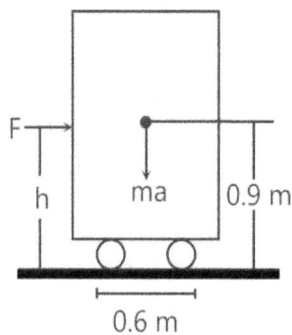

F

h   ma   0.9 m

0.6 m

F = 100 N

$F = ma \Rightarrow a = 5$ m/s²

The cabinet will tip when

$F.h > mg(0.3) + ma(0.9)$

$h > \dfrac{20\times10\times0.3}{100} + \dfrac{20\times5\times0.9}{100}$

$\Rightarrow h > 1.5$ m

and also when

$F.h > m = (0.9) - mg (0.3)$

$\Rightarrow h > 0.3$ m

∴ 0.3 < h < 1.5 m is the range of values of h for which cabinet will not tip.

Sol 22: In the frame of truck,

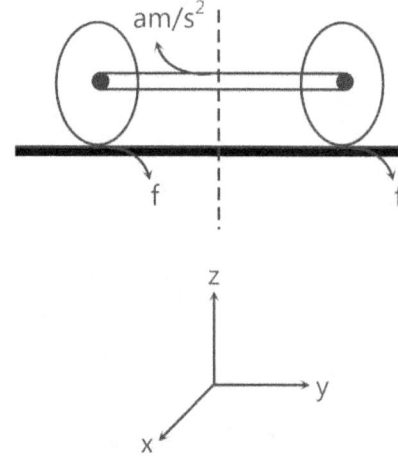

am/s²

f   f

z

y

x

$2M(a) - 2f = 2Ma_1$ (force equation)

$(2f)R = (2T).\alpha$ (Moment equation)

$a^1 = R\alpha$ (pure rolling)

$f = \dfrac{I\alpha}{R} = \dfrac{Ma_1}{2}$

$\Rightarrow a = \dfrac{3a_1}{2} \Rightarrow a_1 = \dfrac{2a}{3} = 6$ m/s²

$$f = \frac{Ma_1}{2} = 6N$$

Frictional torque magnitude about rod is

$$f.R = 0.6 \; Nm$$

Friction torque about O is

$$= \pm \, 0.6 \left(\frac{0.2}{2}\right)(\hat{k}) - 0.6(0.1)\hat{j} = -0.6\,(\hat{j} \pm \hat{k})$$

Sol 23:

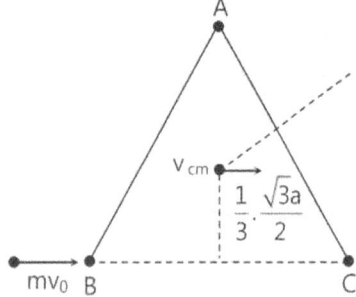

Conserving linear momentum, we get

$$(3m)(v_{cm}) = mv_0$$

$$\Rightarrow v_{cm} = \frac{v_0}{3}$$

Conserving angular momentum about COM we get

$$mv_0 = \frac{a}{2\sqrt{3}} = I.\omega$$

$$\Rightarrow \frac{mv_0 a}{2\sqrt{3}} = 3m\left(\frac{a}{\sqrt{3}}\right)^2.\omega$$

$$\Rightarrow \omega = \frac{v_0}{2\sqrt{3}a}$$

Time taken to complete one revolution $= \dfrac{\pi}{\omega}$

$$= \frac{\sqrt{3}\pi a}{v_0}$$

Displacement of point B will be

$$v_{cm}\, t\, \hat{i} + \left(\frac{2a}{\sqrt{3}} \times \frac{\sqrt{3}}{2}\hat{i} + \frac{2a}{\sqrt{3}}.\frac{1}{2}\hat{j}\right)$$

$$\Rightarrow \left(\frac{2\pi}{\sqrt{3}} + 1\right)(a)\hat{i} + \frac{a}{\sqrt{3}}\hat{j}$$

Exercise 2

Single Correct Choice Type

Sol 1: (D)

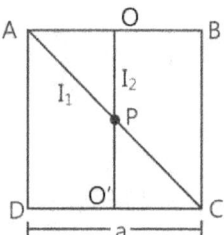

We know the moment of inertia of a square plate along OO' as $\dfrac{Ma^2}{12}$

$$\therefore I_2 = \frac{Ma^2}{12}$$

Also $I_z = I_x + I_y$

(Perpendicular axis theorem)

$$\Rightarrow I_z = 2I_x$$

$(\because I_x = I_y \; \text{fn square})$

$$\Rightarrow I_z = \frac{Ma^2}{6}$$

(I_z is about transverse axis through p) to find I_{APC} or I_1, we take x and y axis as diagonals of square and apply perpendicular axis theorem again

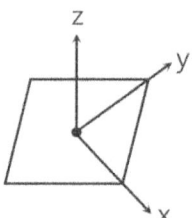

$$\Rightarrow I_x = I_y = \frac{I_z}{2} \; \Rightarrow I_1 = I_x = \frac{Ma^2}{12}$$

$$\frac{I_1}{I_2} = 1$$

Sol 2: (C) Given, moment of inertia of rectangular plate about transverse axis through P as I then the moment of inertia of PQR about P will be greater that $\dfrac{I}{2}$ since mass is distributed away from P unlike in PSR. Since, I depends on distance 'r', the farther the mass, the more the moment of inertia. The moment of inertia of PQR will be less than $\frac{I}{2}$ about R since mass is distributed closes to R

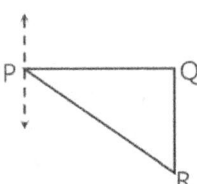

Sol 3: (B) Given, A triangle ABC such AB = BC = a and ∠ACB = 90° of mass M

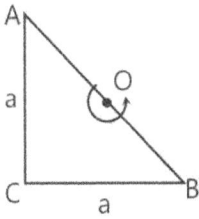

O is the midpoint

Consider a counterpart with same mass such a square is formed

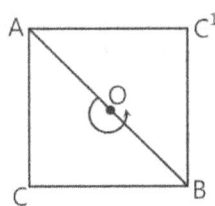

we know the moment of inertia of a square about transverse axis through center as $\frac{Ma^2}{6}$

But, here m = 2M (total mass)

$$\Rightarrow I_{square} = \frac{Ma^2}{3}$$

Since both triangles are symmetric about axis through O they have equal moment of inertia about axis through O.

$$\Rightarrow I_{required} = \frac{I_{square}}{2} = \frac{Ma^2}{6}$$

Sol 4: (A) Given, I is moment of inertia of a uniform square plate about axis parallel to two of its sides and passing through center

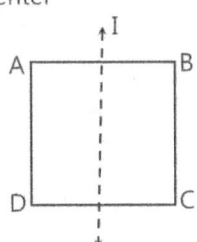

then about any axis as shown below will be

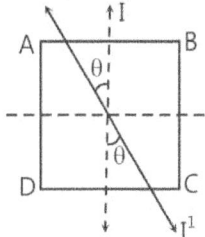

Consider, this axis as x-axis and y-axis perpendicular to a line to it

$(I^1)_{x-axis} = (I^1)_{y-axis}$ |By symmetry|

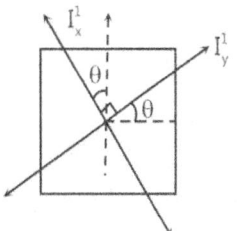

$I_z = I^1_x + I^1_y = 2I^1_x$

$I^1_x = \dfrac{I_z}{2}$

We know that I_z is independent θ since $I_z = \dfrac{Ma^2}{6}$

$\Rightarrow I^1_x$ is also independent of q

$\Rightarrow I^1_x(\theta) = I^1_x(0) = I$

Sol 5: (B) Given, see-saw is out of balance.

$\Rightarrow$ Centre of mass is not at the center of see-saw.

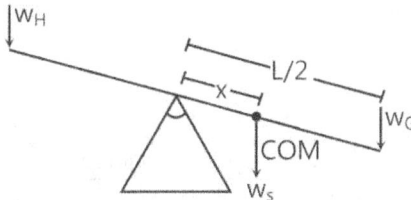

Let 'x' be the distance of COM from center. By moment equilibrium at center

$$[w_H - w_G] \frac{L}{2} = w_s . x \Rightarrow x = \frac{(w_H - w_G)}{w_s} \frac{L}{2}$$

Now if the girl and body move to half of the original

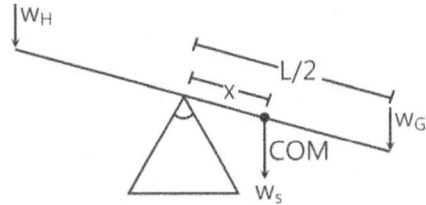

moment due to heavy body and girl is $(w_H - w_G)\dfrac{L}{4}$

(opposite in direction to $w_s.x$)

while

$w_s \cdot x = (w_H - w_G)\dfrac{L}{2}$

$\therefore\ w_s \cdot x > (w_H - w_G)\dfrac{L}{4}$

The side the girl is sitting on will once again tilt downward

Sol 6: (A)

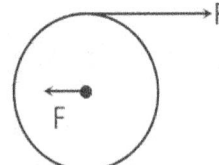

The force on the hinge is same as the force on the thread this can be found by using force equilibrium conditions.

Since, there is a torque always about hinge on pulley, angular velocity increases.

Sol 7: (D)

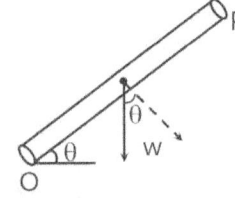

torque acting on the pole due to weight about point O is

$T = \vec{r} \times \vec{F} = \dfrac{L}{2} W\cos\theta$

$T = I\alpha$

$\Rightarrow \alpha = \dfrac{\dfrac{L}{2}W\cos\theta}{\dfrac{ML^2}{3}} = \dfrac{3mg\cos\theta}{2mL} = \dfrac{3g\cos\theta}{2L}$

Acceleration of point P$=L.\alpha = \dfrac{3g\cos\theta}{2}$

Sol 8: (B)

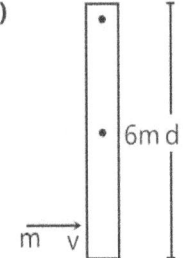

Moment of inertia of rod about point I_{COM}

$= \dfrac{Md^2}{12}$

Angular momentum is conserved as no torque is acting on the system. (while energy is not since a force acts on rod at point)

$\therefore$ Initial angular momentum = final angular momentum

$I_1\omega_1 + m\dfrac{d}{2}v_1 = I_2\omega_2$

$\Rightarrow \dfrac{Md^2}{3}(0) + m\dfrac{dv}{2} = \left(\dfrac{Md^2}{12} + \dfrac{Md^2}{4}\right)\omega^1$

$\Rightarrow \dfrac{md}{2}v = \dfrac{3}{4}md^2\omega^1$

$\Rightarrow \omega^1 = \dfrac{2v}{3d}$

Sol 9: (D)

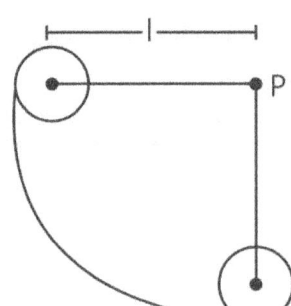

By the principle of conservation of energy

K.E.$_i$ + P.E.$_i$ = K.E.$_f$ + P.E.$_f$

$\Rightarrow 0 + mg(l) = \dfrac{1}{2}mv^2 + \dfrac{1}{2}I\omega^2 + 0$

$\Rightarrow mgl = \dfrac{1}{2}mr^2w^2 + \dfrac{1}{2}\cdot\dfrac{2}{5}mr^2\omega^2$

$\Rightarrow \omega = \sqrt{\dfrac{10}{7}\dfrac{gl}{r^2}}$

Angular momentum of the sphere about P is

$L = I\omega + m.l.v$

$\Rightarrow L = \dfrac{2}{5}mr^2\,\omega + m.\,l.\,r\,\omega$

$\Rightarrow L = m.\sqrt{\dfrac{10}{7}gl}\,.\left[\dfrac{2}{5}r + l\right]$

Sol 10: (C)

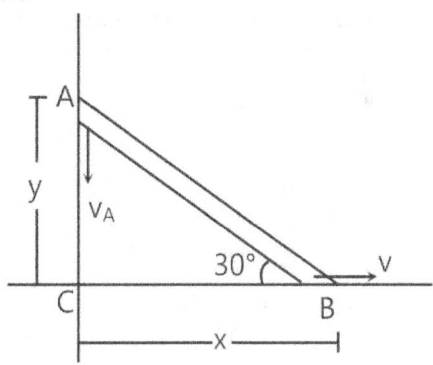

$x^2 + y^2 = l^2$

L = length of ladder = constant

$$\Rightarrow x.\frac{dx}{dt} + y.\frac{dy}{dt} = 0$$

$$\Rightarrow x.v + y.v_A = 0$$

$$\Rightarrow v_A = -\frac{x}{y}.v = -\sqrt{3}\,v$$

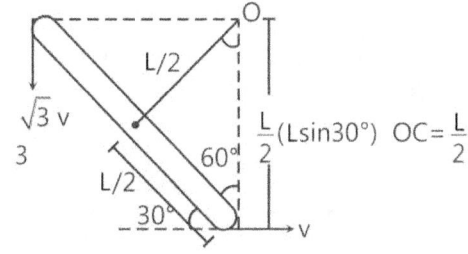

Angular velocity of rod

$$= \frac{|-\sqrt{3}v\hat{j} + v\hat{i}|}{L} = \frac{2v}{L}$$

Velocity of center $= \frac{L}{2}.\omega = v$

Sol 11: (A) If $\frac{dv}{dt} = 0$

$x^2 + y^2 = l^2$

$$\Rightarrow x.\frac{dx}{dt} + y.\frac{dy}{dt} = 0$$

$$\Rightarrow x.v + y\,v_A = 0$$

$$\Rightarrow \frac{dx}{dt}.v + x.\frac{dv}{dt} + \frac{dy}{dt}.v_A + y.\frac{dv_A}{dt} = 0$$

$$\Rightarrow v^2 + v_A^2 + y.\frac{dv_A}{dt}$$

$$\Rightarrow \frac{dv_A}{dt} = \frac{0 - (v^2 + v_A^2)}{y}$$

when $\alpha = 45°$

$v_A = -v$ and $y = \frac{\ell}{\sqrt{2}}$

$$\Rightarrow \frac{dv_A}{dt} = -\frac{2v^2.\sqrt{2}}{\ell}$$

Angular acceleration =

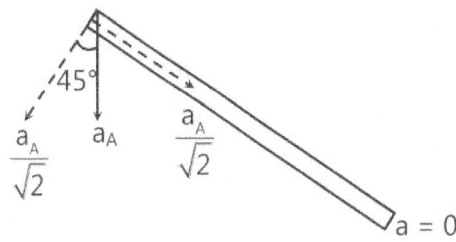

$$\alpha = \frac{\frac{a_A}{\sqrt{2}}}{(L)}$$

$$\alpha = \frac{2v^2}{L^2}$$

Sol 12: (A)

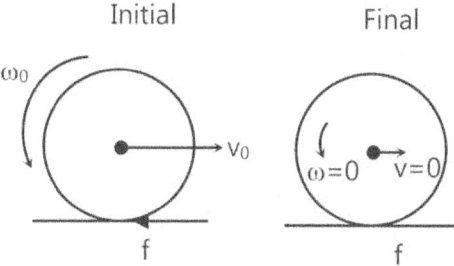

Frictional forces acts to reduce the velocity of bottom most point.

$f = -ma_{cm}$ (By Newton's second law)

$T = I\alpha$

$$\Rightarrow f.R = \frac{MR^2}{2}.a \Rightarrow f = \frac{MR\alpha}{2}$$

After time $t = t$, angular velocity and lineal velocity becomes zero.

$$\Rightarrow 0 - v_0 = -a_{cm}t \text{ and } 0 - \omega_0 = -\alpha t$$

$$\Rightarrow \frac{v_0}{w_0} = \frac{a_{cm}}{\alpha} = \frac{\frac{f}{M}}{\frac{2f}{MR}}$$

$$\Rightarrow \frac{v_0}{\omega_0} = \frac{R}{2} \Rightarrow \frac{v_0}{R\omega_0} = \frac{1}{2}$$

Sol 13: (C) Conserving linear momentum

$2mv - mv = 2m \times v_{cm}$

$v_{cm} = \dfrac{v}{2}$

Initial angular momentum

$= m \times 2v \times \dfrac{b}{2} + mv \times \dfrac{b}{2} = \dfrac{3vbm}{2}$

Final angular momentum

$= \left[m\left(\dfrac{b}{2}\right)^2 + m\left(\dfrac{b}{2}\right)^2 \right] \times \omega = \dfrac{mb^2\omega}{2}$

$\Rightarrow \dfrac{3mvb}{2} = \dfrac{mb^2\omega}{2}$

$\therefore \omega^2 = \dfrac{3v}{b}$

For skater at $x = b/2$

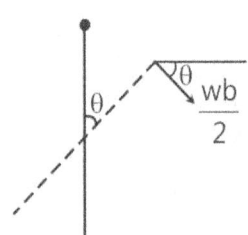

$v_x = v + \dfrac{\omega b}{2}\cos\theta$

$v_y = -\dfrac{\omega b}{2}\sin\theta$

$\theta = \omega t$

$\therefore v_x$ at $t = \dfrac{v}{2}$

$v_x = v + \dfrac{3v}{2}\cos\left(\dfrac{3vt}{b}\right)$

$\therefore x = \int v_x\, dt = \int v + \dfrac{3V}{2}\cos\left(\dfrac{3vt}{b}\right)$

$= vt + \dfrac{3V}{2 \times 3V} \times b \times \sin\left(\dfrac{3vt}{b}\right)$

$= vt + \dfrac{b}{2}\sin\left(\dfrac{3vt}{b}\right)$

$y = \int v_y\, dt = \int -\dfrac{\omega b}{2}\sin(\omega t)$

$= \dfrac{+\omega b}{2 \times \omega}\cos(\omega t) = \dfrac{b}{2}\cos\left(\dfrac{3vt}{b}\right)$

Sol 14: (C) When F_1 is applied, the body moves right and angular acceleration is developed accordingly by friction

when F_3 is applied, the angular acceleration developed it the body move left.

When F_2 is applied the body can move either left of right depending on angle of inclination.

Multiple Correct Choice Type

Sol 15: (B, C)

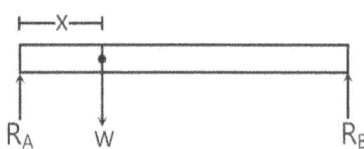

FBD of rod

$R_A + R_B = w$ (force equilibrium)

$R_B.d = w.x$ (torque equilibrium)

$\Rightarrow R_B = \dfrac{wx}{d}$ and $R_A = \dfrac{w(d-x)}{d}$

Sol 16: (A, D)

Sliding condition $= mg\sin\theta > \mu mg\cos\theta$

$\Rightarrow \tan\theta > m$

Toppling condition $= mg\sin\theta.\dfrac{h}{2} > mg\cos\theta.\dfrac{a}{2}$

$\Rightarrow \tan\theta > \dfrac{a}{h}$

If $\mu > \dfrac{a}{h}$

$\tan\theta > \dfrac{a}{h}$ is met earlier than $\tan\theta > m$

$\therefore$ Topples before sliding

If $\mu < \dfrac{a}{h}$

It will slide before toppling

Sol 17: (A, C, D)

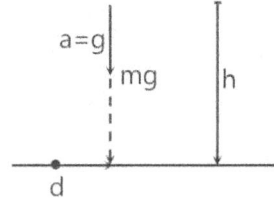

Angular momentum = mvd = mgtd

Torque of gravitational force $= \dfrac{dL}{dt} = mgd$

Moment of inertia $= m(d^2 + h^2)$

where $h = H_0 - \dfrac{1}{2}gt^2$

Angular velocity $= \dfrac{v}{d} = \dfrac{gt}{d}$

Sol 18: (A, B, C) $\vec{T} = \vec{A} \times \vec{L}$

$\Rightarrow \dfrac{d\vec{L}}{dt} = \vec{A} \times \vec{L}$

$\therefore \dfrac{d\vec{L}}{dt} \perp \vec{L}$

Components of $\vec{L}$ on $\vec{A}$ remain unchanged because if $\vec{L}$ component changes the L.H.S changes while R.H.S remains unchanged which is a contradiction. If magnitude of L changes with time, thin L.H.S and R.H.S vary differently with time which is a contradiction.

Suppose $\vec{L} = (x.t) \dfrac{\vec{L}}{|\vec{L}|}$

Then $\dfrac{d\vec{L}}{dt} = \dfrac{x\vec{L}}{|\vec{L}|}$ while $\vec{A} \times \vec{L} = xt \dfrac{\vec{A} \times \vec{L}}{|\vec{L}|}$

Sol 19: (B, C)

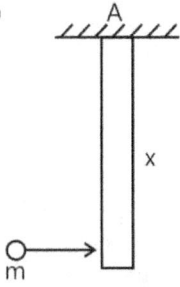

Linear momentum is not conserved because of hinge force angular momentum about A is conserved since torque at A is zero.

Kinetic energy of system before collision is equal to kinetic energy of system just after collision since, the collision is elastic

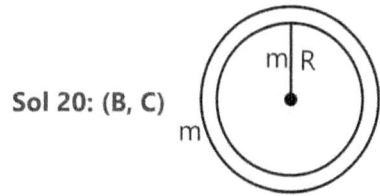

Sol 20: (B, C)

Kinetic energy of the body

$= \dfrac{1}{2}mv^2 + \dfrac{1}{2}mv^2 + \dfrac{1}{2}I\omega^2$

$= \left(1 + \dfrac{1}{2}.\dfrac{2}{3}\right)mv^2 = \dfrac{4}{3}mv^2$ ($\because v = R\omega$)

($\because I = \dfrac{2}{3}mR^2$ only hollow sphere

$\because$ Non-viscous liquid)

Angular momentum about any point on ground

$= 2mRv + \dfrac{2}{3}mR^2\omega = \dfrac{8}{3}mRv$

Sol 21: (B, C)

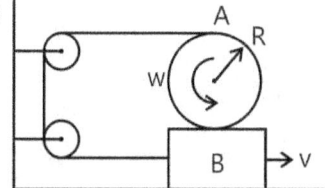

Since the cylinder does not slip

At point B velocity = 0

$\Rightarrow -\vec{V} + \vec{V}_{cm} + \vec{R\omega} = 0$

$\Rightarrow \vec{V}_{cm} = -(v - R\omega)\hat{i}$

At point A, velocity = 0

$\Rightarrow v = R\omega$

$\Rightarrow \vec{V}_{cm} = 0$

Sol 22: (B, C, D) To the right of B, angular acceleration will disappear but linear acceleration will increase since no friction is present angular velocity attained by disc after time T is

$\omega = \alpha T$

and $2\pi = \dfrac{1}{2}\alpha T^2$

Time to complete one rotation

$2\pi = \omega t$

$$\Rightarrow t = \frac{2\pi}{\alpha T} = \frac{2\pi . T}{2\pi . 2} = \frac{T}{2}$$

Sol 23: (C, D) Given,

$180° < Q_f - Q_i > 360°$

$\pi \text{ rad} < Q_f - Q_i < 2\pi \text{ rad}$

Sol 24: (A, B, C, D)

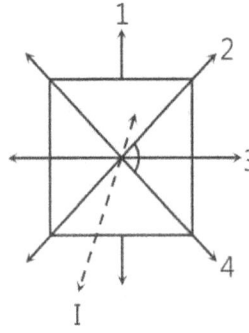

$I = I_1 + I_3$ (Perpendicular axes theorem)

Also $I_1 = I_3$ (by symmetry)

$$\Rightarrow \frac{I}{2} = I_1 = I_3$$

$I = I_2 + I_4$ (perpendicular axis theorem)

Also $I_2 = I_4$ (by symmetry)

$$\Rightarrow \frac{I}{2} = I_2 = I_4$$

Sol 25: (B, C) Option A is incorrect, since the statement indicates a force body system as below.

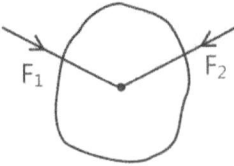

Which is not in equilibrium while B, C are possible

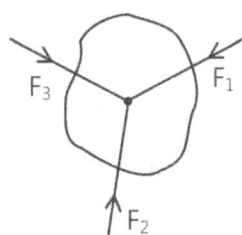

(b) $\vec{F_1} + \vec{F_2} + \vec{F_3} = 0$

(c)

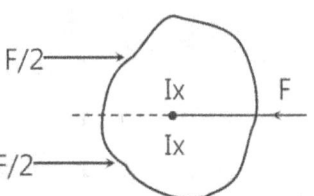

(d) While D is possible

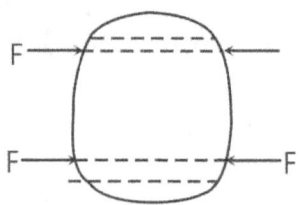

Sol 26: (A, B, D)

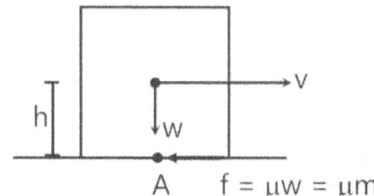

$L = mvh$ (Angular momentum)

$$\text{acceleration} = -\frac{\mu mg}{m} = -\mu g$$

$V_t = v - \mu g t$

$L_t = m(v - \mu g t)h$

$$\frac{dL_t}{dt} = T = -\mu mgh$$

Sol 27: (A, C, D) If Re spreads or curls up his hands, moment of inertia changes, accordingly angular velocity changes too.

If $I\omega = $ Constant, it cant keep

$\frac{1}{2}I\omega^2$ the same, rotational kinetic energy would also change.

Sol 28: (A, C, D)

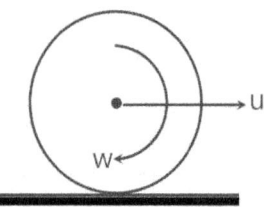

$u = R\omega$ (for pure rolling)

The velocity of bottom most point is $u - R\omega = 0$

the velocity of topmost point is $u + R\omega = 2u$

$\therefore 0 \le v \le 2u$

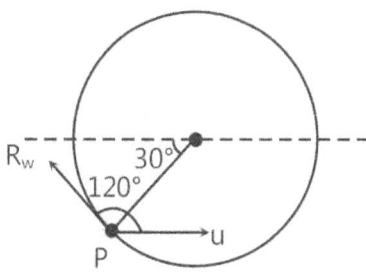

Velocity of P is

$$\vec{V} = u\hat{i} + \left(-\frac{u}{2}\hat{i} + \frac{\sqrt{3}u}{2}\hat{j} \right)$$

$$\Rightarrow \vec{V} = \frac{u}{2}\hat{i} + \frac{\sqrt{3}u}{2}\hat{j}$$

$$\Rightarrow V = u$$

If CR is horizontal

$$\vec{V} = u\hat{i} + u\hat{j}$$

$$V = \sqrt{2}\,u$$

Sol 29: (B, C, D) Angular moment about O is not constant because a component of weight causes torque at point 'O'.

Angular moment about C is zero since weight is parallel axis

About O,

$\vec{L} = m(\vec{r} \times \vec{v})$ gives angular momentum in direction perpendicular to length of thread and velocity. The vertical component never changes direction.

Sol 30: (A, C)

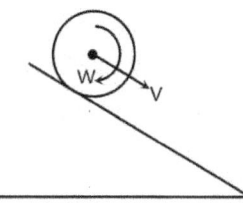

A cylinder rolling down with incline may or may not attain pure rolling. It depends on length of the incline

Sol 31: (B, C) Friction on cylinder under pure rolling depends on the external forces

Sol 32: (A, C)

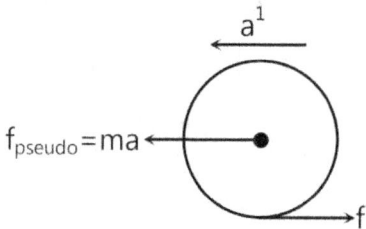

In frame of plank,

$F_{pseudo} - f = ma_1$

$a_1 = a - \dfrac{f}{m}$ where a is acceleration of plank

$a = \dfrac{F - f}{M}$

Total K.E. of system = work done by force F

($\because$ no other external forces is doing work)

Work done on sphere = work done by friction + work done by pseudo force = change in K.E.

Sol 33: (A, B, C)

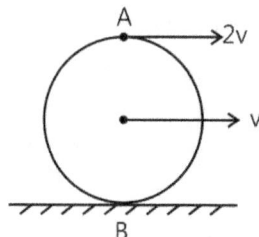

$V_A = V + R\omega = 2V$

$V_B = V - R\omega = 0$

L about B $= mvR + \dfrac{1}{2}mR^2\omega = \dfrac{3mRV}{2}$ clockwise

L about A $= -mvR + \dfrac{1}{2}mR^2\omega = \dfrac{mRV}{2}$ anti-clockwise

Sol 34: (A, B, C)

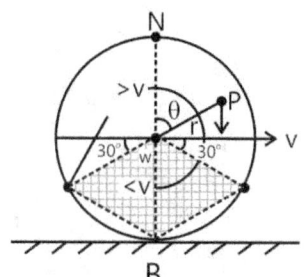

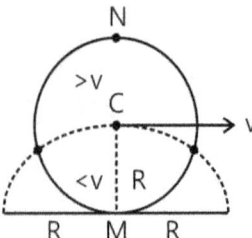

$$\vec{V_P} = (V + r\omega\cos\theta)\,\hat{i} + (r\omega\sin\theta)\,\hat{j}$$

$$|\vec{V_P}|^2 = V^2 + r^2\omega^2 + 2r\omega\cos\theta V$$

$$= (R^2 + r^2 + 2rR\cos\theta)\omega^2$$

$$|\vec{V_P}|^2 = M^2P^2\omega^2$$

$$\Rightarrow |\vec{V_P}| = (\overline{MP}).\omega$$

Using the above relation we draw a circle with radius R to get point with velocities $R.\omega = v$

Sol 35: (A, B, C) As the ring enters frictional force (limiting) acts on sphere increases angular acceleration in clockwise direction and slows down the linear motion.

Sol 36: (B, C, D) Rolling motion starts when the point of contact has zero velocity. Conserving angular momentum about point of ground gives

$$mvR = I\omega + mv'R$$

$$\Rightarrow mvR = mR^2\omega^1 + mR^2\omega^1 \ (\because v^1 = R\omega^1)$$

$$\Rightarrow \omega^1 = \frac{v}{2R}$$

$$v^1 = \frac{v}{2}$$

Time taken to achieve pure rolling is

$$v - u = at \Rightarrow \frac{v}{2} - v = -\mu gt \Rightarrow t = \frac{v}{2\mu g}$$

Sol 37: (A, B, C, D) Distance moved by ring

$$= \frac{v^2 - u^2}{2a} = \frac{3v_0^2}{8\mu g}$$

Work done by friction $= -\mu mg.\dfrac{3v_0^2}{8\mu g}$

$$= -\frac{3mv_0^2}{8}$$

Gain in rotational K.E. $= \dfrac{1}{2}mR^2.\dfrac{v_0^2}{4R^2} = \dfrac{mv_0^2}{8}$

Loss in K.E. $= -\left(-\dfrac{3mv_0^2}{8} + \dfrac{1}{8}mv_0^2\right) = \dfrac{mv_0^2}{4}$

Sol 38: (B, C) By holding a pole horizontally, the moment of inertia is increased leading to slower angular acceleration due to undesired torques.

Also, adjusts center of gravity to be vertically over rope to eliminate torque.

Assertion Reasoning Type

Sol 39: (B) A cyclist always bends inwards to reduce the centrifugal force.

Also, he lowers the center of gravity.

But the reason does not explain the assertion

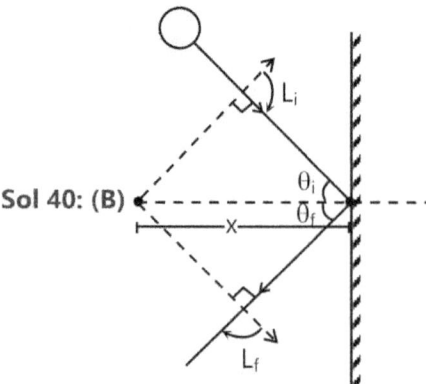

Sol 40: (B)

Statement-I is true because the initial angular moment about any point on xy plane is $L_i = mv_i \sin\theta_i x$ and final angular momentum $L_f = mv_f x\sin\theta_f$

for elastic collision $v_f = v_i,\ \theta_i = \theta_f$

$$\Rightarrow L_i = L_f$$

Statement-II is also correct since the disc is in equilibrium

But II is explanation of I

Sol 41: (A)

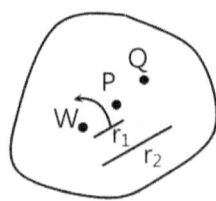

In frame of O

$$\omega_P = \omega_Q = \omega$$

$$v_P = r_P\omega_P\ v_Q = r_Q\omega_Q$$

In frame of P

$$v_0 = -r_P\omega_P\ v_Q = r_Q\omega_Q - r_P\omega_P$$

$$w_0 = -\frac{v_0}{r_P} = -\omega_P = \omega$$

$$w_Q^1 = \frac{(r_Q - r_P)(\omega)}{(r_Q - w_P)} = \omega$$

∴ Statement-I is true and statement-II is correct explanation of I

Sol 42: (B) Statement-I true by parallel axis theorem.

Statement-II is true but doesn't explain parallel axis theorem.

Sol 43: (B) See above

Sol 44: (D)

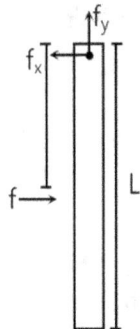

$F_x = F$ (force equilibrium in x-direction)

∴ Assertion is false

While reason is true since

$$T = I\alpha = F.x \Rightarrow \alpha = \frac{F}{I}.x$$

Sol 45: (B) Statement-I is true, which is the condition for pure rolling

Statement-II is also correct by the definition of center of mass but II is not correct explanation of I.

Sol 46: (D) Statement-I is false, because a body can roll if we throw it with property determined linear and angular velocities.

Statement-II is true by definition of pure rolling

Comprehension Type

Sol 47: (C)

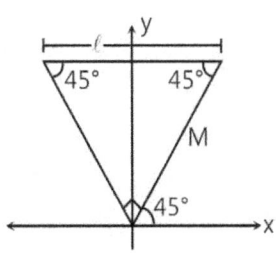

Taking an elemental point dy

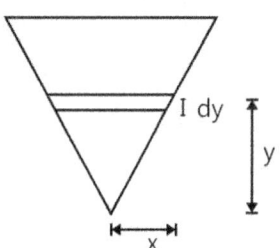

$dm = \rho.(2x)dy$

$= \rho.(2y)dy$ [∵ $y = x$]

$$\rho = \frac{M}{\frac{1}{4}\ell^2} = \frac{4M}{\ell^2}$$

$$dI_2 = \frac{dm(2x)^2}{12} + dm(y)^2$$

$$\left[\frac{m\ell^2}{12} \text{ for rod, parallel axis theorem} \right]$$

$$\Rightarrow \int dI_2 = \int_0^{\ell/2} \left(\frac{y^2}{3} + y^2 \right)(2y)\rho\, dy \; (\because y = x)$$

$$\Rightarrow \int dI_1 = I_2 = \frac{8\rho}{3} \int_0^{\ell/2} y^3.dy = \frac{8\rho}{3}.\left[\frac{y^4}{4} \right]_0^{\ell/2}$$

$$= \frac{2\rho}{3}\left[\frac{\ell^4}{16} \right] = \frac{ML^2}{6}$$

Sol 48: (A) $dI_x = (dm)(y^2)$

$$\Rightarrow \int dI_x = \int_0^{\ell/2} 2\rho y^3 dy$$

$$\Rightarrow I_x = 2\rho \left[\frac{y^4}{4} \right]_0^{\ell/2} = \frac{8M}{\ell^2}.\frac{\ell^4}{64} = \frac{M\ell^2}{8}$$

Sol 49: (C) Moment of inertia about base is

$$dI = dm\left(\frac{L}{2} - y \right)^2$$

$$\Rightarrow \int dI = \int_0^{L/2} \left(\frac{L}{2} - y \right)^2 (2\rho y)\, dy$$

$$\Rightarrow I = 2\rho \int_0^{L/2} y^2\left(\frac{L}{2} - y \right).dy$$

$$\Rightarrow I = 2\rho \left[\frac{L}{2} \cdot \frac{y^3}{3} - \frac{y^4}{4} \right]_0^{L/2}$$

$$\Rightarrow I = 2\rho L^4 \left[\frac{1}{48} - \frac{1}{64} \right] = 8ML^2 \left(\frac{4-3}{64 \times 3} \right) = \frac{ML^2}{24}$$

Sol 50: (C) $I_z = I_x + I_y$ (perpendicular axis theorem)

$$\Rightarrow I_y = \frac{ML^2}{6} - \frac{ML^2}{8} = \frac{ML^2}{24}$$

Sol 51: (B) $d\vec{L} = dm\,(\vec{r} \times \vec{v})$

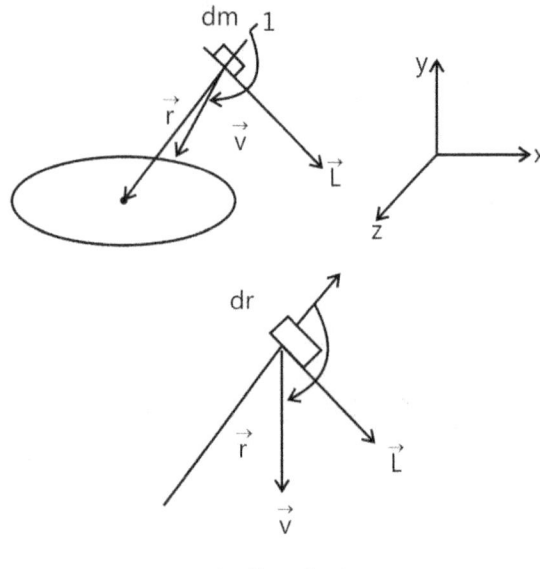

(z direction)

∴ Angular momentum is down at 20° to horizontal

Sol 52: (B) There is a torque since angular momentum is always changing direction.

Sol 53: (A)

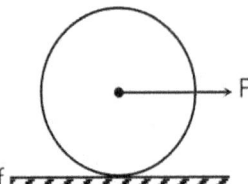

Friction reduces linear acceleration and increases angular velocity

Sol 54: (A) Same as previous

So 55: (D)

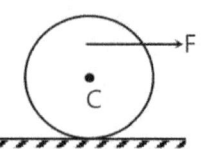

Frictional force in this depends on distance between application of force and center

Sol 56: (B)

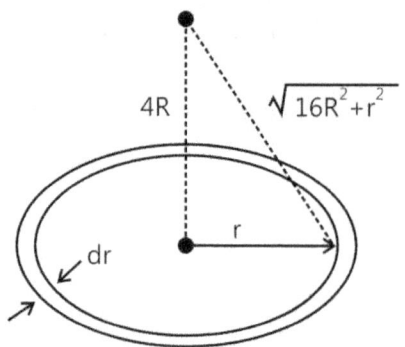

By basic FBD's we can understand that friction acts in forward direction

∴ Option B is correct

And the torque is acting horizontally, since the horizontal component of angular momentum is only changing.

Previous Years' Questions

Sol 1: (A) $W = \Delta U = U_f - U_i = U_\infty - U_P$

$$= -U_P = -mV_P = -V_P \text{ (as } m = 1\text{)}$$

Potential at point P will be obtained by integration as given below.

Let dM be the mass small ring as shown

$$dM = \frac{M}{\pi(4R)^2 - \pi(3R)^2}(2\pi r)dr = \frac{2Mrdr}{7R^2}$$

$$dV_P = -\frac{G\,dM}{\sqrt{16R^2 + r^2}}$$

$$= -\frac{2GM}{7R^2} \int_{3R}^{4R} \frac{r}{\sqrt{16R^2 + r^2}} \, dr = -\frac{2GM}{7R}(4\sqrt{2} - 5)$$

$$\therefore W = +\frac{2GM}{7R}(4\sqrt{2} - 5)$$

Sol 2: (A) $\frac{2}{5}MR^2 = \frac{1}{2}Mr^2 + Mr^2$

Or $\frac{2}{5}MR^2 = \frac{3}{2}Mr^2$

$$\therefore r = \frac{2}{\sqrt{15}}R$$

Sol 3: (B) Condition of sliding is

$ms \sin \theta > \mu\, mg \cos \theta$ or $\tan \theta > \mu$

or $\tan \theta > \sqrt{3}$ ... (i)

Condition of toppling is

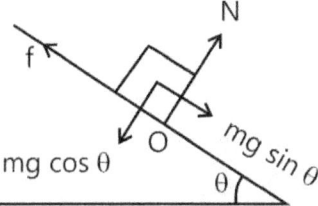

Torque of mg sin θ about O>torque of mg about

$\therefore (mg \sin \theta)\left(\dfrac{15}{2}\right) > (mg \cos \theta)\left(\dfrac{10}{2}\right)$

or $\tan \theta > \dfrac{2}{3}$ ... (ii)

With increase in value of θ, condition of sliding is satisfied first.

Sol 4: (A) $I_{remaining} = I_{whole} - I_{removed}$

or $I = \dfrac{1}{9}(9M)(R)^2 - \left[\dfrac{1}{2}m\left(\dfrac{R}{3}\right)^2 + \dfrac{1}{2}m\left(\dfrac{2R}{3}\right)^2\right]$ ....(i)

Here, $m = \dfrac{9M}{\pi R^2} \times \pi\left(\dfrac{R}{3}\right)^2 = M$

Substituting in Eq. (i), we have $I = 4MR^2$

Sol 5: (A) A'B' $\perp$ AB and C'D' $\perp$ CD

From symmetry $I_{AB} = I_{A'B'}$ and $I_{CD} = I_{C'D'}$. From theorem of perpendicular axes,

$I_{zz} = I_{AB} + I_{A'B'} = I_{CD} + I_{C'D'} = 2I_{AB} = 2I_{CD}$

$I_{AB} = I_{CD}$

Alternate The relation between I_{AB} and I_{CD} should be true for all values of Nθ

At $\theta = 0$, $I_{CD} = I_{AB}$

Similarly, at $\theta = \pi/2$, $I_{CD} = I_{AB}$

(By symmetry)

Keeping these things in mind, only option (a) is correct.

Sol 6: (D) In case of pure rolling,

$$f = \frac{mg\sin\theta}{1 + \dfrac{mR^2}{I}} \quad \text{(Upwards)}$$

$\therefore f \propto \sin\theta$

Therefore, as θ decreases force of friction will also decrease.

Sol 7: (A) On smooth part BC, due to zero torque, angular velocity and hence the rotational kinetic energy remains constant. While moving from B to C translational kinetic energy converts into gravitational potential energy.

Sol 8: (B) From conservation of angular momentum ($I\omega$ = constant), angular velocity will remains half. As

$$K = \frac{1}{2}I\omega^2$$

The rotational kinetic energy will become half. Hence, the correct option is (b).

Sol 9: (A) Let ω be the angular velocity of the rod. Applying angular impulse = change in angular momentum about center of mass of the system

$$J.\frac{L}{2} = I_C\omega$$

$$\therefore (Mv)\left(\frac{L}{2}\right) = (2)\left(\frac{ML^2}{4}\right)\omega \therefore \omega = \frac{v}{L}$$

Sol 10: (A) In case of pure rolling bottom most point is the instantaneous center of zero velocity.

Velocity of any point on the disc,, where r is the distance of point from O.

$r_Q > r_C > r_P$

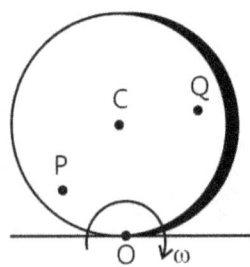

$$\therefore \quad v_Q > v_C > v_P$$

Paragraph 1

Sol 11: (C) $\dfrac{1}{2}I(2\omega)^2 = \dfrac{1}{2}kx_1^2$ ...(i)

$$\dfrac{1}{2}(2I)(\omega)^2 = \dfrac{1}{2}kx_2^2 \qquad \text{..... (ii)}$$

From Eqs. (i) and (ii), we have $\dfrac{x_1}{x_2} = \sqrt{2}$

Sol 12: (A) Let ω' be the common velocity. Then from conservation of angular momentum, we have

$$(I + 2I)\omega' = I(2\omega) + 2I(\omega)$$

$$\omega' = \dfrac{4}{3}\omega$$

From the equation,

Angular impulse = change in angular momentum, for any of the disc, we have

$$\tau . t = I(2\omega) - I\left(\dfrac{4}{3}\omega\right) = \dfrac{2I\omega}{3}$$

$$\therefore \quad \tau = \dfrac{2I\omega}{3t}$$

Sol 13: (B) Loss of kinetic energy = $K_i - K_f$

$$= \left\{\dfrac{1}{2}I(2\omega)^2 + \dfrac{1}{2}(2I)(\omega)^2\right\} - \dfrac{1}{2}(3I)\left(\dfrac{4}{3}\omega\right)^2 = \dfrac{1}{3}I\omega^2$$

Sol 14: (D) $\dfrac{1}{2}mv^2 + \dfrac{1}{2}I\left(\dfrac{v}{R}\right)^2 = mg\left(\dfrac{3v^2}{4g}\right) \therefore I = \dfrac{1}{2}mR^2$

$\therefore$ Body is disc.

Paragraph 2

Sol 15: (D) $a = R\alpha$

$$\therefore \quad \dfrac{2kx - f}{M} = R\left[\dfrac{fR}{\dfrac{1}{2}MR^2}\right]$$

Solving this equation, we get

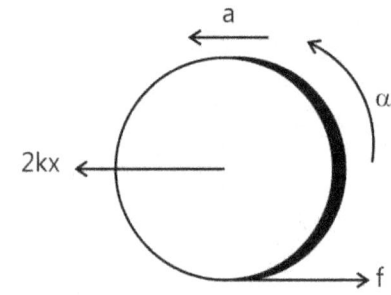

$$f = \dfrac{2kx}{3}$$

$$\therefore \quad |F_{net}| = 2kx - f = 2kx - \dfrac{2kx}{3} = \dfrac{4kx}{3}$$

This is opposite to displacement.

$$\therefore \quad F_{net} = -\dfrac{4kx}{3}$$

Sol 16: (D) $F_{net} = -\left(\dfrac{4kx}{3}\right)x$

$$\therefore \quad a = \dfrac{F_{net}}{M} = -\left(\dfrac{4k}{3M}\right)x = -\omega^2 x$$

$$\therefore \quad \omega = \sqrt{\dfrac{4k}{3M}}$$

Sol 17: (C) In case of pure rolling mechanical energy will remains conserved.

$$\therefore \quad \dfrac{1}{2}Mv_0^2 + \dfrac{1}{2}\left(\dfrac{1}{2}MR^2\right)\left(\dfrac{v_0}{R}\right)^2 = 2\left[\dfrac{1}{2}kx_{max}^2\right]$$

$$\therefore \quad x_{max} = \sqrt{\dfrac{3M}{4k}}v_0$$

As $f = \dfrac{2kx}{3}$

$$\therefore \quad F_{max} = \mu Mg = \dfrac{2kx_{max}}{3} = \dfrac{2k}{3}\sqrt{\dfrac{3M}{4k}}v_0$$

$$\therefore \quad v_0 = \mu g \sqrt{\dfrac{3M}{k}}$$

Sol 18: (B) Angular momentum about rotational axis

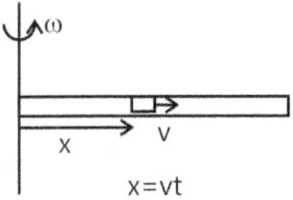

$$x = vt$$

$$L_{(t)} = \left[I + m(vt)^2 \right] \omega$$

$$\frac{dL_t}{dt} = 2mv^2 t \omega$$

Torque $\tau = \left(2mv^2 \omega \right) t$

Sol 19: (C)

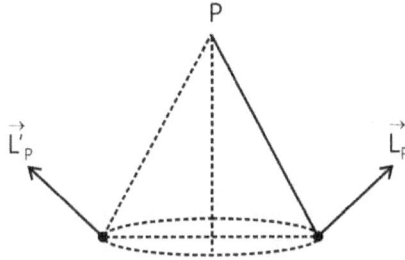

$$\vec{L}_0 = \vec{r}_0 \times \vec{p}$$

$\vec{L}_0$ is always directed along the axis & its magnitude is constant.

Sol 20: (C)

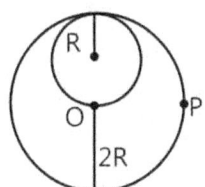

Let mass of original disc = m

The mass of disc removed $= \dfrac{m}{\pi\left(4R^2\right)} \times \pi R^2 = \dfrac{m}{4}$

So M.O.I of remaining section about axis passing

through "O" $I_O = \dfrac{m(2R)^2}{2} - \left[\dfrac{m}{4}\dfrac{R^2}{(2)} + \dfrac{m}{4}R^2 \right]$

$\Rightarrow 2mR^2 - \left[\dfrac{mR^2 + 2mR^2}{8} \right] \Rightarrow \left[2 - \dfrac{3}{8} \right]mR^2 \Rightarrow \dfrac{13}{8}\,mR^2$

MOI of remaining section about "P"

$$I_P = \left[\frac{m(2R)^2}{2} + m(2R)^2 \right] - \left[\frac{m}{4}\frac{R^2}{(2)} + \frac{m}{4}5R^2 \right]$$

$$\Rightarrow \left[2mR^2 + 4mR^2 \right] - \left[\frac{mR^2}{8} + \frac{5mR^2}{4} \right]$$

$$\Rightarrow 6mR^2 - \frac{11}{8}mR^2 \Rightarrow \frac{37}{8}\,mR^2$$

$$\frac{I_P}{I_O} = \frac{37}{8} \times \frac{8}{13} \approx 3$$

Sol 21: (A)

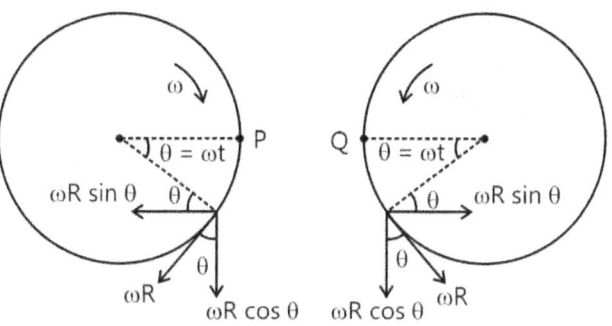

So, $v_r = 2\omega R \sin(\omega t)$

At $t = T/2$, $v_r = 0$

So two half cycles will take place.

Sol 22: (C, D)

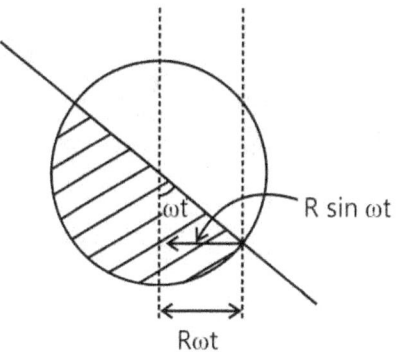

According to problem particle is to land on disc.

If one consider a time 't' then x component of disc is $R\omega t$

$R \sin\omega t < R\omega t$

This particle 'P' land on unshaded region. For "Q" x-component is very small and y-component equal to P it will also land in unshaded region.

Now repeat same thing when right part is shaded then correct answer is "C" or "D"

3.117

Sol 23: (A) In both the cases, the instantaneous axis will be along z-axis i.e. along vertical direction.

Sol 24: (D) w.r.t. centre of mass only pure rotation of disc will be seen. So in both the cases, angular speed about instantaneous axis will be "ω".

Sol 25: (A, B)

$$V_p = R\omega\hat{i} + \frac{\omega}{2}(-\hat{j}) \times \left(R\cos 30°\hat{i} + R\sin 30°\hat{k}\right)$$

$$= 3R\omega\hat{i} + \sqrt{3}\frac{\omega}{4}R\hat{k} - \frac{\omega}{4}R\hat{i} = \frac{11}{4}R\omega\hat{i} + \frac{\sqrt{3}}{4}R\omega\hat{k}$$

Sol 26: (D)

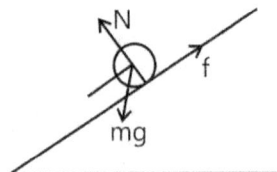

Translation motion:

$$mg\sin\theta - f = ma_{cm} \qquad \ldots(i)$$

Rotational motion

$$fR = I_{cm}\alpha \qquad \ldots(ii)$$

Rolling without slipping

$$\alpha R = a_{cm} \qquad \ldots(iii)$$

From (ii) & (iii)

$$f = \frac{I_{cm}a_{cm}}{R^2}$$

Put this in (i)

$$mg\sin\theta - \frac{I_{cm}a_{cm}}{R^2} = ma_{cm}$$

$$a_{cm} = \frac{mg\sin\theta}{\left(\dfrac{I_{cm}}{R^2} + m\right)}$$

As $I_P > I_Q$

Sol 27: (8) Conservation of angular momentum about vertical axis of disc

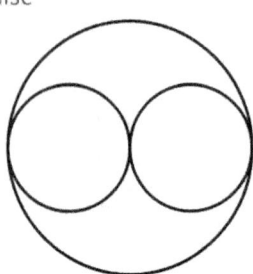

$$\frac{50(0.4)^2}{2} \times 10 = \left[\frac{50(0.4)^2}{2} + 4(6.25)(0.2)^2\right]\omega$$

$$\omega = 8\,rad/sec$$

Sol 28: (C, D) Condition of translational equilibrium

$$N_1 = \mu_2 N_2$$

$$N_2 + \mu_1 N_1 = Mg$$

Solving $N_2 = \dfrac{mg}{1 + \mu_1\mu_2}$

$$N_1 = \frac{\mu_2 mg}{1 + \mu_1\mu_2}$$

Applying torque equation about corner (left) point on the floor

$$mg\frac{\ell}{2}\cos\theta = N_1\,\ell\sin\theta + \mu_1 N_1\,\ell\cos\theta$$

Solving $\tan\theta = \dfrac{1 - \mu_1\mu_2}{2\mu_2}$

Sol 29: (2)

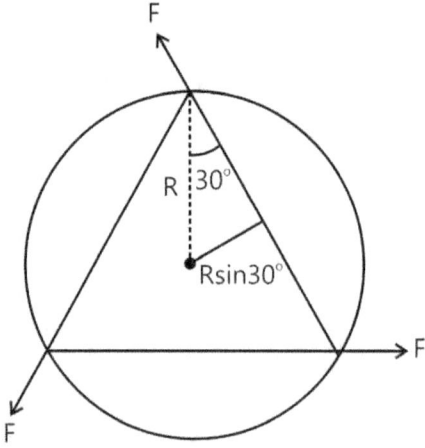

$$\tau = I\alpha$$

$$3FR\sin 30° = I\alpha$$

$$I = \frac{MR^2}{2}$$

$$\alpha = 2$$

$$\omega = \omega_0 + \alpha t$$

$$\omega = 2\,rad/s$$

Sol 30: (4) Since net torque about centre of rotation is zero, so we can apply conservation of angular momentum of the system about center of disc

$L_i = L_f$

$0 = I\omega + 2mv(r/2)$; comparing magnitude

$\therefore \left(\dfrac{0.45 \times 0.5 \times 0.5}{2}\right)\omega = 0.05 \times 9 \times \dfrac{0.5}{2} \times 2$

$\therefore \omega = 4$

Sol 31: (7) Kinetic energy of a pure rolling disc having

velocity of centre of mass $v = \dfrac{1}{2}mv^2 + \dfrac{1}{2}\left(\dfrac{mR^2}{2}\right)\dfrac{v^2}{R^2} = \dfrac{3}{4}mv^2$

So,

$\dfrac{3}{4}m(3)^2 + mg(30) = \dfrac{3}{4}m(v_2)^2 + mg(27) \therefore v_2 = 7\ m/s$

Sol 32: (D) Using conservation of angular momentum

$mR^2\omega$

$= \left(mR^2 \times \dfrac{8\omega}{9}\right) + \left(\dfrac{m}{8} \times \dfrac{9R^2}{25} \times \dfrac{8\omega}{9}\right) + \left(\dfrac{m}{8} \times x^2 \times \dfrac{8\omega}{9}\right)$

$\Rightarrow x = \dfrac{4R}{5}$

Sol 33: $I = \int \dfrac{2}{3}\rho 4\pi r^2 r^2\, dr$

$I_A \propto \int (r)(r^2)(r^2)\, dr$

$I_B \propto \int (r^5)(r^2)(r^2)\, dr$

$\therefore \dfrac{I_B}{I_A} = \dfrac{6}{10}$

Sol 34: (D) Force balance

N sin 30°

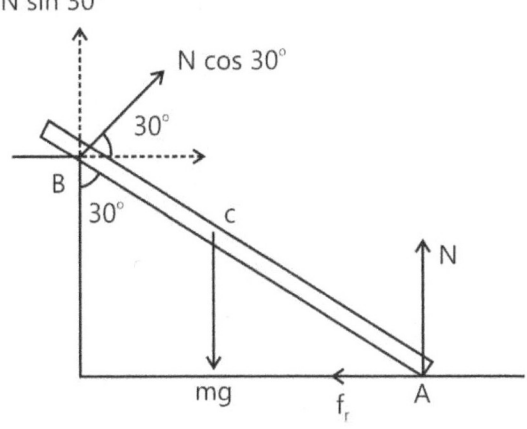

N cos 30°

30°

B

30°

c

N

mg

f_r

A

$N + N\ \sin 30° = mg$

$\dfrac{3}{2}N = mg$

$\boxed{N = \dfrac{2}{3}mg}$

$f_r = \dfrac{mg}{\sqrt{3}} = \dfrac{16}{\sqrt{3}} = \dfrac{16\sqrt{3}}{3}$

Torque balance (about A)

$N \times \dfrac{h}{\cos 30°} = mg \times \dfrac{L}{2}\sin 30°$

$\dfrac{2}{3}mg \times \dfrac{2h}{\sqrt{3}} = mg \times \dfrac{L}{4}$

$\boxed{\dfrac{h}{L} = \dfrac{3\sqrt{3}}{16}}$

Sol 35: (A, B, D)

$\vec{v} = \dfrac{d\vec{r}(t)}{dt} = 3\alpha t^2\hat{i} + 2\beta t\hat{j}, \vec{a} = \dfrac{d\vec{v}}{dt} = 6\alpha t\hat{i} + 2\beta\hat{j}$

At t = 1, $\vec{v} = \left(10\hat{i} + 10\hat{j}\right)\ ms^{-1}$

$\vec{a} = 20\hat{i} + 10\hat{j}\ ms^{-2}$

$\vec{r} = \dfrac{10}{3}\hat{i} + 5\hat{j}\ m$

$\vec{L}_0 = \vec{r} \times m\vec{v} = \left(-\dfrac{5}{3}\hat{k}\right)\ N\,m\,s$

$\vec{F} = m\dfrac{d\vec{v}}{dt} = \left(2\hat{i} + \hat{j}\right)N$

$\vec{\tau} = \vec{r} \times \vec{F} = \vec{r} \times m\vec{a} = \left(-\dfrac{20}{3}\hat{k}\right)N\,m$

Sol 36: (A, D) $\omega_z = \dfrac{\omega a}{\ell}\cos\theta = \omega/5$